D1592720

Biomaterials Science and Biocompatibility

Springer
New York
Berlin
Heidelberg
Barcelona
Hong Kong
London
Milan
Paris
Singapore
Tokyo

Biomaterials Science and Biocompatibility

Frederick H. Silver
David L. Christiansen

With 166 Illustrations

Frederick H. Silver
Department of Pathology
UMDNJ—Robert Wood Johnson Medical School
Piscataway, NJ 08854
USA
silverfr@umdnj.edu

David L. Christiansen
Department of Pathology
UMDNJ—Robert Wood Johnson Medical School
Piscataway, NJ 08854
USA
christia@umdnj.edu

Cover illustration: Joan Greenfield. Cover photo: "The Thinker" by Auguste Rodin, Musee Rodin, Paris, France/Bernard Bansse/Superstock.

Library of Congress Cataloging-in-Publication Data
Silver, Frederick H., 1949–
Biomaterials science and biocompatibility/Frederick H. Silver, David L. Christiansen.
p. cm.
Includes bibliographical references and index.
ISBN 0-387-98711-8 (hbk.: alk. paper)
1. Biomedical materials. 2. Biocompatibility. I. Christiansen, David L. II. Title.
R857.M3S553 1999
610'.28—dc21 98-51768

Printed on acid-free paper.

Production coordinated by Impressions Book and Journal Services, Inc., and managed by Francine McNeill; manufacturing supervised by Joe Quatela.
Typeset by Impressions Book and Journal Services, Inc., Madison, WI.
Printed and bound by Hamilton Printing Co., Rensselaer, NY.
Printed in the United States of America.

9 8 7 6 5 4 3 2 1

ISBN 0-387-98711-8 Springer-Verlag New York Berlin Heidelberg SPIN 10707743

Preface

Although many advances have been made in the 1990s concerning the biocompatibility of implants, we still are only beginning to understand how foreign materials affect biological systems. The purpose of this book is to introduce the study of biocompatibility of implants by integrating background material, including information on the structure and properties of individual biological macromolecules, macromolecular packing, histology of tissues and organs, immunology, wound healing, and pathobiology. Although each of these topics is extensive enough to warrant its own course, we attempt to provide enough information and continuity among these topics so that the graduate student or biomaterials scientist can get a solid overview of the relationships among these topics without having to read more than one text. The information contained in this text was developed from course notes covering graduate courses in the Joint

Biomedical Engineering Program between Rutgers University and the University of Medicine and Dentistry of New Jersey entitled "Biopolymers" and "Pathobiology" during the past 15 years.

It is our hope that this book will provide the reader with all the information necessary to understand the complexity of the biological reactions that are set into motion by implantation of a material or a device. We hope that this book will provide a framework for thinking about implant interactions with biological systems. Although the field of studying pathobiological responses to implants is still in its infancy, we are now more aware of acute and chronic conditions that generate inflammatory responses as a result of wear debris, activation of complement, and acute hypersensitivity. As we learn more concerning these responses, it is hoped that our ability to design implants will also improve. We encourage readers to send to us any suggestions of additional topics that they would like to see covered in our book.

Frederick H. Silver
David L. Christiansen
Piscataway, New Jersey

Contents

Preface v

1 Introduction to Biomaterials Science and Biocompatibility 1

Scope of Text 1
Structures of Soft and Hard Tissues 3
Properties of Tissues 16
Structure of Synthetic Materials 19
Properties of Synthetic Materials 21
Structures and Properties of Cells 22
Cellular and Tissue Responses to Materials 24
Summary 25

2 Introduction to Structure and Properties of Biological Tissues 27

Introduction to Principles of Stereochemistry 27
Basic Building Blocks 30
Stereochemistry of Polymer Chains 33
Primary and Secondary Structures of Biological Macromolecules 53
Higher-Order Structures 61
Structure of Extracellular Matrix Macromolecules 62
Cell Membrane Polymers 71
Other Polymeric Materials 78
Summary 83

3 Introduction to Structure and Properties of Polymers, Metals, and Ceramics 87

Introduction to Synthetic Materials 87
Polymer Structures 89
Polymer Mechanical Properties 95
Structure of Metals 106
Structure of Ceramics 114
Structure of Composites 117
Materials Degradation 119
Summary 120

4 Microscopic and Macroscopic Structure of Tissue 121

Introduction to Methods for Cellular and Tissue Analysis 121
Surface and Internal Linings 125
Conduit and Holding Structures 133
Parenchymal or Organ-Supporting Structures 139
Skeletal Structures 139
Summary 144

5 Determination of Physical Structure and Modeling 147

Introduction 147
Viscosity 148
Light Scattering 152
Quasi-Elastic Light Scattering 154
Ultracentrifugation 157
Electron Microscopy 160
Determination of Physical Parameters for Biological Macromolecules 161
Summary 162

6 Assembly of Biological Macromolecules 165

Introduction 165
Methods for Studying Self-Assembly Processes 169
Collagen Self-Assembly 174
Assembly of Cytoskeletal Components 176
Actin–Myosin Interaction 179
Fibrinogen 182
Summary 182

7 Mechanical Properties of Tissues 187

Introduction to Analysis of Tissue Mechanical Properties 187
Mechanical Properties of Collagenous Tissues 196
Mechanical Properties of Hard Tissue 207
Cellular Biomechanics 208
Summary 209

8 Pathobiology and Response to Tissue Injury 213

Introduction 213
Cellular Components 214
Cell Attachment, Proliferation, and Differentiation 223
Cellular Adaptation 228
Cell Injury 234
Summary 238

9 Wound Healing 241

Introduction 241
Biological Cascades Involved in Healing 242
Cells Involved in Wound Healing 250
Inflammatory and Immunological Aspects of Wound Healing 252
Cell Adhesion Molecules Involved in Wound Healing 261
Growth Factors Involved in Wound Healing 263
General Wound-Healing Process: Inflammatory Phase 264
Capsule Formation Around Implants 269
Cartilage Wound Healing 269
Tendon and Ligament Healing 272
Peripheral Nerve Repair 272
Hard Tissue Repair 273
Cardiovascular Wound Healing 274
Summary 275

10 Pathobiological Responses to Implants 279

Introduction 279
Pathobiological Complications with Silicone Implants 280
Inflammation Induced by Wear Particles 286
Complement Activation by Biomaterial Surfaces 289
Restenosis After Vascular Stenting 292
Thrombosis and Small-Diameter Vascular Grafts 296
Ectopic Mineralization of Implants 298
Summary 300

11 Tissue Engineering 305

Introduction to Tissue Engineering 305
Immunology of Cell and Tissue Transplantation 307
Role of Matrixes and Scaffolds in Cell Transplantation 310
Tissue Engineering of External Lining Structures 312
Skeletal Tissue Engineering 316
Tissue Engineering of Specialized Organs 320
Summary 321

12 Future Considerations 327

Introduction 327
New Directions in Tissue Engineering 328
Design of New Polymeric Materials 328
Design of Interfaces 328
Vascular Device Design 329
Summary 329

Index 331

1

Introduction to Biomaterials Science and Biocompatibility

Scope of Text

The field of biomaterials science dates back centuries to the ancient Greeks and Chinese, who used natural materials to ameliorate the effects of diseases. However, not until late in the twentieth century did the design and use of medical devices using synthetic and natural materials advance rapidly. The largely empirical problem-solving strategies, such as trial-and-error materials selection procedures, have evolved into a multidisciplinary field that requires in-depth knowledge of biochemistry, anatomy, structural biology, immunology, histology, pathobiology, engineering, and materials science.

The modern era in biomaterials science led to progress in the treatment of cardiovascular disease, thus extending the lives of patients with coronary artery disease, dissection or aneurysm of a major vessel, abnormal electrical conduction of the heart, or a weak heart. The first vascular prosthetics were introduced in the 1950s when parts of the aorta were replaced with vessels made of synthetic fabrics. At about the same time, the first pacemakers were introduced to prevent defibrillation. These advances resulted in decreased mortality from cardiovascular diseases and increased collaboration among scientists in the fields of materials sciences, engineering, and medicine.

It was also about 1950 when researchers discovered that soft and hard tissues are composed of collagen, a rodlike molecule, which, together with hydroxyapatite, is responsible for maintaining tissue shape and providing rigidity to the skeleton. Since that time, our knowledge of vertebrate and human tissue structure has exploded, yielding an information base that can require a lifetime to master. In this information base is contained the product of millions of years of evolutionary design. Studying materials such as collagen in the skin of prehistoric woolly mammoths, cuticles of worms, and the skins of fish and other invertebrates helps us understand collagens found in human tissue. This wealth of information in biological structure and architecture serves as nature's guide to artificial implant design.

In this book, we will attempt to understand biomaterials and their interaction with tissues by studying normal pathobiological processes. This approach relies on defining fundamental principles that relate the structure of natural tissues and their interaction with foreign materials. To do this, it is necessary to first define what is foreign and how it is recognized as foreign by the host. In addition, we will explain what goes wrong when the host treats its own tissues as foreign.

Perhaps the most exciting part of biomaterials science is the complexity of macromolecular structures that have been designed by nature to perform specific functions. The molecular structures are rich in interesting features, such as the ability of large molecules to change their sizes and shapes as a result of applied loads or chemical signals. These structures have evolved ordered packing patterns that not only have short-range order but form repeated patterns that continue for distances as long as meters. The tendons in the hand that control the movement of the fingers are controlled by insertions into the biceps muscle in the arm. The tendons are composed of collagen fibrils and bundles of fibrils that are composed of molecules only 0.00003 m long. The packing of millions of these molecules into a regular array allows for the precise movement of the fingers. It is this precise movement and the macromolecular packing patterns that have evolved in nature that are of interest to us in this book. As engineers, we are interested in how much force this precise array of molecules can generate without failing. This knowledge can then be applied to the design of synthetic materials to replace those designed by nature.

This is an exciting area of study. We take structures apart, analyze their components, and crash test them—we pull and pound on these structures to see how strong they are. Much of what we learn in biomaterials science is a direct result of our inquisitiveness. It is essential to *think* and *feel* biological structures to get a better idea of what specific physical parameters mean and how they might function in biological settings. A good example of applying these physical principles is the question of how large macromolecular chains deform. Like a twisted rubber band, macromolecules extend by untwisting when a tensile force is applied to the ends. You can perform this experiment at home by twisting a rubber band and pulling on the ends; a twisted rubber band uncoils when a tensile force is applied at the ends. What happens when a tensile load is applied to a macromolecule is a little more complex; however, the response observed is still dictated by the physical principle that deformation occurs at the least rigid element. This will be discussed later when we address the mechanical properties of tissues.

Structures of Soft and Hard Tissues

The biomaterials scientist is responsible for developing replacement for tissues and organs and must be familiar with the structure and function of tissues and organs to be replaced. From a simple, engineering point of view, mammalian tissues are classified as either soft or nonmineralized and hard or mineralized; the mechanical properties of soft and hard tissues are widely divergent. More accurately we can classify body tissues as (1) surface and internal lining tissues, (2) conduit or fluid transport tissues, (3) parenchymal or organ-supporting structures, and (4) skeletal structures.

Surface and internal lining structures are similar; they are composed of an epithelial layer in contact with the external environment or a mesothelial layer in contact with internal organs that are supported by connective tissue containing collagen, proteoglycans, elastin, glycoproteins, cells, and water. These structures are found at external or internal interfaces and are composed of layers of cells and extracellular matrix and typically, with the exception of the cornea, contain blood vessels for nutrition. Examples of these structures are given in Table 1.1 and include external linings such as found in cornea, skin, oral mucosa, vagina, and uterus that provide mechanical, chemical, and microbiological barriers. These linings keep external substances out of the body, and internal

Table 1.1 Surface and internal lining structures

Structure	Composition of lining	Function
Alveoli	Squamous epithelium	Allows oxygen transport
Cornea	Stratified squamous epithelium	Protects eye from injury
Mouth	Stratified squamous epithelium	Protects oral tissues
Peritoneum	Mesothelial cells	Protects stomach organs
Pleura	Mesothelial cells	Protects chest organs
Skin	Stratified squamous epithelium	Protects body surface
Uterus	Columnar epithelium	Protects internal surfaces
Vagina	Stratified squamous epithelium	Protects internal surfaces

Table 1.2 Conduit and holding structures

Structure	Composition	Function
Bladder	Mucosa, muscularis, serosa	Holds urine
Blood vessels	Intima, media, adventitia	Transports blood
Bronchiole	Mucosa, smooth muscle, adventitia	Distributes air
Bronchus	Mucosa, submucosa, adventitia	Distributes air
Esophagus	Mucosa, submucosa, muscularis, fibrosa	Collect air
Large intestine (colon)	Mucosa, submucosa, muscularis externa, serosa	Transports food
Rectum	Mucosa, submucosa, muscularis externa, serosa	Transports waste
Stomach	Mucosa, submucosa, muscularis serosa	Hydrolyzes food
Small intestine (duodenum)	Mucosa, submucosa, muscularis externa, serosa	Adsorbs food
Trachea	Mucosa, submucosa, fibrocartilage, fibrosa	Distributes air
Ureter	Mucosa, muscularis, fibrosa	Transports liquid waste

Table 1.3 Dental and skeletal structures

Structure	Composition	Function
Articular cartilage	Superficial, intermediate, deep zones	Absorbs shock
Compact bone	Circumferential, concentric lamellar bone	Prevents bending of long bone
Cruciate ligaments	Collagen, proteoglycans	Stabilizes knee
Intervertebral disc	Nuclear pulposis, annular fibrosa	Supports spine
Muscular tissue	Smooth muscle	Constricts tubular walls
	Skeletal muscle	Allows for locomotion
Periodontal ligament	Collagen, proteoglycans	Connects tooth to bone
Spongy bone	Circumferential, concentric lamellar bone with cavities	Stores blood cell precursors

linings such as the pleural and peritoneal membranes protect organs in the thoracic and abdominal cavities from injury. A second classification of tissues includes conduit and holding structures (Table 1.2), which are tubes and containers that are characteristically composed of three layers. For example, blood vessels contain an intimal or cell layer in contact with flowing blood, a muscular or medial layer, and an adventitia that blends into the surrounding tissues. Other tubular structures are similar in that they consist of an internal cellular layer (mucosa), surrounded by a muscular layer (muscularis), and an external connective-tissue-containing layer (serosa).

The final classification of tissues is dental and skeletal tissues (Table 1.3). These tissues include oral tissues and joint and spine structures. Organ-supporting structures in the parenchyma of different tissues, such as the nephron in the kidney, are not discussed in detail in this book; however, their physiological function is described in the next section.

Structure of Surface and Internal Lining Tissues

Epithelial cells line the surface of external and internal tissues in mammals. They are derived from two of the primary layers in the embryo, the *ectoderm* and the *endoderm* (Figure 1.1). The epidermis of the skin and the epithelia of the cornea that together cover the external surfaces of the body develop from the ectoderm. The glandular appendages of the skin, including the sebaceous glands, which

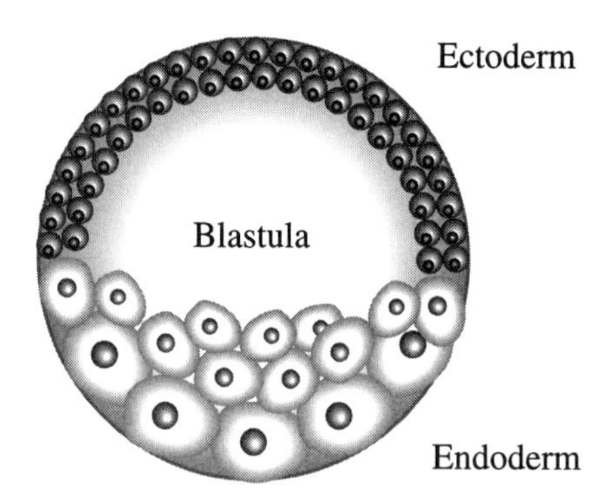

Figure 1.1 Embryonic development. Diagram showing the derivation of epithelial cells from the ectoderm and the endoderm in the embryo. The epidermis of the skin and the epithelia of the cornea develop from the ectoderm. Epithelium of the digestive system derives from the endoderm.

secrete an oily substance, and the mammary glands, are formed by folding and proliferation of the outer epithelia. The digestive system is lined by epithelia that derive from the endoderm, and its associated glands, including liver, pancreas, stomach, and intestines, arise by folding and outgrowth of primitive gut tissue. In addition to the epithelial structures that develop from the ectoderm and endoderm are several internal lining layers that are composed of epithelium derived from mesoderm. Examples include the epithelia of the kidney and the male and female reproductive tracts.

The linings of the blood and lymph vessels are all derivatives of the mesenchyme. The epithelial cell linings of blood and lymph vessels are usually referred to as *endothelium*, and the linings of the body cavities (peritoneum and pleura) are referred to as *mesothelium*. However, all these cells are considered epithelia.

All epithelia sit on connective tissue that contains collagen, the major structural protein; proteoglycans, sugar polymers that maintain tissue hydration; and glycoproteins, which act as connecting elements between epithelia and other cells and the extracellular matrix.

Types of Epithelia

Epithelia are classified according to the number of cell layers and the cell shape. A surface or lining with one cell layer is *simple* and one with two or more layers is *stratified*. The outermost or superficial layer of cells can be described as squamous, cuboidal, or columnar. Squamous epithelia are characterized by elongation of the cell cytoplasm along the surface on which the cells lie. Cell shape is described as cuboidal or columnar; cuboidal cells are cubelike, and columnar cells are approximated by a right parallelepiped (six-sided rectangle). In the case of skin, keratin accumulates in the cell cytoplasm, and the cells in the outermost layer are devitalized.

Examples of epithelia include simple squamous, simple cuboidal, simple columnar, stratified squamous, stratified columnar, pseudostratified columnar, and transitional cells (Figure 1.2). Simple squamous epithelia consist of thin platelike cells arranged in a sheet with each cell adhered closely to one another. From a surface view, individual cells have a polygonal or hexagonal outline, and each contains a nucleus. The side view shows a thin cell profile with an elongated nucleus. Examples of simple squamous epithelia include the mesothelia that form the surface of the linings of the pleura and peritoneum and the endothelia that line the blood vessels and the lymphatics.

Simple cuboidal epithelia from a top view appear as hexagonal polygons that form a continuous sheet; from the side view, they appear as a line of square or rectangular profiles. Secreting epithelia of many glands can be placed in this class.

Simple columnar epithelia look from the top view much like other epithelia except the polygonal outlines are smaller in size. From a side view, the surface looks similar to that of the epithelium except that the cells are a little higher and the aspect ratio (height/width ratio) is greater. In addition, the nuclei are all at about the same vertical level. Simple columnar epithelia line the surface of the digestive tract from the stomach to the anus and are also common in the excretory ducts of many glands. Ciliated simple columnar epithelia are found in the uterus, in the small bronchi of the lung, and in the central canal of the spinal cord. Cilia are motile cell extensions that propel fluid or mucous films over the surface of the epithelia by dynamic processes.

Stratified squamous epithelia consist of a thick sheet containing more than one cell layer with cells varying in shape; from the top, they are elongated along the surface to the bottom, where they are rounded. The bottom cell layer rests on a membrane, the *basement membrane.* These epithelia are found in the epidermis (the top layer of skin), esophagus (windpipe), cornea, vagina, and part of the female urethra. When stratified squamous epithelia are found at the outer surface of the body, the cells are keratinized.

Stratified columnar epithelia consist of more than one layer of columnar cells and are rarely observed. Pseudostratified columnar epithelia consist of

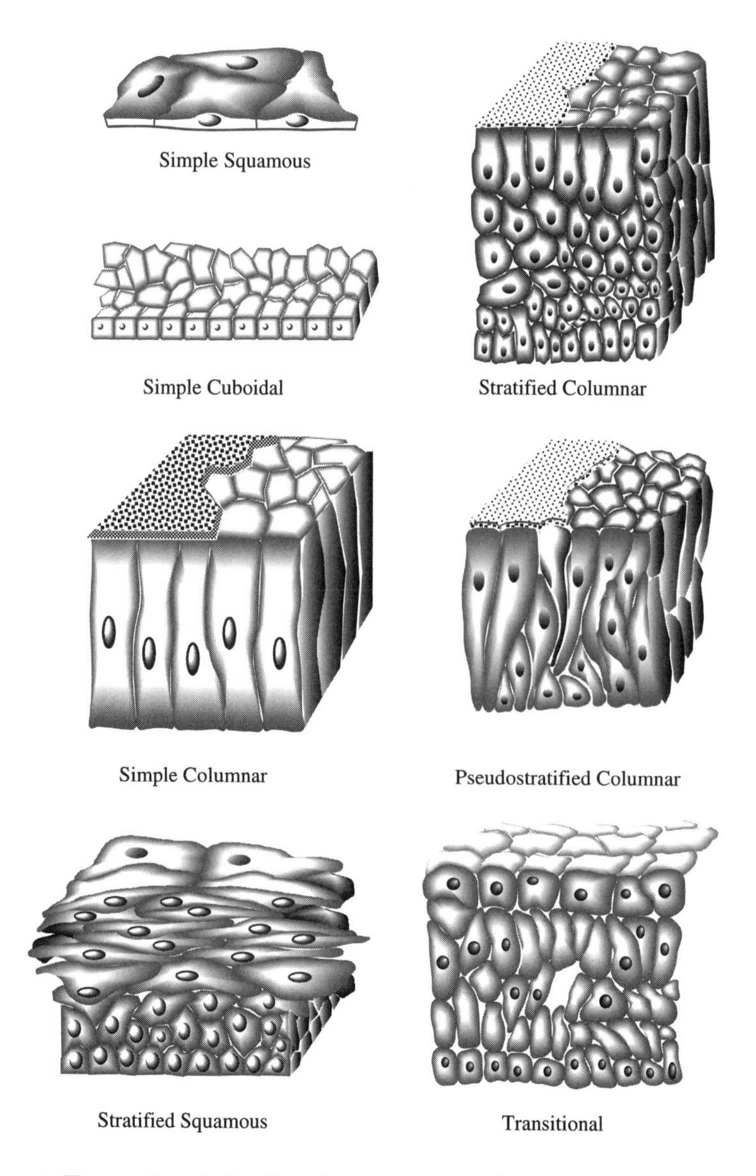

Figure 1.2 Types of epithelia. Simple squamous epithelia consist of thin, platelike cells arranged in a sheet. From the top view, each cell appears either polygonal or hexagonal in shape. Simple cubodial epithelia appear from the top as hexagonal polygons and are rectangular from the side. Simple columnar epithelia from the front view have a higher length/width ratio than do cuboidal cells. Stratified epithelia are found in layers, whereas pseudostratified epithelia are found in groups, with some of the cells in layers and others not. Transitional epithelia, as the name indicates, are structurally somewhere between cuboidal and columnar cells.

columnar cells that are on top of a basement membrane; however, not all the cells reach the surface. The respiratory tract is lined with ciliated pseudostratified epithelia.

Connective Tissue

Connective tissue is the primary structural tissue in vertebrates and serves to maintain tissue shape (skin), transmit and absorb loads (tendons and ligaments), prevent premature mechanical failure (skin and blood vessels), partition cells and tissues into functional units (fascia), and act as a scaffold that defines tissue and organ architecture (parenchyma). Connective tissue is composed of 19 types of collagen (primarily the fibril-forming collagens), at least three categories of proteoglycans, elastic fibers, cells, cell attachment factors, water, and ions. The material outside of the cell is the *extracellular matrix*, which is the major component of connective tissue.

The role of each of these components in the behavior of connective tissue is complex. Collagen fibers are the elements that provide tensile strength and limit creep of tissues under constant loads. Elastic fibers provide recovery to skin and cardiovascular tissue under the constant mechanical loads. Proteoglycans are found in high concentrations in tissues such as cartilage that undergo large compressive forces during locomotion. Cells in these tissues act as force transducers that sense shear loading and convert it into chemical stimulators that signal for the production of additional matrix components. Cell attachment factors are important in providing continuity between the cell cytoskeleton and the extracellular matrix.

Collagens

The collagens are a family of proteins that are the major components of vertebrate tissues. To date, at least 19 different types of vertebrate collagens have been identified. Additional invertebrate collagens have been isolated, but in this book we will not discuss nonvertebrate tissues. The common structural feature found in all collagens is a triple helix that consists of three left-handed helixes that are wound into a right-handed triple helix. Individual α-chains, which are the basic units of collagen, contain one or more polypeptide sequences (Gly-X-Y) that form the triple helix with one or more nontriple helical modules (Figure 1.3). The α-chains vary in size from 600 to 3,000 amino acids and have been classified into fibrillar, nonfibrillar, and novel collagens. The fibrillar collagens include types I, II, III, V, and XI, which form cross-striated fibrils, and all share a triple helical region containing about 1,000 amino acids per chain, which

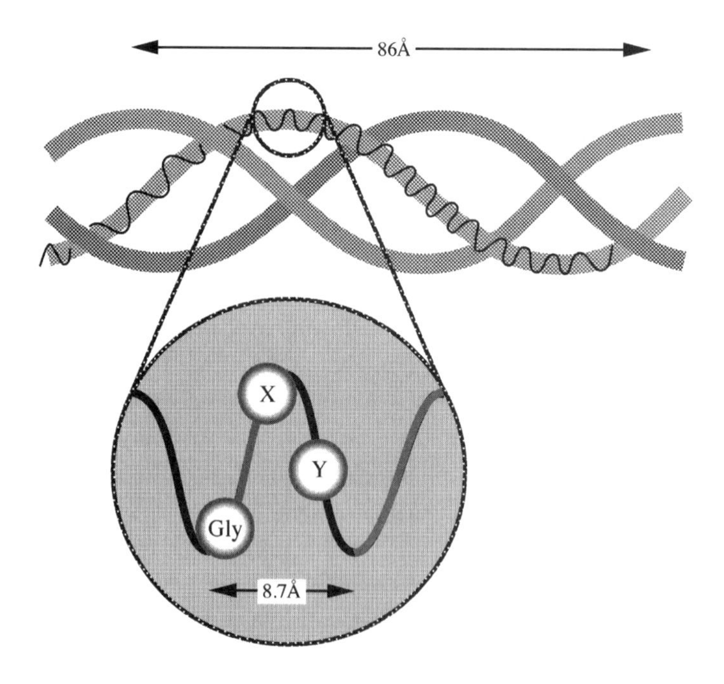

Figure 1.3 Collagen triple helix. Diagram showing section of collagen triple helix that in fibrillar collagens goes on for about 300 nm (3,000 Å) and in nonfibrillar collagens is connected to other modules. The triple helix requires glycine to be present every third residue (Gly-X-Y) because there is no room for any other amino acid residue at the center of the superhelix. The X and Y positions are frequently occupied by proline and hydroxyproline because these residues stabilize the structure.

has a length of about 300 nm (Figure 1.4) Nonfibrillar collagens may associate with fibrillar collagens or form separate networks of microfibrils. These collagens contain triple helical segments of varying lengths interrupted by sequences containing larger segments of noncollagenous sequences. The noncollagenous sequences include modules containing sequences found in von Willebrand factor (binds to and protects factor VIII, which is necessary for blood coagulation), collagen type IV (found in basement membranes), and fibronectin (a cell-adhesive glycoprotein). The novel collagens are similar to the nonfibril-forming collagens because they consist of triple helical regions separated by nontriple helical regions. The discovery of novel collagens is based on preliminary classification derived from complementary DNA sequencing, and they have not been characterized in their tissue forms.

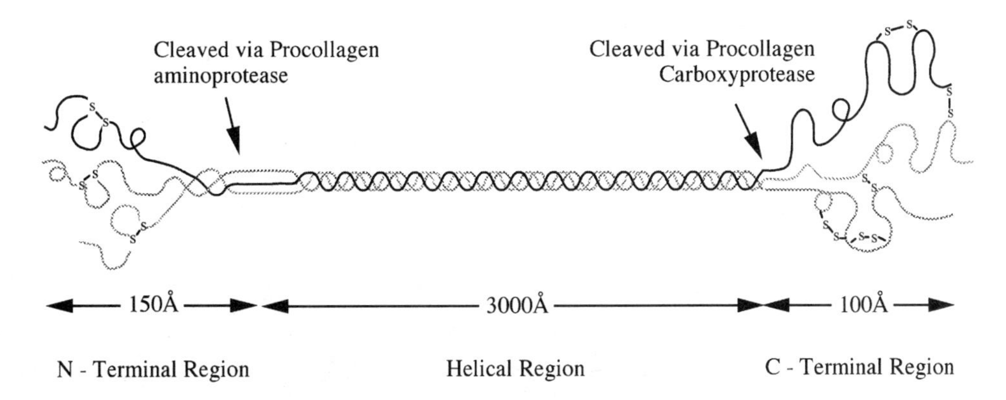

Figure 1.4 Diagram of procollagen molecule. Collagen is formed from procollagen molecules that are synthesized on the rough endoplasmic reticulum found within the cell. The N-terminal and C-terminal nontriple helical ends are removed to make striated fibers in the extracellular matrix. Fibrillar collagens have a triple helix that is about 300 nm (3,000 Å) long.

Collagen type I contains two α_1(I) chains and one α_2(I) chain and is the major component of cross-striated fibrils in tissues. It is found associated with types V and III collagens, with type V as the core and type III as an outer layer. Cartilage contains striated fibrils containing types II and XI collagens.

Nonfibrillar collagens form microfibrillar meshes (type VI), short filaments (type VII), or sheetlike structures (types IV, VIII, and X). Type VI forms microfibrils by end-to-end assembly with 100-nm periodicity (repeat pattern) with alternating globular and filamentous structures. Type VII forms anchoring fibrils that connect the basement membrane to underlying structures in the dermal component of skin. Type IV collagen forms networks by dimerization of type IV collagen molecules via the C-terminal globular regions and by the formation of tetramers via parallel and antiparallel alignment along the N-terminal triple helical segment, which leads to the formation of infinite networks.

All striated collagens are cross-linked into networks via formation of lysine-, histidine-, and hydroxylysine-derived cross-links that are catalyzed by lysyl oxidase. These cross-links are fundamental to collagen's role in transmitting and absorbing force.

Proteoglycans

Proteoglycans are a heterogeneous family of macromolecules found in most tissues. They are located intracellularly, at the cell surface, and in the extracellular

Table 1.4 Repeat units found in glycosaminoglycans

Glycosaminoglycan	Hexuronic acid	Hexosamine	Sulfate
Hyaluronan	D-Glucuronic acid	D-Glucosamine	—
Chondroitin-4-sulfate	D-Glucuronic acid	D-Galactosamine	O-Sulfate
Chondroitin-6-sulfate	D-Glucuronic acid	D-Galactosamine	O-Sulfate
Dermatan sulfate	D-Glucuronic acid or L-iduronic acid	D-Galactosamine	O-Sulfate
Keratan sulfate	D-Galactosamine	D-Glucosamine	O-Sulfate
Heparan sulfate	D-Glucuronic acid or L-iduronic acid	D-Glucosamine	O-Sulfate or *N*-sulfate
Heparin	D-Glucuronic acid or L-iduronic acid	D-Glucosamine	O-Sulfate or *N*-sulfate

matrix. Proteoglycans are complex macromolecules that consist of a core protein to which one or more glycosaminoglycan (GAG) side chains are attached. The one exception is hyaluronan (HA), which is found free in the extracellular matrix. The GAG side chains are polymers of an amino sugar and a uronic acid, as shown in Table 1.4. The GAGs found in proteoglycans include HA, chondroitin sulfate (CS), dermatan sulfate (DS), heparin (H), heparan sulfate (HS), and keratan sulfate (KS). Proteoglycans are found extracellularly and intracellularly in connective tissue. Some proteoglycans carry one or more than one type of GAG. For example, aggrecan, a cartilage proteoglycan, contains CS and KS, and syndecan, a cell membrane proteoglycan, contains HS and CS.

Specialized proteoglycans, including large (aggrecan and versican), small (biglycan, decorin, and fibromodulin), and basement membrane (fibroglycan, glypican, neurocan, and syndecan) proteoglycans, have been identified. Aggrecan, a large CS-containing proteoglycan that also contains KS, forms aggregates with HA and is believed to provide resistance to compressive deformation in weight-bearing joints. This molecule is composed of three domains: G1, G2, and G3. Between regions G2 and G3 is the GAG attachment region to the core protein. The G2 domain is the site for attachment of link protein that provides attachment to HA. As many as 100 aggrecan and link protein molecules can bind to a single molecule of HA, forming a complex with a molecular weight as high as 2×10^8. Each aggrecan molecule has a molecular weight of about 10^6. Versican (synthesized by fibroblasts) has a domain structure similar to aggrecan, although some differences exist, and it can also bind to hyaluronan.

Biglycan, decorin, and fibromodulin are small proteoglycans that contain a leucine-rich region. Their molecular weights are 70,000, 100,000, and 50,000, respectively. They all have similar core proteins with molecular weights of about

37,000; however, the number of GAG side chains varies from one (decorin) to four (fibromodulin). Biglycan contains two short GAG chains. The distribution of biglycan has led to the proposition that this proteoglycan may have a more general function related to cell regulation. In contrast, decorin and fibromodulin, which bind to the surface of collagen fibrils, are believed to be involved in possible regulation of fibril diameters.

The final classification of proteoglycans is derived from their association with cell membranes through a hydrophobic segment of their core proteins. They are associated with hepatocytes, epithelial cells, fibroblasts, marrow stromal cells, glial precursor cells, and Schwann cells. Syndecan is found in adult epithelial tissues and has been studied most extensively. It behaves as a matrix receptor that can interact, via its HS side chains, to a number of extracellular matrix molecules, including fibrillar collagens (types I, III, and V), fibronectin, and thrombospondin, to link the cell cytoskeleton to the extracellular matrix. It has been proposed that syndecan anchors epithelial cells to the matrix and stabilizes the formation of epithelial sheets. The role of syndecan is similar to that of other HS chains that bind to growth factors (promoting migration and proliferation of cells) and cytokines (releasing effector molecules from white blood cells); as such, they can modulate changes in cell migration and proliferation. Recently, it has been proposed that proteoglycans form complexes with growth factors that regulate their activity. Fibroblast growth factor binds to HS proteoglycan, platelet factor 4 binds to CS proteoglycan, and transforming growth factor β binds to decorin.

Elastic Fibers

Elastic fibers are the elements that provide reversible recovery of tissue shape during small deformations. These fibers provide retraction of skin stretched during movement as well as reversible expansion of aortic tissue when the aortic valve opens and closes. This component is observed after staining with special stains such as orcein. It consists of microfibrillar and amorphous components. The microfibrillar component consists of a ring of microfibrils 10 to 12 nm in diameter composed of microfibrillar proteins (one that has been identified is fibrillin) and is surrounded by a protein core composed of elastin. The microfibrils are believed to serve as sites for elastin deposition.

Cell Attachment Factors and Cell Surface Proteins

Entactin, fibronectin, laminin, tenascin, thrombospondin, and vitronectin are several of the noncollagenous extracellular proteins that play important roles in many cell surface interactions. Entactin binds tightly to laminin and mediates cell adhesion; fibronectin (found in plasma and extracellular matrix) helps mediate cell adhesion, cell migration, and wound healing; laminin (found in basement membranes) has been implicated in cell adhesion, migration, and differentiation; tenascin (found in remodeling tissues and during embryonic development) mediates both adhesive and repulsive interactions and binds to certain proteoglycans and fibronectin; thrombospondin (a product released from activated platelets) mediates or inhibits cell proliferation of smooth muscle cells; and vitronectin (found in plasma and extracellular matrix) mediates cell adhesion and protects cells from destruction by activated complement. Cells are bound to extracellular matrix collagen by one or more cell attachment factors that have specific receptors for extracellular matrix molecules such as collagen and fibrinogen and other cell surface receptors.

Integrins are a family of cell surface proteins that mediate cell-to-extracellular matrix adhesion and cell-to-cell adhesion. In the classic model, proteins of the extracellular matrix, including collagen, laminin, and fibronectin, bind to integrins in the extracellular region of a plasma membrane receptor. Integrins are a superfamily of integral membrane proteins that are made up of α and β subunits. Many integrins have affinity for the amino acid sequence arginine-glycine-aspartic acid (Arg-Gly-Asp) in the extracellular ligand portion. The integrins have small cytoplasmic regions that bind to elements of the cytoskeleton (actin filaments) and to actin-binding proteins. Specific examples of cell adhesion via integrins include adhesion of platelets and epithelial cells to type I collagen fibrils via the $\alpha_2\beta_1$ receptor.

Specialized Organs

Specialized organs in the body include the blood and lymph system, thymus, lymph nodes, spleen, pituitary gland, thyroid, adrenal glands, parathyroid, skin, oral cavity, teeth, esophagus and stomach, intestines, liver, pancreas, respiratory system, urinary system, reproductive system, mammary gland, eye, and ear. Each specialized organ consists of an extracellular matrix either in the form of tubes or sheets and a parenchyma or scaffold that defines the tissue architecture. For example, the thymus is the site in which stem cells differentiate into lymphocytes that on contact with antigens further differentiate into T cells involved in cell-mediated immunity. Table 1.5 summarizes the functions of these specialized organs.

Table 1.5 Functions of specialized organs

Organ	Function
Blood vessels	Distribute blood to peripheral tissues and organs
Esophagus	Conveys food to stomach
Eye	Transduces light into electrical signals
Intestines	Absorb digested food into blood and lymphatics
Liver	Metabolize or transform absorbed products and returns them to blood
Lymphatics	Recirculate extravascular fluid to cardiovascular system
Lymph nodes	Filters lymphatic fluid and initiates immune reaction to antigens
Mammary gland	Provides for release of milk protein
Pancreas	Elaborates digestive juice and controls carbohydrate metabolism
Reproductive system	Provides for release and movement of cells and embryo
Respiratory system	Collects and absorbs oxygen and releases carbon dioxide
Spleen	Filters of blood particulates and dead cells; triggers immune response to blood-borne antigens
Stomach	Stores and digests food
Thymus	Serves as a supply of stem cells that differentiate into T cells
Urinary system	Filters waste products from blood

Muscular Tissue

Muscular tissue is responsible for locomotion and for movements of the various parts of the body with respect to each other. The fundamental unit that provides for contractility is the muscle fiber. There are two types of muscle fibers: smooth and striated. Striated muscle exhibits regularly spaced transverse bands along the length of the fiber. Smooth muscle is composed of individual cellular units that are enervated by the autonomic nervous system and are not subject to voluntary control. Striated muscle is composed of cardiac and skeletal muscle. Skeletal muscle fibers are enervated by a system of nerves controlled by the central nervous system. Cardiac muscle is made up of individual fibers and is involuntarily controlled. Muscle fibers are cells that contain filaments of actin, myosin, and other proteins responsible for muscle contraction. In striated muscle, thin filaments of actin interact with thick filaments of myosin, causing contraction and sliding of actin filaments over myosin. This process is associated with the conversion of adenosine triphosphate to adenosine diphosphate and provides the contractile force necessary to cause movement of a joint.

Structure of Hard Tissue

The structure of hard tissue is similar to that of soft tissue. Hard tissue contains collagen (type I) that is formed into lamellar bone or osteonic bone. Lamellar bone consists of sheets of aligned collagen fibrils separated by plates of a mineral containing calcium, phosphate, and hydroxyl groups (hydroxyapatite). Osteonic bone consists of collagen fibrils containing hydroxyapatite that are rolled into concentric cylinders. Long bone is a combination of lamellar and osteonic bone and contains regions with little free space (cortical bone) or a lot of free space (trabecular bone). The center of long bone is filled with trabecular bone, and the pores are filled with cells containing bone marrow that are precursors of the cells that circulate in the blood. Bone contains a number of specialized factors involved in bone mineralization.

Properties of Tissues

The science of biomaterials concerns the physical properties of tissues. The optical and mechanical properties of replacement tissues must be matched, and the physical aspects of the design of engineered tissues must be considered. This section discusses the mechanical and optical properties of tissues and replacement materials.

Although collagen, proteoglycans, water, and cells make up the bulk of the human body, the arrangement of these components and the addition of mineral in the form of hydroxyapatite markedly affect the physical properties. The optical and mechanical properties of tissues vary from soft and transparent (cornea) to hard and opaque (bone). The variations in these materials have evolved over millions of years and can be simply described.

From a physical point of view, a material is transparent if the size of the crystalline unit within the material is small (5%) compared to the wavelength of light. Plexiglass is clear because it is an amorphous (structureless) material, whereas crystalline polyethylene is opaque because the crystallite size is greater than about 30 nm. In the human body, the cornea is transparent because the interfibrillar distance is similar to the fibril diameter, and the fibrils are regularly spaced in three dimensions. What is amazing is that very similar collagen fibrils are found in tendon, but they are opaque because the large fibrils are closely packed together. Even though the cornea is transparent, the regular arrangement of the collagen fibrils provides toughness to the material in the plane of the cornea. Typically, mechanical injury to the cornea other than scrapes (which are painful because the epithelia slough off) involves penetration of the cornea perpendicular to the plane of the collagen fibrils.

The skin is similar to the cornea because the collagen fibrils are found within a plane almost parallel to the surface. This is required so that skin can stretch in either direction within the plane to dissipate energy and maintain its shape. It has been estimated that skin must stretch about 100% to accommodate joint movement. Without this flexibility, the skin would permanently deform over the joints, leading to tearing and eventual scarring. Third-degree burns sometimes lead to abnormal collagen organization and mechanical properties. Scar tissue and contractures form around joints, indicating the significance of normal tissue architecture and skin mobility. In Chapters 4 and 7 we go into more depth concerning the relationship between connective tissue architecture and mechanical properties.

The mechanical properties of tissues depend on a number of factors. Because tissues are viscoelastic, their properties are somewhere between those of a viscous liquid and those of an elastic solid. The consequence of this is that the force per unit area required to deform a tissue to a fixed extension varies depending on how fast the deformation is applied. Another way of looking at this is that the mechanical properties are time dependent or strain-rate dependent. Figure 1.5 is a plot of force per unit area versus strain as the change in length over original length for skin, bone, and tendon. Readers who have studied mechanics will note that strain is incorrectly used in the biomedical literature, and we use it here only to be consistent with the literature. Engineering strain is defined as the logarithm of the change in length over the original length.

The stress at failure, *ultimate tensile strength* (UTS), varies from about 5 MPa to 150 MPa. This range is extraordinary, because use of the same collagen fibers that are present in skin and tendon with the addition of hydroxyapatite drastically changes the mechanical properties. The difference in the UTS between skin and tendon is primarily a result of the difference in orientation of the col-

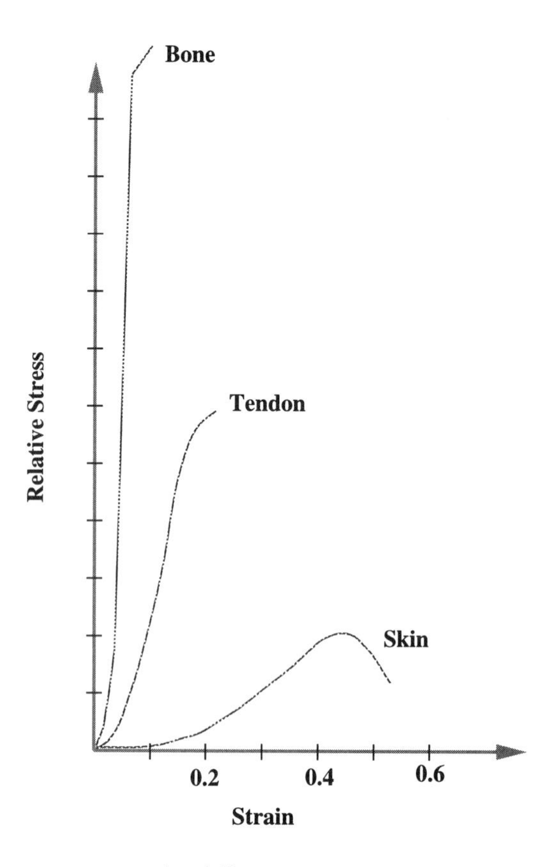

Figure 1.5 Stress–strain curve for different tissues. Plot of stress (force per unit area) versus strain (change in length divided by original length) for skin, tendon, and bone. The following definitions are used to refer to the mechanical properties: stress at failure is called the *ultimate tensile strength* (UTS), slope of the stress–strain curve is called the *modulus*, and the strain at failure is called the *ultimate strain*.

lagen fibers. However, in addition to the high UTS values is the very low deformability (strain at failure) of bone compared to skin and tendon. The strain at failure drops from about 100% for skin to about 1% for bone. The resistance to deformation (the slope of the stress–strain curve) is called the *modulus* or *tangent* modulus. The modulus for tissue depends on the extent of deformation, a property of tissue viscoelasticity, and varies from about 50 MPa (1 MPa equals 10^6 N/m^2) for skin to about 20 GPa (1 GPa equals 10^9 N/m^2) for bone.

An important concept to remember is that mechanical properties change drastically depending on the orientation of collagen fibers in tissue and on the amount of hydroxyapatite. Ectopic calcification, deposition of hydroxyapatite where it normally is not found, is a problem associated with cardiovascular implants, especially heart valves, because it drastically changes the mechanical properties of valvular tissue. It is important to understand the relationship among composition, structure, and mechanical properties; in many cases, this relationship will determine the success of a device being designed. The mechanical properties of a device being designed must be matched with those of the surrounding tissues, especially for cardiovascular devices. Failure to do so will result in stress concentrations, excessive tissue production at the interface (intimal hyperplasia), and device failure.

Structure of Synthetic Materials

The previous sections discussed the components of tissues, including cells, macromolecules, water, ions, and minerals. In this section, the components of implant materials will be considered. Implants are largely formed from macromolecules, also called polymers, that are synthesized from raw materials such as oil. These polymers include polyethylene and poly(tetrafluoroethylene) that are used in orthopedic and vascular implants. Throughout this book, we will refer to a number of polymers used in implants; in addition to polyethylene and poly(tetrafluoroethylene), polyurethanes and silicones are the most widely used. Other macromolecules from natural sources are also used in medicine, such as hyaluronic acid, synthetic peptides such as the attachment peptide Arg-Gly-Asp, growth factors either synthetically derived or made in bacteria, derivatives of cellulose such as carboxymethylcellulose, hydroxypropylmethylcellulose, chitin and its derivatives, alginates, and a variety of others.

The polymers themselves are usually not detrimental to the surrounding tissues; however, low molecular weight compounds added into the polymers as well as breakdown products of the implants can cause allergic and inflammatory responses. Inflammation in the form of redness, swelling, and pain is associated with the release of breakdown products that are either hydrolytically or enzymatically liberated from polymers. Fortunately, implants composed of polyethylene, poly(ethylene terephthalate), poly(tetrafluoroethylene), silicone, and stable polyurethane are well tolerated by the human body. However, because a certain percentage of patients will have an allergic reaction to any implant, depending on their immune systems, it is important to understand the potential systemic and local complications associated with all implant materials.

Natural polymers such as polypeptide growth factors or cells that have been transplanted or passaged in tissue culture can be somewhat more complex in terms of their response. Most foreign proteins will stimulate inflammatory and immune responses that will lead to hydrolysis of the implant and a cytotoxic response to cellular material. However, this response can be minimized by chemical cross-linking of foreign proteins and encapsulation of foreign cells. Ultimately, it is the balance between activation of the biological responses by the implant and performance of the intended function that will determine whether the implant is useful. However, much attention must be placed on evaluation of the response and function in humans, because this cannot be effectively determined from animal studies.

A number of synthetic polymers have been developed that are dissolved either by hydrolysis or by enzymatic action. These include poly(lactic acid), poly(glycolic acid), and other derivatives. In addition, other attempts have been directed at use of polymers that mimic proteins and other naturally occurring polymers. These materials have found application in biodegradable suture material and in bone pins where it is desired to transmit the stress to host tissue after a repair response. The response to these materials is transient and is associated with the degradation of the polymer chain. However, the release of lactic acid or other physiologically active material from a biodegradable polymer may limit their utility.

Metallic implants have received much attention as a result of Charnley's pioneering work with joint replacement using stainless steel total joint replacements. Stainless steels are used in vascular stents and in orthopedic and dental applications, where wire is used to stabilize a device. Cobalt base alloys and titanium base alloys are used in orthopedic implants and have better corrosion resistance than do stainless steels. The compatibility of metallic implants is generally related to their stability, like polymers, although there are individuals who are allergic to ions released from these implants. The major pathological response to metallic implants is associated with wear particles from the implant surface, leading to inflammation and eventual implant failure as a result of loosening. Implant failure can occur in the absence of implant wear debris as a result of the mechanical mismatch between bone and metal or poor force transfer between the implant and the bone. In this case, failure of the bone at the interface results in inflammation and eventual implant failure. This is an example of how mismatching of the mechanical properties between an implant and tissue result in implant failure. In the case of orthopedic implants, implant failure is often the result of a discontinuity across the implant-to-tissue interface.

Ceramic materials used in implants include aluminum oxide, silica glass, and the carbons. These materials are used in cochlear implants (silica glass), dental and orthopedic implants (aluminum oxide), and heart valves (pyrolytic carbons). The advantage of these materials is that they are very stiff and chemically stable. The disadvantage is that they tend to be brittle and hard to process.

The last group of materials used in implants are referred to as composites. They traditionally refer to a group of materials that are composed of stiff fibers in a polymeric matrix that is ductile. Composite materials have been envisioned to replace metallic implants where stiffness in combination with light weight is required.

Properties of Synthetic Materials

The properties of implant materials are quite diverse and reflect the diversity of structures. Most polymers used in devices are in a crystalline form; therefore, their mechanical properties reflect those of crystalline polymers in general. The general behavior of crystalline polymers is that below their crystalline melting temperatures their stiffness ranges from 1 GPa to about 10 GPa; their UTS values range from about 100 to 200 MPa. Typical stress-to-strain curves for crystalline polymers are shown in Figure 1.6

In comparison to polymers, metallic implants are stiffer and stronger but less deformable. Typical stress–strain curves for stainless steel, cobalt-based, and titanium-based implants are shown in Figure 1.7. Stainless steels have UTS values between 480 and 860 MPa and ultimate strains between 12 and 40%. In metals, the amount of stress required to cause the material to deform by 2% is referred to as the *yield strength at 2% offset,* and it is used for characterization. The yield at 2% for stainless steel varies from 170 to 690 MPa, depending on if they are processed by cold working. Cobalt-based alloys exhibit higher moduli and much higher strengths compared to stainless steels. However, they have lower ductility and cannot deform as much before failure. Their UTS values range from about 700 to 1,500 MPa; yield at failure and ultimate strain are between 276 and 1,500 MPa and 8 to 45%, respectively. Titanium-based alloys have moduli that are similar to that of stainless steel and cobalt-based alloys and lower ductility; they have UTS values that approach those of stainless steel and cobalt-based alloys. UTS values range from 240 to about 890 MPa, yield at 2% of 170 to 827 MPa, and ultimate strain of 10 to 24%, respectively.

Ceramic materials typically are known for their high moduli. They range from 28 GPa for pyrolytic carbon to more than 300 GPa for aluminum oxide. Compressive strengths for these materials range from 517 (pyrolytic carbon) to

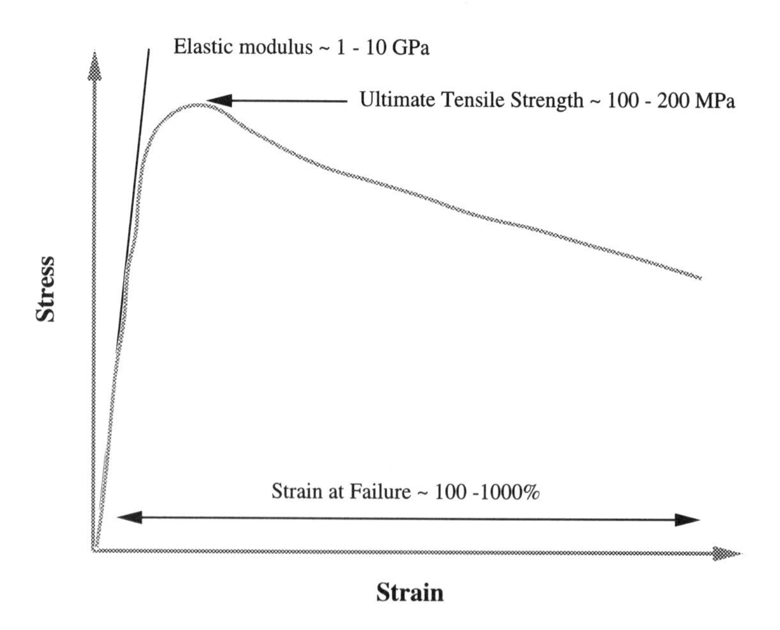

Figure 1.6 Stress–strain properties for crystalline polymers. A typical stress–strain curve for crystalline polymers below their crystalline melting temperature is shown. The modulus for crystalline polymers varies from 1 to 10 GPa, the UTS varies from 100 to 200 MPa, and the strain at failure varies from 100 to 1,000%.

more than 1,000 GPa for aluminum oxide. The modulus and compressive strength for silica glasses are about 100 MPa.

Structures and Properties of Cells

Cells are functional units within tissues and organs that serve to regulate the internal and external environments, synthesize macromolecules and small molecules for internal and external use, and replicate through cell division. The cell consists of a plasma membrane that contains integral proteins, such as integrins, histocompatibility markers (human leukocyte antigens that identify self versus foreign cells), and syndecans, which provide attachment, identify foreign cells, and act as receptors for triggering cellular processes. The membrane controls ion movement in and out of the cell and separates the inside of the cell from the extracellular compartment. The cell membrane also has specialized attachments by which it adheres to other cells. Inside of the cell is the cytosol, which contains

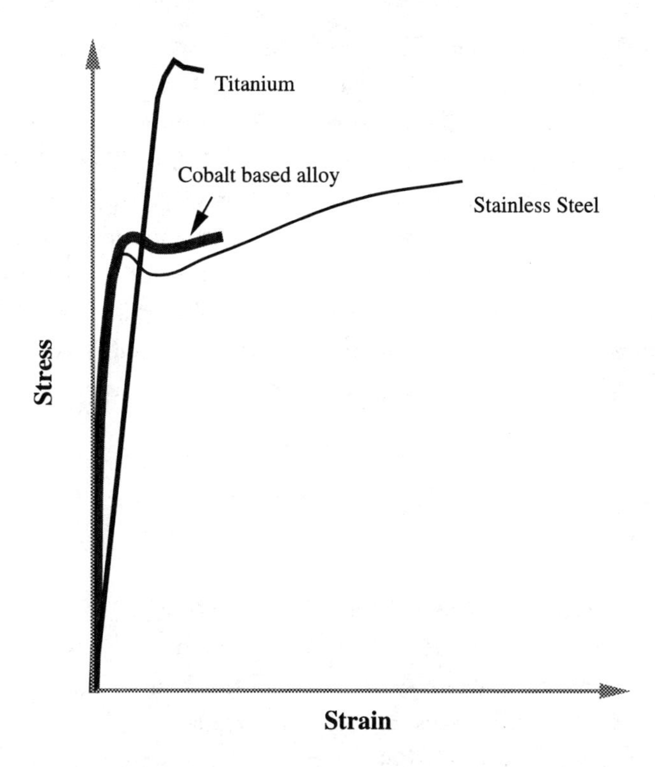

Figure 1.7 Stress–strain properties for metals. Typical stress–strain curves for metals, including titanium alloys, cobalt-based alloys, and stainless steel. Note that the UTS of metals exceeds that for crystalline polymers, whereas the strain at failure is typically reduced.

numerous organelles, including endoplasmic reticulum (transport of proteins extracellularly), mitochondria (generation of energy), ribosomes (synthesis of proteins), Golgi apparatus (packing of proteins for extracellular transport), lysosomes (hydrolysis of intracellular and extracellular material), and nucleus (synthesis of chromosomes and ribonucleic acids).

Although these components are found in all cells, there are differences in morphology and the actual spectrum of products synthesized by a cell. Some cells have the potential to synthesize a variety of products depending on the stimulating signals, whereas other cells are differentiated, meaning that they can only synthesize a subset of all proteins. The morphology of a cell can change depending on its environment and what is being synthesized. Fibroblasts are found in the extracellular matrix and change their morphology depending on their biosynthetic capabilities. They are highly elongated (long and thin) when they are depositing collagen and extracellular matrix macromolecules.

The mechanical properties of cells have been estimated based on the deformation of the red cell membrane during blood flow. The red cell stiffness has been estimated to be about 1 MPa, which is very small compared to the stiffness of collagen fibers and bone. It is similar to the stiffness of proteoglycans and elastic fibers. Therefore, cells that are located in tissues subject to tensile or compressive forces must not be directly loaded by collagen fibers and mineralized collagen fibers. Instead, cells in these tissues must be in parallel to collagen fibers and deform by shear when collagen fibers are loaded in tension.

Cellular and Tissue Responses to Materials

There are a number of responses that can be observed when an implant contacts cells and tissues. Cellular responses include adaptation, such as changes in the amount and types of proteins synthesized; changes in cell size and number; changes in cell type and characteristics; and cellular injury. Although cellular injury is not necessarily irreversible, under certain instances such as ischemic injury (loss of blood supply and oxygen), cellular injury becomes irreversible if the insult is not removed. These responses can occur as a result of mechanical trauma (pressure), infectious agents (bacteria and viruses), and chemicals (low molecular weight compounds released from implants) or from inflammation, immune responses, and activation of blood clotting, fibrinolysis, complement, or other biological cascades.

Implants that are in contact with the surface of the body are less likely to result in cellular activation and therefore require fewer and less stringent tests to assess their activation of inflammation, immune, and other biological processes. In general, materials to be used in contact with skin or other surface linings must not cause irritation of the surface or cellular cytotoxicity. For materials to be permanently implanted and placed in contact with blood, extensive testing must be done before an implant is used in humans. The required tests include assessment of the inflammatory potential, immunity, and ability to cause red cell hemolysis and blood clotting. All permanent implants made out of non-degradable materials must not induce tumors in animals during a long-term implantation (2 years in rats). Beyond this, it is now believed that materials that deplete any component of a biological cascade, such as complement or a blood clotting factor, may have negative consequences if used in permanent implants. These are some of the considerations examined in more detail in Chapter 9 on the implant-tissue response.

Summary

Biomaterials science and biocompatibility is a complex interdisciplinary subject. The anatomy and physiology of a wide variety of tissues and organs must be understood as well as the engineering properties and microstructure down to the level of the individual macromolecule. Most research on implant design has been largely empirical, based on the availability of a limited number of materials, including polymers, metals, ceramics, and composites; however, we are just beginning to understand how different these materials are in structure and properties compared to that of the host. The purpose of this book is to provide the student of biomaterials with enough information to understand the structure and complexity of biological materials and systems, which have evolved for millions of years, so that future materials will be designed that more closely mimic natural tissues. Although the number of synthetic materials is currently limited, it is our hope that advances in tissue engineering and transplantation biology will provide the biomaterials scientist of the next millennium with limitless possibilities for duplicating replacements for host tissues.

Suggested Reading

Bernfield M., Kokenyesi R., Kato M., Hinkes M.T., Spring J., Gallo R.L., and Lose E.J., Biology of the Syndecans: A Family of Transmembrane Heparan Sulfate Proteoglycans, Annu. Rev. Cell Biol. 8, 365, 1992.

Black J., Orthopaedic Biomaterials in Research and Practice, Churchill Livingstone, New York, chapter 7, 1988.

Bloom W. and Fawcett D.W., A Textbook of Histology, W.B. Saunders Company, Philadelphia, chapter 3, 1965.

Brown J.C. and Timpl R., The Collagen Superfamily, Int. Arch. Allergy Immunol. 107, 484, 1995.

Cowin S.C., The Mechanical Properties of Cortical Bone Tissue, in Bone Mechanics, edited by S.C. Cowin, CRC Press, Boca Raton, FL, chapter 6, pp. 98–127, 1989.

Goetinck P., Proteoglycans in Development, Curr. Topics Dev. Biol. 25, 111, 1991.

Heidemann S.R., A New Twist on Integrins and the Cytoskeleton, Science 260, 1080, 1993.

Ruoslahti E. and Yamaguchi Y., Proteoglycans as Modulators of Growth Factor Activities, Cell 64, 867, 1991.

Silver F.H., Kato Y.P., Ohno M., and Wasserman A.J., Analysis of Mammalian Connective Tissue: Relationship Between Hierarchical Structures and Mechanical Properties, J. Long-Term Effects Med. Implants 2, 165, 1992.

Yamada K.M., Fibronectin and Other Cell Interactive Glycoproteins, in Cell Biology of Extracellular Matrix, second edition, edited by E.D. Hay, Plenum Press, New York, chapter 4, pp. 111–146, 1991.

2

Introduction to Structure and Properties of Biological Tissues

Introduction to Principles of Stereochemistry

Macromolecules and polymers are the principle building blocks of tissues. Without these large molecules, life as we know it would not be possible, because these moieties are responsible for the completion of most biological processes. Biological macromolecules are classified into four groups of large molecules: proteins, polysaccharides (sugar polymers), nucleic acids, and lipids. These classes are differentiated by their repeat units, the chemical structure that is repeated over and over again to make a large chain. The properties of long chains of repeat units linked together are dependent on the chemistry of the chain. The physical properties of long-chained molecules also depend on the rotational freedom around the backbone, as diagrammed in Figure 2.1. Regardless of the exact chemistry of a macromolecule's backbone, the physical behavior is fixed. The modulus or resistance of a polymer to deformation is independent of the backbone chemistry, but the temperature at which a particular behavior is observed is dependent on the backbone chemistry. At some temperature, all polymers behave like a rubber band, stretching easily and reversibly. This temperature, at which a polymer behaves like a rubbery material, is the *glass-transition temperature*. The glass-transition temperature is affected by the chemistry of the repeat unit and by how it affects the backbone flexibility. The relationship between the chemistry of the backbone of a polymer and its rubberiness is more complex than just analysis of the backbone rotational freedom; this is discussed later in this chapter.

Consider a chain of carbon atoms similar to that observed in lipids and polyethylene. How does this chain exist in three dimensions? We know from general chemistry that a chain of atoms covalently bonded together must have a fixed bond length and bond angle. Therefore, a carbon chain must adhere to these principles, and all carbon-to-carbon bonds must have a bond length of about 1.5 Å and a bond angle of 108°. In a carbon chain with fixed bond angles, the bond angle easily forms a planar zigzag in three dimensions (see Figure 2.1). The hydrogen atoms are attached to the backbone, but how are they arranged in space?

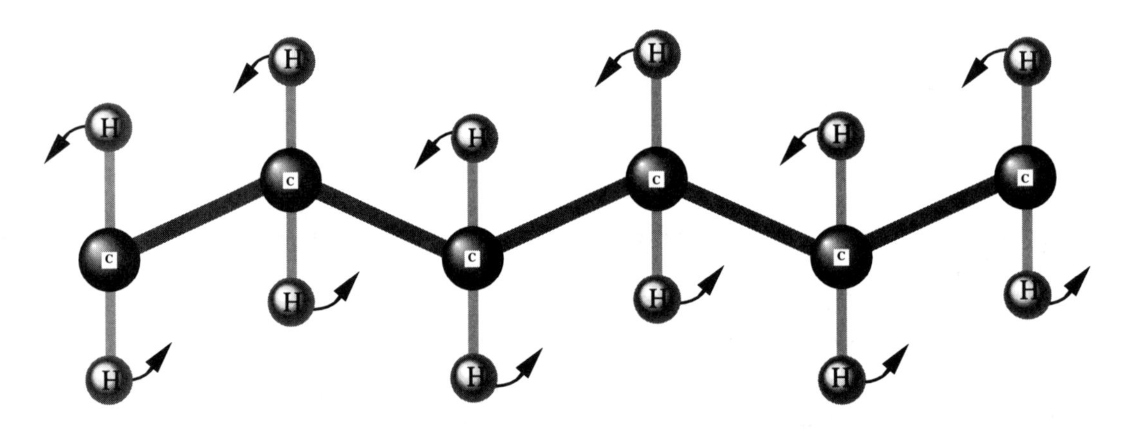

Figure 2.1 Mobility of hydrocarbon chain. The diagram shows mobility of a polymer chain composed of carbon atoms attached by single bonds with hydrogen side chains. Rotational freedom of the single carbon-to-carbon bonds allows hydrogen atoms to rotate freely about the backbone.

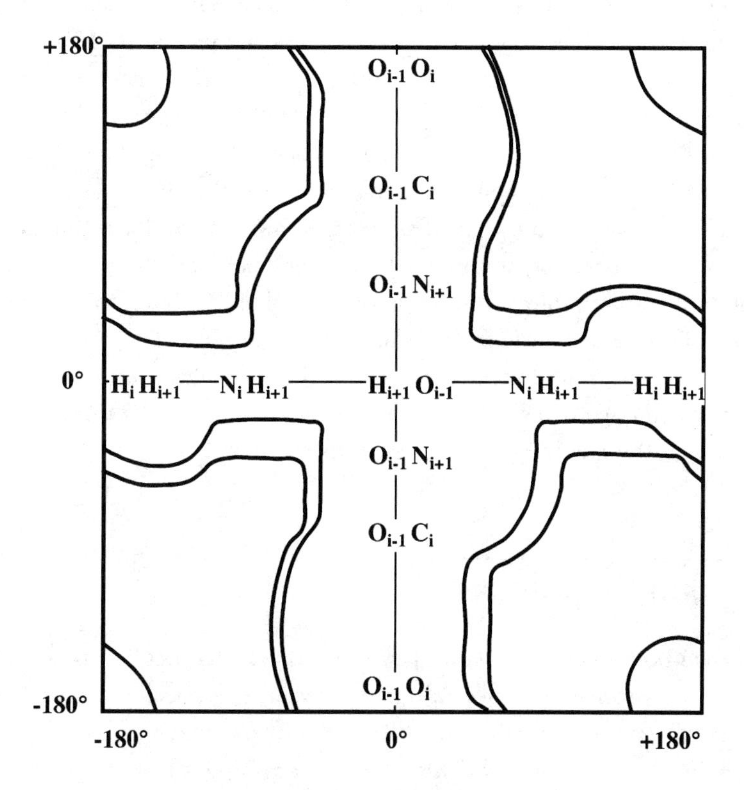

Figure 2.2 Allowed conformations for a dipeptide of glycine. Plot of fully (solid lines) and partially (dotted lines) allowed combinations of ψ (vertical axis) and ϕ calculated using a hard sphere model with normal (solid lines) and minimum (dotted lines) atomic contact distances. Dihedral angles (ϕ,ψ) in center of plot (0°,0°; 0°,180° and 180°,180°) are unallowed because of contacts between backbone atoms in the neighboring (*i*th, *i*th + 1 and *i*th − 1) peptide units. The contacts between nitrogen (N), hydrogen (H), oxygen (O), and carbonyl groups (CO_2H) prevent allowed conformations in the center of this diagram. This diagram can be constructed using coordinates of the atoms within the peptide unit and interatomic distances (see Tables 2.2 and 2.3). To construct this diagram, a standard dipeptide unit (see Figure 2.7) is formed by translating the first peptide unit along the line between the α carbons and then flipping the second unit into the *trans* configuration. The values of ϕ and ψ were varied using matrix multiplication, and the interatomic distances were checked for all nonbonded atoms. Pairs of ϕ and ψ that have allowable interatomic distances are found *within* the dotted and solid lines shown as the allowable conformations.

The allowable conformations of a polymer chain can be determined by considering atomic bond angles and bond lengths. Nonbonded atoms must not be closer than the sum of the van der Waals radii between two atoms. Each conformation, which is a set of dihedral angles within the backbone, can be assessed by a computer model that knows the atomic coordinates of all atoms in the repeat unit. The model then translates and rotates a repeat unit through the available combinations of dihedral angles. This can be done using vector mathematics, and the resulting conformational plot gives the allowable values of the backbone angles for a polymer. This plot, known as the Ramachandran plot (Figure 2.2) for proteins, is modified by interactions among side chains and among polymer molecules. Conformational freedom is related to flexibility and mechanical properties through the relationship between the number of available conformations and the Boltzmann constant (k) for chains that deform without interactions among chains. Chain flexibility and stiffness under these conditions are related to the number of available conformations.

Basic Building Blocks

The human body is constructed of proteins, polysaccharides, lipids, and nucleic acids. Proteins form the structural materials of the extracellular matrix, in the form of collagen and elastin, and they function as enzymes and cell surface markers. In this book, we will discuss the role of proteins as structural materials. Some protein molecules of significance include collagens, myosin, actin, tubulins, integrins, and class I and II histocompatibility markers. The basic repeat unit in proteins, which are also called *polypeptides,* is the peptide unit formed when two amino acids condense to form a dipeptide and water (Figure 2.3). Proteins are synthesized for transport extracellularly on the endoplasmic reticulum and then transported through the Golgi apparatus for release extracellularly. Proteins are also synthesized on free ribosomes within the cell cytoplasm, which consist of large and small subunits that translate the genetic code of the cell into a sequence of repeat units.

Polysaccharides are found as sugar polymers that are components of the extracellular matrix (hyaluronan), as carbohydrates in vegetables (starch), and as energy stores in humans (glycogen). There are numerous repeat units of polysaccharides, but many are derivatives of the simple sugars glucose and galactose, which are six-membered ring structures (Figures 2.4 and 2.5). The repeat unit of polysaccharides consists of a six-membered ring linked via an oxygen molecule

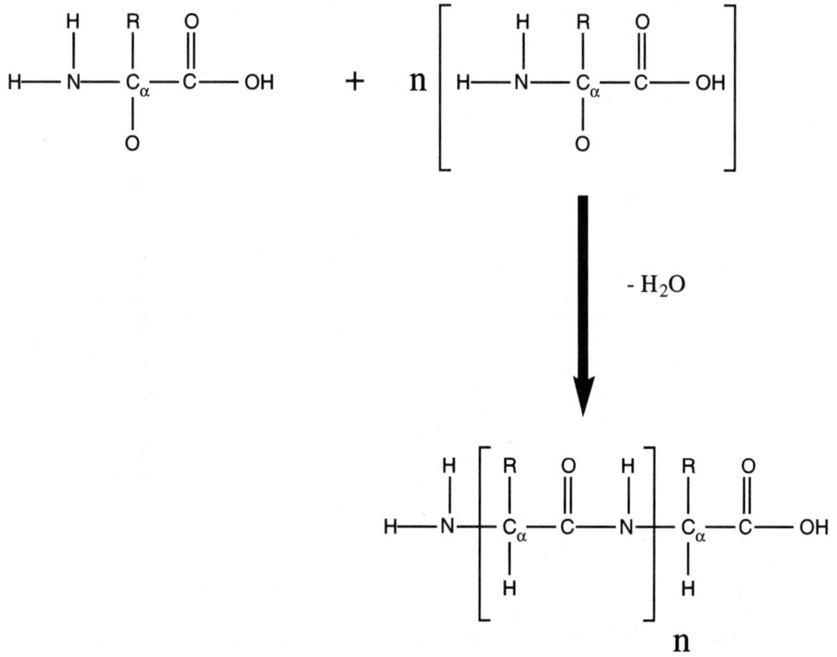

Figure 2.3 Formation of a polypeptide. Polypeptides are formed when a condensation reaction occurs between amino acids, releasing water when the amino and acid end groups react.

to the next six-membered ring. Precursors of polysaccharide molecules are synthesized in the cell cytosol, and polysaccharides are assembled at the cell membrane or inside the cell, depending on whether the polysaccharide is to be released extracellularly or used intracellularly.

Lipids are a heterogeneous group of long hydrocarbon chains that can have a free carboxyl group (fatty acids), an ester group (neutral fats), or derivatives of fatty acids (membrane phospholipids). Derivatives of phosphatidic acid (phosphatidyl choline, phosphatidyl serine, and phosphatidyl inositol) are membrane components that are modified hydrocarbon chains. Rotation around the carbon-to-carbon bonds in the backbone gives these molecules great flexibility. The membrane of mammalian cells contains a wide array of polymer structures, including lipids, polysaccharides, and proteins. The function of these polymers depends on their size, shape, and flexibility. Polymer molecules of importance in the cell membrane are phospholipids, hyaluronic acid, syndecan, class I and II histocompatibility markers, and integrins.

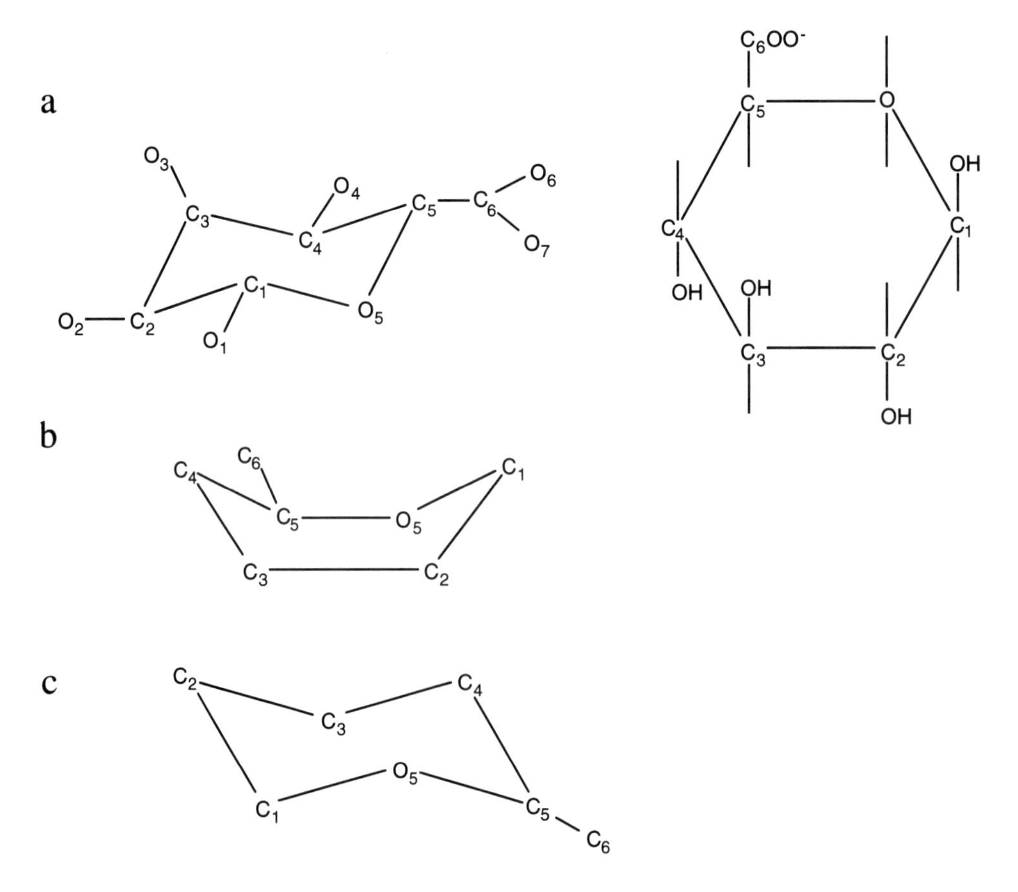

Figure 2.4 Boat and chair forms of β-D-glucuronic acid. (**a**) C_1 chair conformation (left) and planar projection (right); (**b**) Boat conformation; and (**c**) ^1C chair conformation of β-D-glucuronic acid. The latter two conformations are energetically less stable than the C_1 chair conformation.

Nucleic acids are polymers composed of a nitrogenous base (either a purine or a pyrimidine), a five-membered sugar ring (a pentose), and phosphoric acid. Deoxyribonucleic acid (DNA) contains the purines adenine and guanine, the pyrimidines cytosine and thymine, 2-deoxyribose, and phosphoric acid. Deoxyribonucleic acid is found in the cell nucleus in double-stranded form in the chromosomes. Ribonucleic acid (RNA) contains the purines adenine and guanine, the pyrimidines cytosine and uracil, ribose, and phosphoric acid. Ribonucleic acid is found in the small and large subunits of ribosomes (rRNA), as a copy of the genetic material (messenger [m]RNA), and for adding amino acids to a

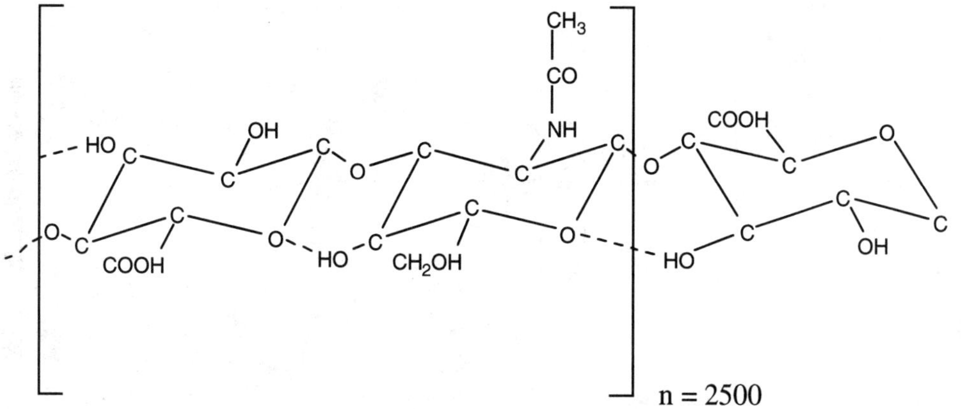

Figure 2.5 Repeat disaccharide of hyaluronan. Hyaluronan is composed of β-(1–3) linkage of glucuronic acid to D-*N*-acetyl glucosamine that is linked β-(1–4) to D-glucuronic acid. Both β-D-glucuronic acid and β-D-*N*-acetyl glucosamine are in the C_1 chair conformation. The 0°,0° conformation shown occurs when the atoms attached to the carbons connected to the oxygen linking the sugar units are along the axis of the unit (axially) and eclipse each other when viewed along the chain.

growing protein chain (transfer [t]RNA). The repeat units and structures of DNA and RNA are shown in Figure 2.6.

Stereochemistry of Polymer Chains

The stereochemistry of several types of polymer chains has been studied, including proteins, hydrocarbon chains, and polysaccharides. If we know the chemistry of a polymer backbone, we can predict the chain flexibility. In addition, from a stereochemical plot and the degree of chain extension, it is possible to predict qualitatively how much resistance to deformation (stiffness) is associated with a particular structure. Proteins have been studied extensively by *X-ray diffraction* and by direct visualization with *electron microscopy.*

Stereochemistry of Polypeptides

In proteins, amino acids are the building blocks that are synthesized into long polymer chains. When amino acids are added together, they form peptide units, which are contained within a single plane except for the side chain or R group

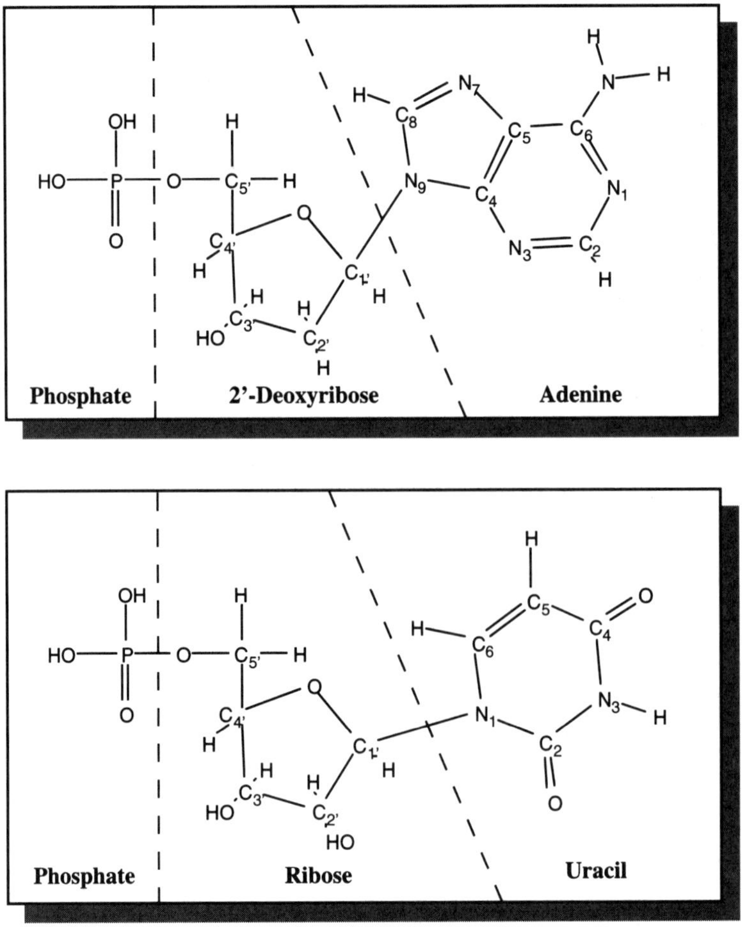

Figure 2.6 Structure of nucleic acids. The diagram shows the linkage between phosphate, sugars, and a purine or pyrimidine for DNA (top) and RNA (bottom).

(Figure 2.7). The sequence of amino acids is important in dictating the manner in which polymer chains behave, because the sequence dictates whether a chain can fold into a specific 3-D structure or whether the polymer chain does not fold. For example, because of its primary sequence of amino acids, keratin folds into a series of α helixes that pack together to form twisted ropes with very high modulus and tensile strength values. Elastin forms short segments with random chain structure, which results in a material with a much lower modulus and ultimate tensile strength compared to keratin.

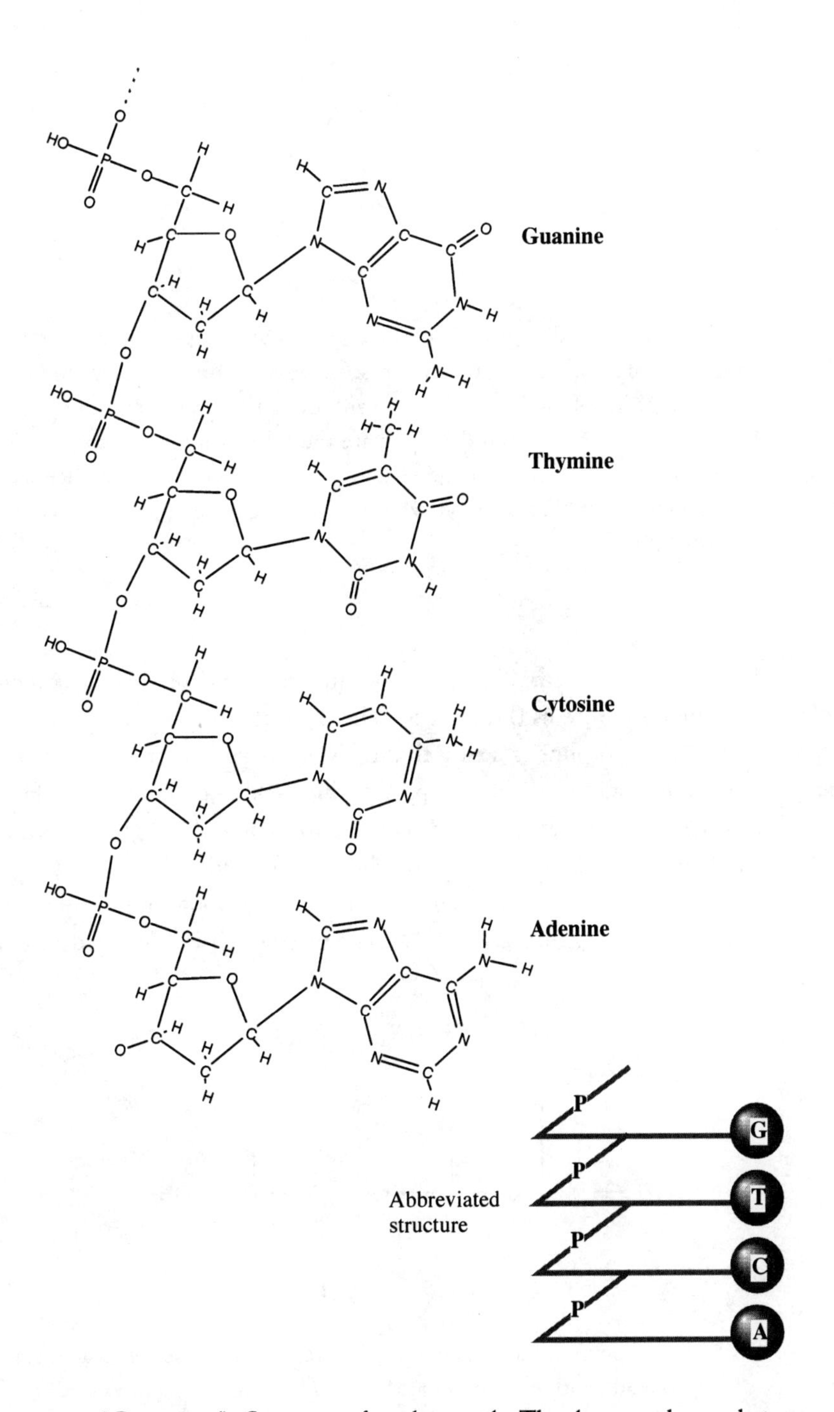

Figure 2.6 (*Continued*) Structure of nucleic acids. The diagram shows chain structure of DNA.

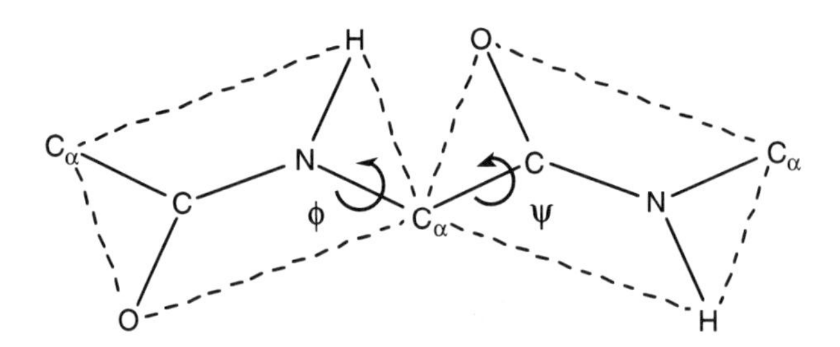

Figure 2.7 Diagram of a dipeptide. The diagram shows a dipeptide of glycine in the standard conformation 180°,180°. To a first approximation, all atoms seen in the dipeptide backbone are found within the plane of the paper. For a dipeptide of glycine, the only atoms not within the plane are the side chain hydrogens. The unit begins at the first Cα and goes to the second Cα. Both ϕ and ψ are defined as positive for clockwise rotation when viewed from the nitrogen and carbonyl carbon positions toward the Cα.

The general representation of amino acids holds true for 19 of 20 common side chains found in proteins (Figure 2.8). There are 19 different amino acid side chains, or R groups; proline is a ring that does not have two angles of rotation because the amino acid side chain is part of the backbone, as shown in Figure 2.9. The presence of proline and a hydroxylated form of proline, hydroxyproline, is characteristic of collagen and collagen-like polypeptides that form a triple-helical structure. The primary consequence of the incorporation of proline into the backbone is that it stiffens the 3-D structure, which is explained in more detail later.

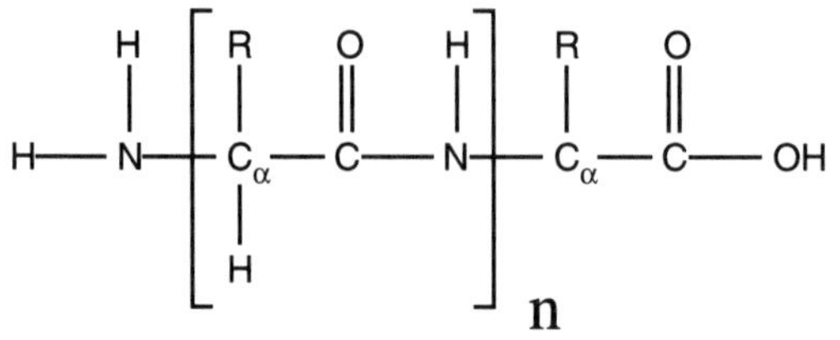

$$\mathrm{H{-}N(H)}\left[\mathrm{C_{\alpha}(R)(H){-}C({=}O){-}N(H)}\right]_{n}\mathrm{C_{\alpha}(R){-}C({=}O){-}OH}$$

Figure 2.8 Chain structure of a polypeptide. The repeat unit of a polypeptide chain consists of amino acid residues, each characterized by a specific amino acid residue designated by R. The size of the polypeptide chain is dictated by n, the degree of polymerization.

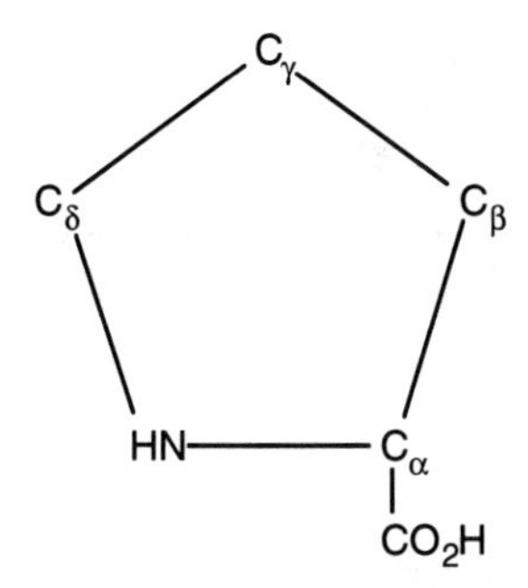

Figure 2.9 Chemical structure of proline. The structure of proline differs from that of other amino acids because a ring is part of the polypeptide backbone. The chain backbone includes the amide nitrogen (N), Cα, and the carbonyl carbon (CO_2H).

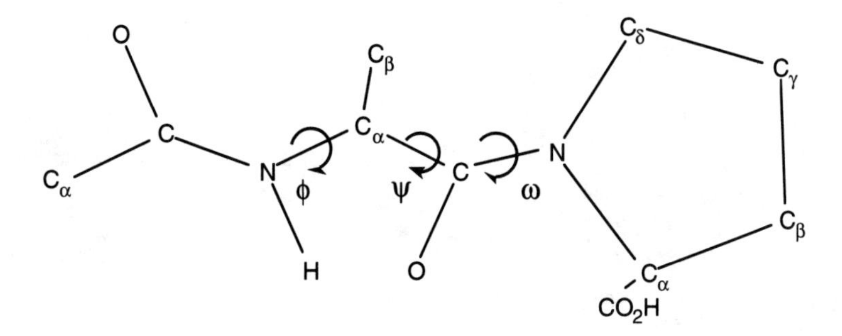

Figure 2.10 Polypeptide chain structure containing proline. Chains containing proline lack the flexibility of other peptides because the proline ring has only one available angle for backbone rotation. Rotation occurs around the angles ϕ, ψ, and ω.

There are 20 standard amino acids commonly found in proteins (Table 2.1). The sequence of these amino acids dictates the shape of the resulting chain, and the presence of large R groups limits mobility of the chain backbone by preventing rotation (Figure 2.10). Also interfering with rotation around the backbone is the form of the amino acid. Each amino acid is in the D or L form because the carbon to which the R group is attached, the α-carbon (Cα), is asymmetrical; it has four different chemical groups attached to it. Therefore, there are two different chemical forms of amino acids that rotate a plane of polarized light differently (the L form rotates the plane to the left and the D form to the right). The structure and properties of the resulting protein synthesized differ with all D- or L-amino acids. The L form is predominantly found in proteins of higher

Table 2.1 Standard amino acids

Amino acid	Abbreviation	R group
Alanine	Ala	$-CH_3$
Arginine	Arg	$-CH_2-CH_2-CH_2-NH-CH(NH_2)_2$
Asparagine	Asn	$-CH_2-C(=O)-NH_2$
Aspartic acid	Asp	$-CH_2-C(=O)-O^-$
Cysteine	Cys	$-CH_2-SH$
Glutamic acid	Glu	$-CH_2-CH_2-C(=O)-O^-$
Glutamine	Gln	$-CH_2-CH_2-C(=O)-NH_2$
Glycine	Gly	$-H$
Histidine	His	$-CH_2-C$ (ring: $C(H)$, N, CH, $N(H)$)
Isoleucine	Ile	$-CH_2-CH_2-CH(CH_3)-CH_3$

Table 2.1 (*Continued*) Standard amino acids

Amino Acid	Abbreviation	R Group
Leucine	Leu	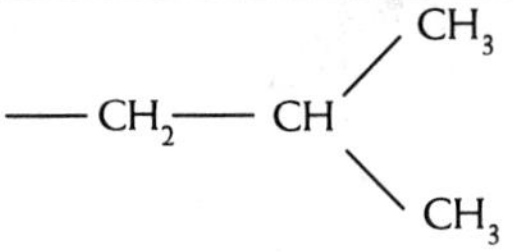 $-CH_2-CH(CH_3)_2$
Lysine	Lys	$-CH_2-CH_2-CH_2-CH_2-NH_3^{+}$
Methionine	Met	$-CH_2-CH_2-S-CH_3$
Phenylalanine	Phe	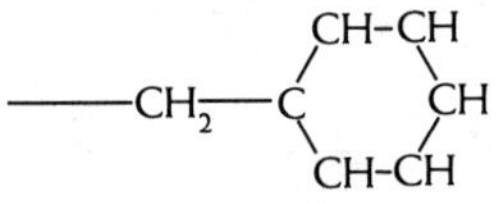 $-CH_2-C(CH\text{-}CH)_2CH$ (ring: C, CH-CH, CH, CH-CH)
Proline	Pro	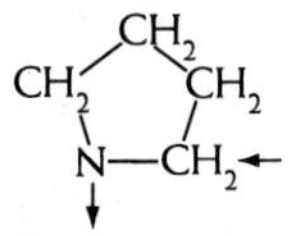 ring $CH_2-CH_2-CH_2-CH_2-N$ (attachment at N↓ and at CH_2←)
Serine	Ser	$-CH_2OH$
Threonine	Thr	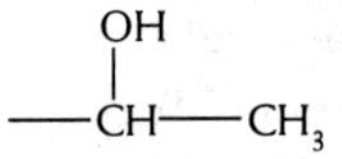 $-CH(OH)-CH_3$
Tryptopfan	Trp	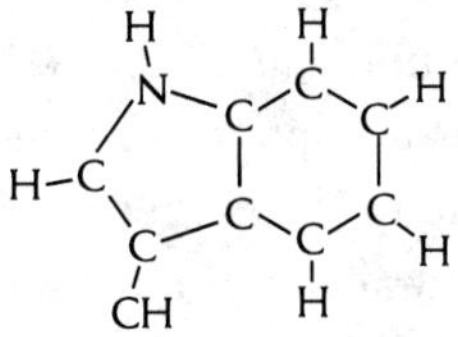 indole ring: N(H), C(H), C–CH, benzene ring C(H), C(H), C(H), C(H)
Tyrosine	Tyr	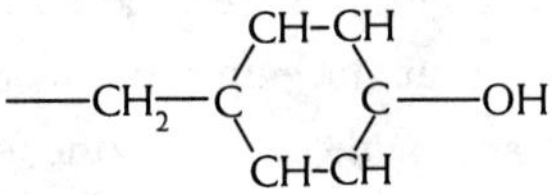 $-CH_2-C(CH\text{-}CH)_2C-OH$
Valine	Val	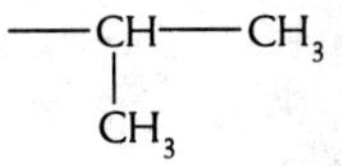 $-CH(CH_3)-CH_3$

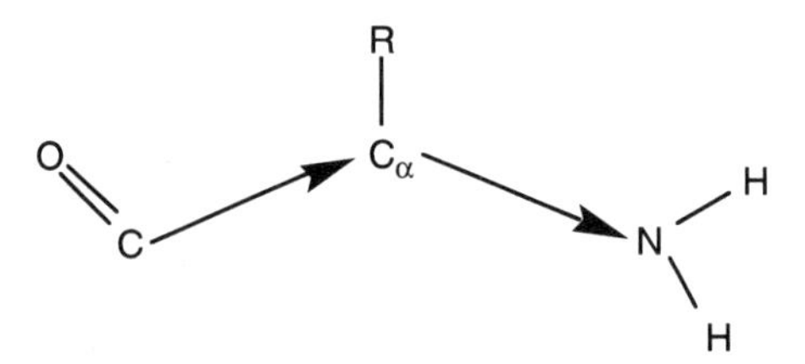

Figure 2.11 Difference between L and D amino acids. The two naturally occurring forms of amino acids differ in the position of the R group with respect to the backbone. An L-amino acid has the R group on the left if viewed along the chain from the free carbonyl end to the free amino end as shown. If the R group is on the right, it is defined as the D form. The predominant form found in proteins is the L form, although some amino acids in the D form are present in proteins.

life species and can be deciphered from the D form by the position of the R group with respect to the backbone. Because the Cα is attached to three different groups of atoms, the L form cannot be rotated around the carbonyl carbon–Cα bond (oxygen attaches to the carbonyl carbon). The L form is obtained by moving along the peptide chain, beginning at the free COOH end of the chain (the free-carboxyl-acid-containing group) and moving toward the free amino end, with the Cα as the point of reference. The amino acid side chain (R group) is on the left, as shown in Figure 2.11. When the side chain is on the right, the D form is present. This is important because the position of the side chain dictates how a chain can fold in three dimensions.

From a structural viewpoint, a polypeptide is composed of planar peptide units (see Figure 2.7). The usefulness of considering the peptide unit instead of the amino acid is that the peptide unit is almost planar, whereas the amino acid has atoms that are in more than one plane. The coordinates of atoms in the peptide unit are given in Table 2.2. Note that all the atoms from the first Cα to the second Cα do not have a z coordinate. These coordinates come from X-ray diffraction studies on proteins and represent the average coordinates found among many proteins. The 3-D structure of a dipeptide is shown in Figure 2.7 in the standard conformation.

The standard conformation is obtained by taking the first peptide unit and rotating it around the line between C-1α and C-2α by 180°, translating the rotated peptide unit along the line between C-1α and C-2α by the distance between C-1α and C-2α, and then rotating counterclockwise along the extension of the line C-1α–C-2α by 33° (Figures 2.12 and 2.13). If we define the

Table 2.2 Coordinates of atoms in peptide unit

Atom	**Coordinates (Å) x, y, z**
C_α (first amino acid)	0, 0, 0
Carbonyl carbon	1.42, 0.58, 0
Oxygen	1.61, 1.79, 0
Nitrogen	2.37, −0.33, 0
Hydrogen	2.19, −1.31, 0
Cα (second amino acid)	3.70, 0, 0
C_β (first carbon of side chain)	0.51, 0.72, 1.25

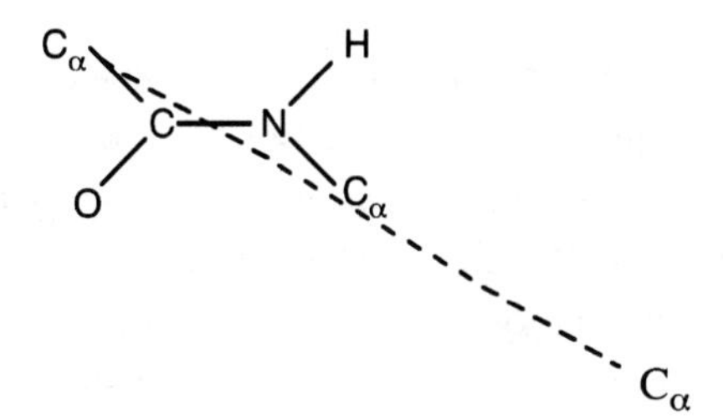

Figure 2.12 Location of the second peptide unit. Dipeptides are constructed by knowing the coordinates of atoms in the first peptide unit (using X-ray diffraction) and then translating along the line between the first and second Cα by a distance of 0.37 nm.

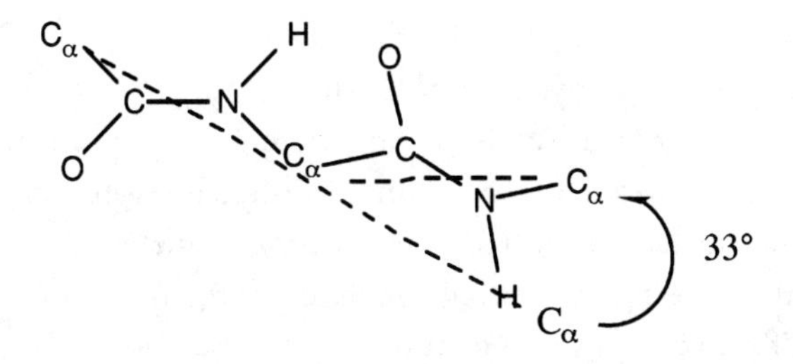

Figure 2.13 Graphical construction of a dipeptide. Once the second peptide unit is located (see Figure 2.12), it is rotated through an angle of 33° clockwise to generate a dipeptide in the standard conformation, with $\phi = 180°$ and $\psi = 180°$.

Table 2.3 Minimum interatomic distances for nonbonded atoms

Nonbonded atom pairs	Contact distances normal (minimum), Å
Carbon to carbon	3.20 (3.00)
Carbon to nitrogen	2.90 (2.80)
Nitrogen to nitrogen	2.70 (2.60)
Carbon to oxygen	2.80 (2.70)
Nitrogen to oxygen	2.70 (2.60)
Oxygen to oxygen	2.70 (2.60)
Carbon to hydrogen	2.40 (2.20)
Hydrogen to nitrogen	2.40 (2.20)
Hydrogen to oxygen	2.40 (2.20)
Hydrogen to hydrogen	2.00 (1.90)

angle of rotation of the bond containing the atoms N–Cα as ϕ and the angle of rotation of bond containing Cα and C (carbonyl) as ψ, the conformation shown is arbitrarily defined as $\phi = 180°$ and $\psi = 180°$. Using a computer and matrix multiplication techniques, the first and second peptide unit can be rotated through all possible combinations of values of ϕ and ψ. For each possible set of these dihedral angles, the distances between all pairs of nonbonded atoms can be compared using a set of minimum interatomic contact distances (Table 2.3). If the distance between any set of atoms is smaller than the minimum contact distance, then that conformation is not allowed: two atoms cannot be closer than the sum of the van der Waals radii of each atom or electron repulsion occurs. Therefore, by determining the values of ϕ and ψ that result in conformations that are not prevented by electron repulsion, we can determine the number of available conformations (combinations of ϕ and ψ that are allowed). Conformational maps show values of ϕ and ψ that fall into two regions. The first region contains points that are always allowable, and the second region contains points that are sometimes allowed. As shown in Figure 2.2, all points within the solid lines are always allowed, and those within the dotted lines are sometimes allowed. The area within the solid or dotted lines on a conformational map (Figure 2.14) is proportional to the number of allowable conformations of a dipeptide unit. We define the flexibility of a polypeptide as the natural logarithm of the number of allowable conformations (#) times the Boltzmann constant (k) (equation 2.1).

$$S = \text{entropy or chain flexibility} = k \ln (\#) \qquad (2.1)$$

Figure 2.2 demonstrates a conformational map for a dipeptide of glycine (the side chain in glycine is small—just a hydrogen) showing mostly allowed or partially allowed conformations; therefore, polyglycine is flexible. How do we know that the conformational plot for a polypeptide is the same as for a dipeptide? Because the side chain points away from the backbone for most conformations, the atoms in the side chains are separated by more than the sum of the van der Waals radii. However, we will later discuss several highly observed conformations of proteins in which the conformational map is an overestimation of the flexibility because of interactions between atoms more than two peptide units apart in space.

Polypeptides are made up of sequences of amino acids that are not identical and have side chains bigger than glycine. A chain of carbon atoms bonded together is ideally flexible for the same reason that polyglycine is flexible. In fact, this is a practical conclusion of viewing the conformational plot because polyethylene and lipids are composed of carbon chains.

Polypeptides, however, are composed of amino acids with side chains that are longer; therefore, the area of allowed conformations is reduced when an alanine (Figure 2.14), aspartic acid (Figure 2.15), or a proline (Figure 2.16) is added to the second peptide unit. The conformational map for a dipeptide of proline–hydroxyproline is dramatically reduced. Rings in the backbone of any polymer reduce the ability of the polymer backbone to adopt numerous conformations, thereby stiffening the structure.

For polypeptides, the addition of a methyl or a slightly longer side chain reduces the number of available conformations to the extent that two major regions of allowable conformations stand out (Figure 2.17). When conformations that are most frequently observed in proteins are tabulated, they fall into these two regions. These conformations include the α helix (Figure 2.18) (which differs from the α-chain helix in collagen), β sheet (Figure 2.19), and the collagen triple helix. The regions that these conformations fall into are ϕ between $-45°$ and $-120°$ and ψ between $150°$ and $-90°$. The β sheet and the collagen triple helix are highly extended structures and are very rigid and stable. The α helix can be extended into a β sheet and is very deformable. This will be discussed in more detail later.

The mechanical properties of polymer chains that do not exhibit interaction between either the side chains and the backbone or one part of the backbone and another part of the backbone are related to the number of available conformations and the chain entropy. The stiffness of a polymer chain that does not exhibit bonding with other parts of the chain is related to the change in the

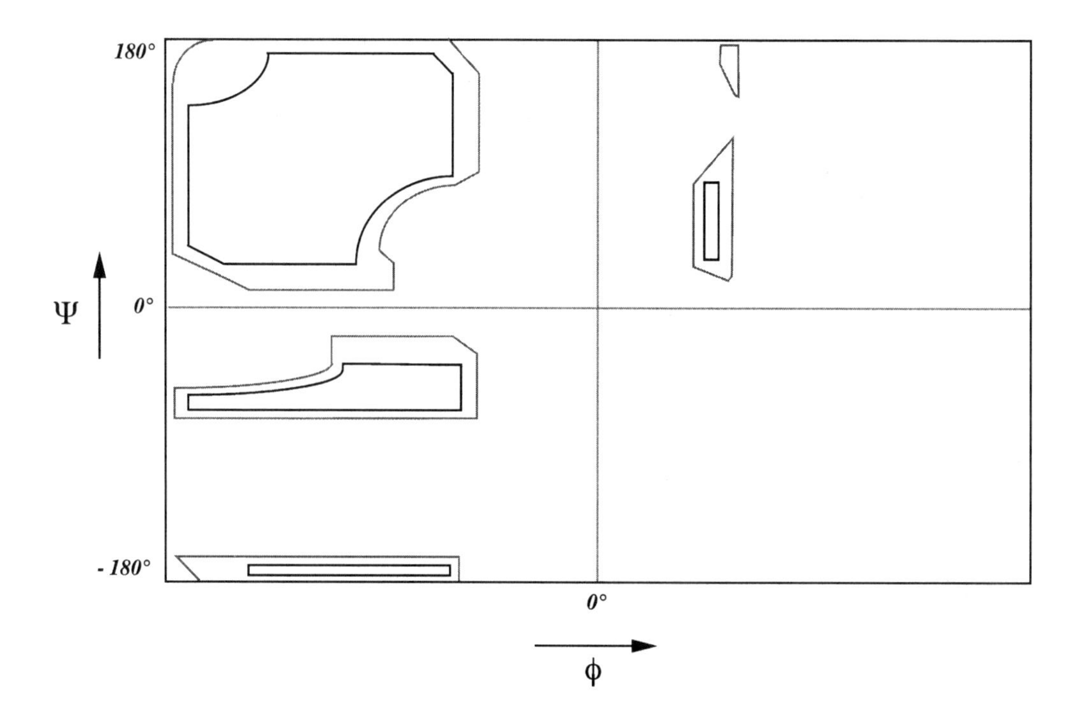

Figure 2.14 Conformational plot for glycine-alanine. Plot of allowable angles for peptides containing a repeat unit of glycine and alanine showing totally (solid lines) and partially (dotted lines) allowed conformations determined from normal and minimum interatomic distances.

number of available conformations. Random chain polymers of elastin, polyethylene at high temperatures, and natural rubber are discussed later. As a polymer chain is stretched, the number of conformations and the resulting flexibility are reduced, thereby increasing the stiffness and resistance to deformation. When a random chain polymer is unloaded, the mobility and flexing of the backbone result in a return of the chain to its original range of rotational motions.

The behavior of proteins is generally more complex. The α helix, β sheet, and collagen conformations are found repeated in different structures in mammals. As we saw earlier, a large polymer molecule that has small side chains and no rings in the backbone has the potential to constantly rearrange itself, like an eel moving through water. A single conformation is a single set of dihedral angles that characterize the rotational state of each peptide unit that makes up the backbone of the molecule. As a molecule moves by diffusion, it changes its conformation by changing the set of dihedral angles that characterize each dipeptide unit. Folding of a polypeptide chain into the most commonly observed

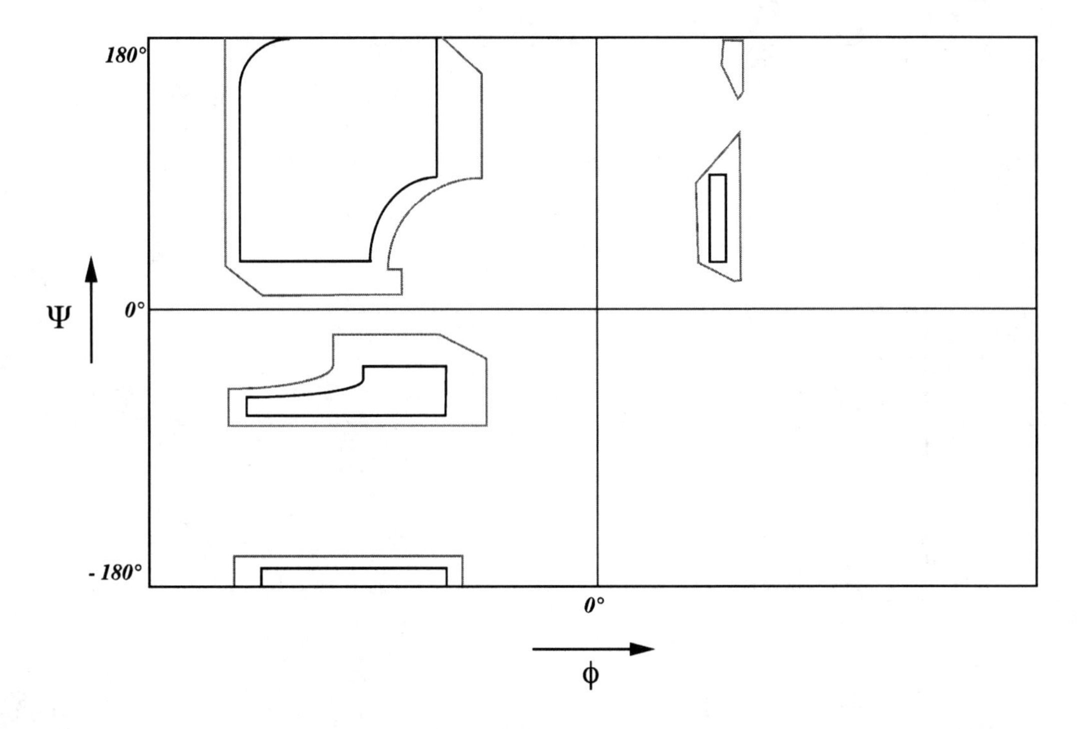

Figure 2.15 Conformational plot for glycine-aspartic acid. The allowable conformations are only slightly reduced compared to the plot in Figure 2.14 because the side chain length is increased in going from alanine to aspartic acid.

conformations—the α helix, β sheet, and collagen triple helix—occurs because of the inherent flexibility of peptide chains; however, once it folds into a particular structure, it is held in place by secondary and other forces.

An interesting perspective is to look at the free energy change associated with transition from a flexible chain into a folded conformation. From thermodynamics we know that the free energy of the transition must be negative for it to occur spontaneously. The Gibbs free energy (G) is a measure of the total energy and entropy (flexibility) that a system of macromolecules has. Thus, for any process to occur spontaneously (within our lifetime), the change in free energy (ΔG) must be equal to the change in enthalpy (ΔH), minus the temperature times the change in entropy ($T\Delta S$) (equation 2.2). The enthalpy (H) of a macromolecule is related to the number of covalent (P_c), dispersive (P_d), and electrostatic bonds (P_e) that are formed as well as the translational (E_t), and rotational kinetic energy (E_r) of the chains (equation 2.3).

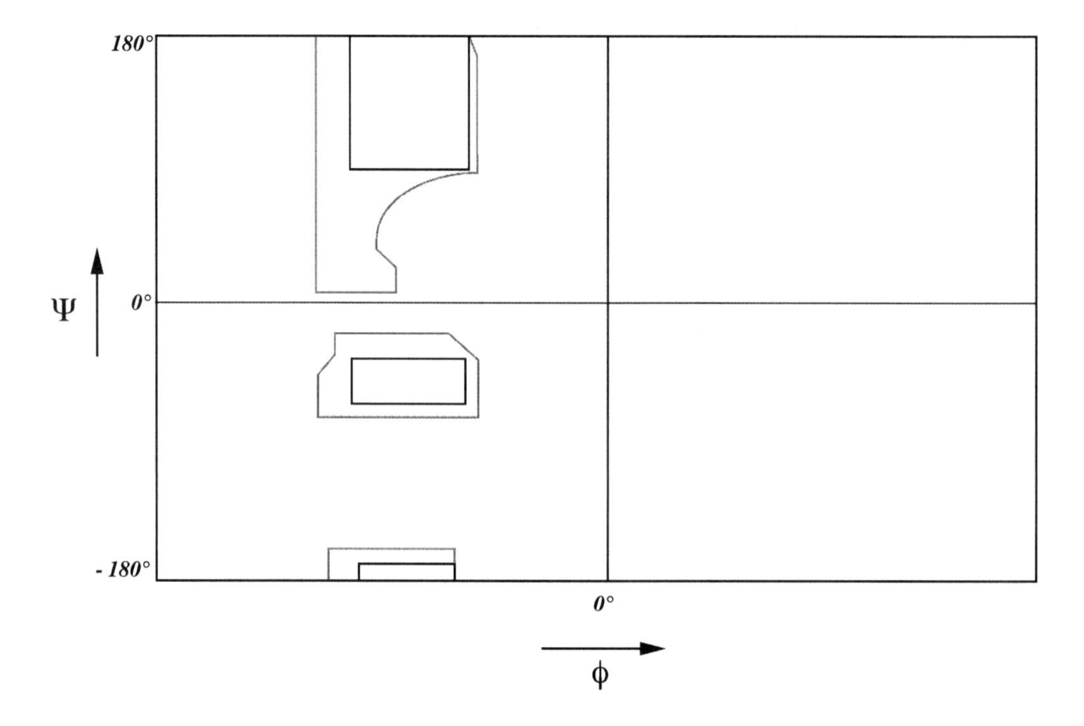

Figure 2.16 Conformational plot for glycine-proline. Addition of proline to a dipeptide further reduces the number of allowable conformations when compared to Figures 2.14 and 2.15.

$$\Delta G = \Delta H - T\Delta S \tag{2.2}$$

$$H = -(P_c + P_d + P_e) + E_t + E_r \tag{2.3}$$

Therefore, the change in enthalpy associated with a transition from a flexible chain to a folded chain involves a change in the number of bonds (the P term) and a change in the flexibility of the chain (the S term). We can calculate the change in the S term by a change in the area of allowable conformations (ideally this is k times natural logarithm of the area under Figure 2.17 divided by 1 because there is one final conformation). For the process to be spontaneous, the change in P must be positive (it must form bonds). Formation of a covalent bond lowers the enthalpy by 100 kcal/mol; hydrogen and electrostatic bonds lower it by between 1 and 5 kcal/mol. Van der Waals or dispersive forces lower it by 0.01 to 0.2 kcal/mol. Electrostatic and hydrogen bonds occur between atoms with partial charges and can be quantitatively assessed using Coulomb's

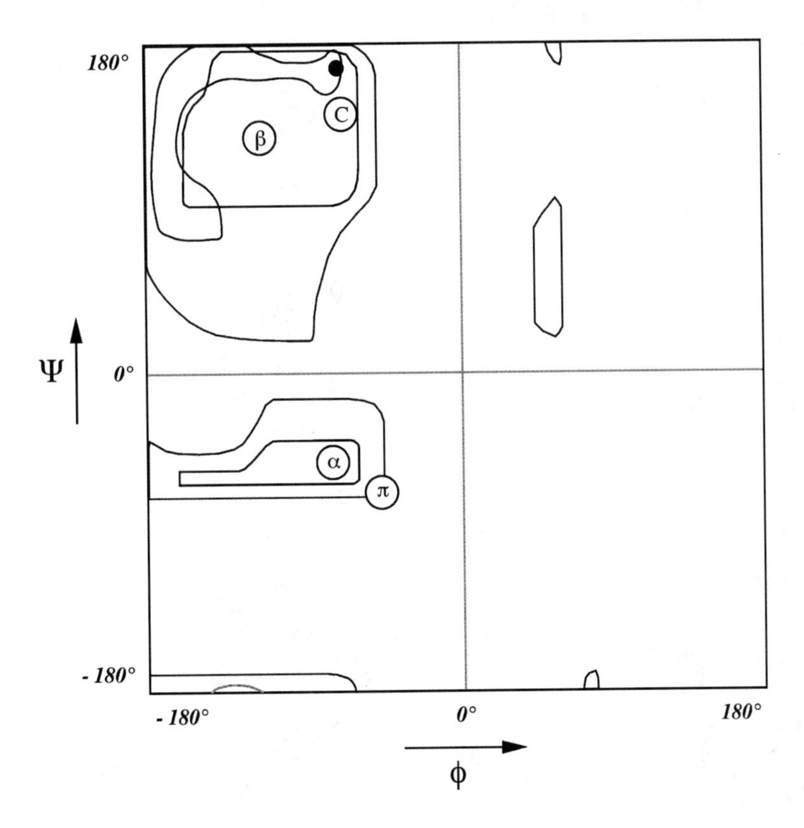

Figure 2.17 Conformational plot showing location of α helix, β sheet, and collagen triple helix. The plot shows the localization of the predominant chain structures found in proteins, including the α helix (α), β sheet (β), and collagen triple helix (C).

law (equation 2.4), where q_1 and q_2 are the partial charges on the atoms involved, D is the dielectric constant of the medium between the charges, and r_{ij} is the separation distance between charges. Dispersive energy (P_d) is calculated using the Lennard-Jones 6–12 potential energy function, where A and B are two constants specific to the two atoms involved (equation 2.5).

$$P_e = q_1 q_2/(4\pi D r_{ij}) \tag{2.4}$$

$$P_d = (A/r_{ij}^{12}) - (B/r_{ij}^{6}) \tag{2.5}$$

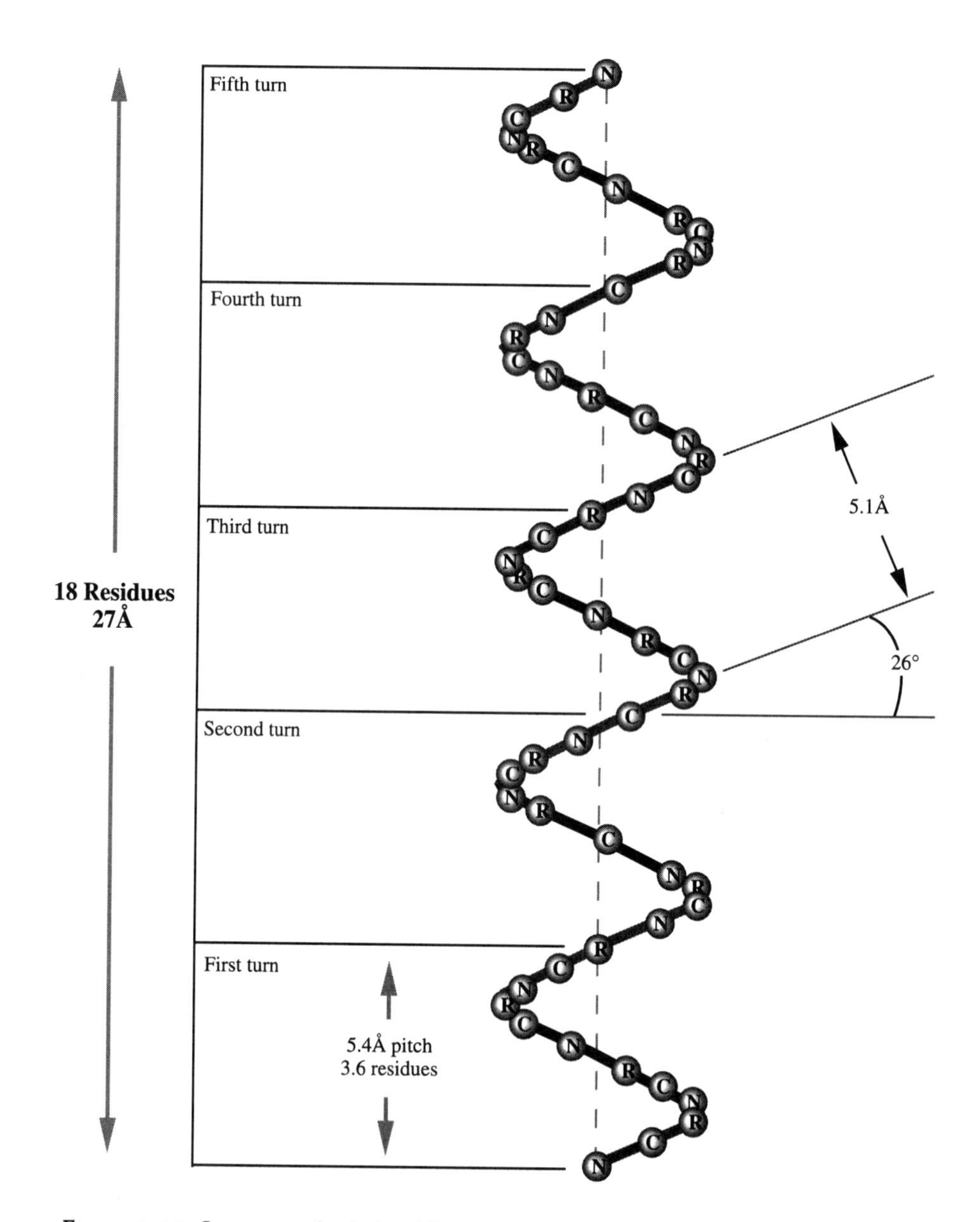

Figure 2.18 Structure of a helix. The structure of the α helix is characterized by 3.6 amino acid residues (R groups) per turn of the helix over an axial distance of 0.54 nm. This is consistent with 18 residues per five full turns of the helix.

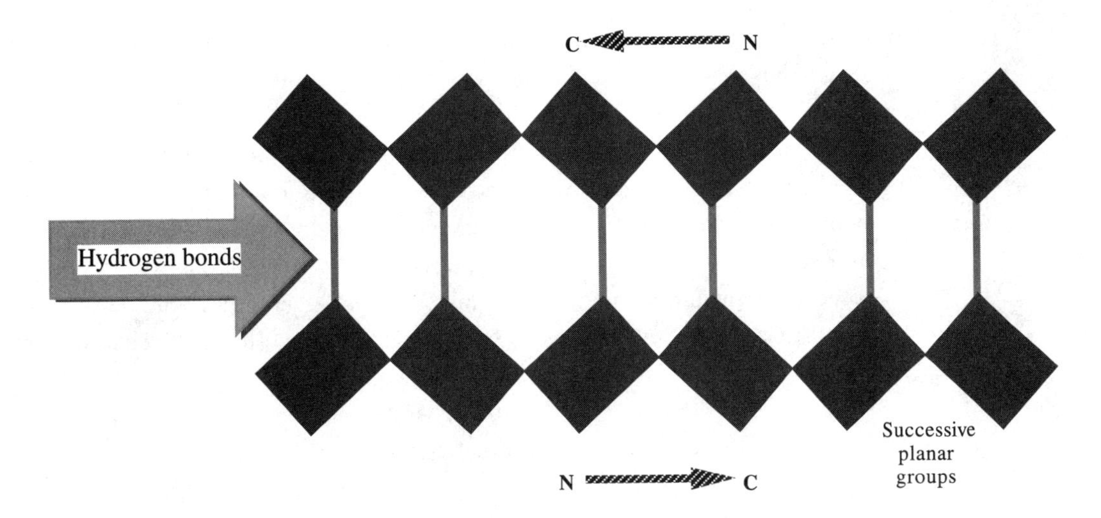

Figure 2.19 Hydrogen bonding in antiparallel β sheet. Antiparallel hydrogen bonding between carbonyl and amide groups within the peptide unit stabilizes the β extended conformation.

In the section on primary and secondary structures of macromolecules, we will see that hydrogen bonds are the primary type of bonds that stabilize helical and extended polypeptide conformations. The difference in these structures is dependent on the different hydrogen bond patterns that occur. Therefore, it is the hydrogen bond pattern that stabilizes folding of polypeptide chains.

Stereochemistry of Polysaccharides

The stereochemistry of polysaccharides is found using procedures similar to those used with polypeptides, except the repeat unit is a sugar and not a peptide unit. Polysaccharides are found as free molecules, such as glycogen, starch, and hyaluronan, and as side chains on molecules such as proteoglycans. There are three possible conformations of the basic glucose repeat unit: $^4C_1$ or C_1 chair conformation, $^1C_4$ chair conformation, or the boat conformation (see Figure 2.4). The most observed conformation of glucose and its derivatives is the C_1 chair conformation, also known as the $^4C_1$ conformation. Glucose repeat units can be linked through oxygen atoms that are either up (α linkage) or down (β linkage).

Stereochemical calculations are similar to those used for polypeptides to determine the allowable conformations for different repeat units. Figure 2.5 shows a repeat unit containing β-D-glucuronic acid and β-D-*N*-acetyl glucosamine (hyaluronan). Both units are derivatives of glucose and are connected via oxygen linkages. A β linkage is one where the position of the linkage is oriented equatorially; the bonds of the side chains are not perpendicular to the chain backbone. An α linkage is one where the bonds can be perpendicular to the backbone. If the linkage involves the carbon at the 1st and 3rd position, it is an α- or β-(1–3) linkage. If it involves the carbon at the 1st and 4th position, it is an α- or β-(1–4) linkage. The repeat disaccharide shown in Figure 2.5 is composed of D-glucuronic acid β-(1–3)-linked to D-*N*-acetyl glucosamine β-(1–4)-linked to D-glucuronic acid. Both sugar units are shown in the C_1 chair conformations. The 0°,0° conformation occurs when the atoms attached to the 1st and 3rd or 1st and 4th positions (i.e., the axially oriented side chains) eclipse each other when viewed from a plane perpendicular to the bonds. The stereochemical plot that is produced when the dihedral angles are rotated through 360° is shown in Figure 2.20. Compared to the stereochemical plots for proteins, the plot for hyaluronan shows a limited amount of allowable conformations, about 4% of the theoretical total. Hyaluronan behaves to a first approximation at high shear as a flexible molecule as compared to collagen, which has a single rigid stable conformation; this will be discussed further later. Flexibility is associated with the continuous range of allowable conformations around the 0°,0° position. Hyaluronan is flexible even though the conformational plot has a limited number of allowable conformations that center around 0°,0°.

Polymers of glucose and glucose derivatives, such as cellulose, have inherent chain flexibility. These polymers are also thixotropic; at low shear they are rigid and at high shear they are flexible. The thixotropic properties of high molecular weight polysaccharides are a reflection of their stereochemistry. In the ice cream manufacturing industry, cellulose derivatives are often used to thicken the ice cream. Because of the property of thixotropy, the manufacturer can pump the mixture of ingredients easily before the ice crystals are formed. Incidentally, most of the cellulose derivatives in food products come from low-quality wood pulp, so enjoy your ice cream!

Stereochemistry of Lipids

The stereochemistry of lipids is similar to that of polyethylene chains and proteins with small side chains. A freely rotating polymer chain (lipids without double bonds in the backbone) has a stereochemical plot that is similar to that shown in Figure 2.2. Therefore, most hydrocarbon chains are quite flexible and

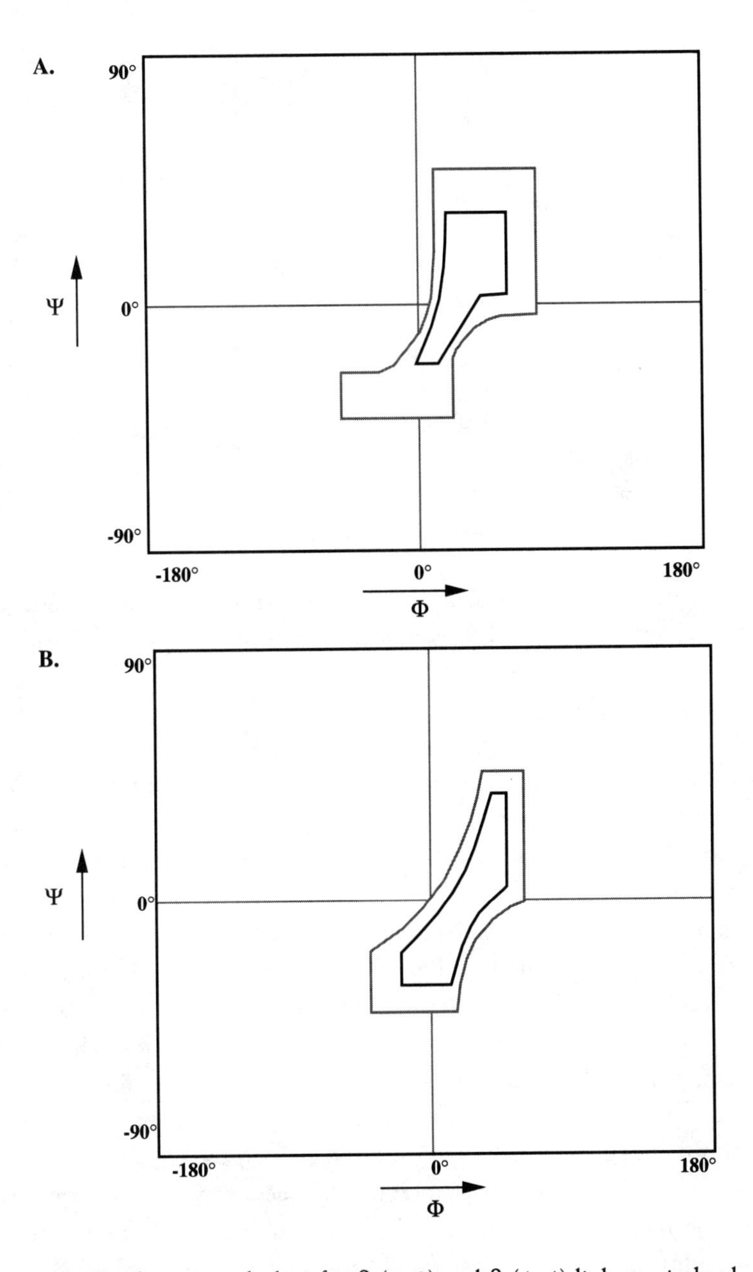

Figure 2.20 Conformational plots for β-(1–3) and β-(1–4) linkages in hyaluronan. Plots of fully allowed (solid lines) and partially allowed (dotted lines) conformations of ϕ and ψ for (**A**) D-glucuronic acid, which is β-(1–3)-linked to *N*-acetyl glucosamine, and (**B**) *N*-acetyl glucosamine, which is β-(1–4)-linked to D-glucuronic acid. Note that the allowed conformations center around 0°,0° and show that hyaluronan has flexibility.

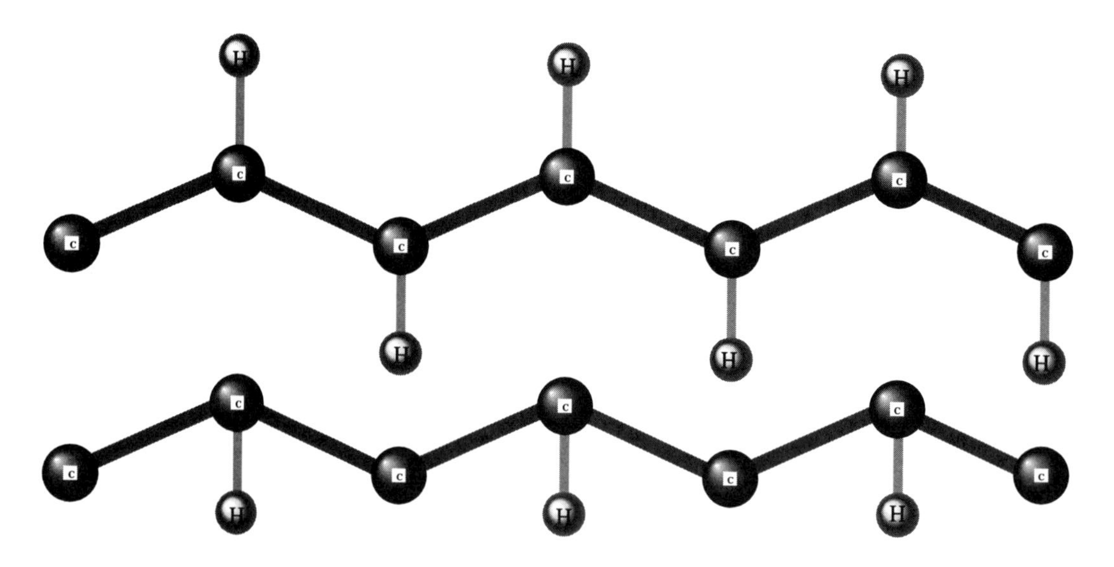

Figure 2.21 Packing of hydrocarbon chains. This diagram illustrates the planar zigzag packing pattern of hydrocarbon chains. Note the formation of van der Waals forces between hydrogens of the side chains and the backbone carbons.

adopt a number of conformations. In order to pack hydrocarbon chains efficiently in the cell membrane, the hydrocarbon component is in a planar zigzag. The hydrocarbon component is only about 14 carbon atoms long in cell membranes. Longer chains would crystallize as a result of the van der Waals bonds that form between hydrogen atoms between the chains, which is what happens with polyethylene when it is polymerized into chains with molecular weights exceeding 100. This is easily illustrated by examining the physical form of hexane, a low molecular weight polyethylene. At room temperature, hexane is a liquid. As the chain length of the hydrocarbon is increased, the material is first a wax; it will flow under pressure and heat like paraffin. At higher molecular weights, the material becomes a solid, although if heated enough the solid will melt. Therefore, when the chains get long enough, the secondary forces between them prevent the molecules from being flexible and allow the chains to crystallize. The flexibility of the chains can be demonstrated by noting the viscosity of gasoline, which is very low compared to that of a 2% solution of carboxy methylcellulose (CMC), the ice cream additive. Gasoline is composed mostly of hydrocarbon chains (gasoline contains octane, a hydrocarbon) and has almost no viscosity because the chains are flexible and slide by each other easily. If this were not so, then it would be difficult in cold weather for gasoline to flow into fuel injectors. In contrast, it would be undesirable for CMC to flow easily; ice cream would have no viscosity and a texture of flavored milk. Manufacturers use cellulose derivatives to replace much of the fat (hydrocarbon chains) that historically has made ice cream thick. The polymer chains in cellulose derivatives are less flexible than hydrocarbon chains and therefore are more efficient in increasing the viscosity.

In lipids, the hydrocarbon chain is composed of a carbon backbone with hydrogen side chains. The small nature of the hydrogen atoms attached to the backbone and the similarity in size allow these polymer chains to pack efficiently, because of the stereochemistry of the side chains. In a drawing of a hydrocarbon chain in a zigzag conformation, the hydrogens as viewed from the side of the chain are staggered and alternately fall between the other hydrogens (Figure 2.21). They can also be positioned to overlap the other hydrogen (eclipse each other). As mentioned in the discussion of equation 2.2, the Gibbs free energy should be minimized in the most favorable conformation. Figure 2.22 is a diagram of the Gibbs free energy as a function of the hydrogen rotational angle. The free energy is minimized when the hydrogens are staggered with respect to the backbone of the hydrocarbon chain.

Stereochemistry of Nucleic Acids

Examples of the repeat units in DNA and RNA are shown in Figure 2.6. The stereochemistry is similar to that of polysaccharides in that a sugar unit is linked through an oxygen to another sugar unit. The difference is that the oxygen linkage has several other atoms attached to it between the sugars, providing additional flexibility to the chain backbone. Therefore, the stereochemical plot for nucleic acid polymers will be similar to that of polysaccharides, except there will be additional allowed conformations. This flexibility is necessary for DNA so that it can fold into a double helix, as discussed later.

Primary and Secondary Structures of Biological Macromolecules

The stereochemistry of macromolecules and the concept of repeat units in macromolecules of biological relevance were introduced in the previous section. In this section, the relationship between repeat sequences of biological macro-

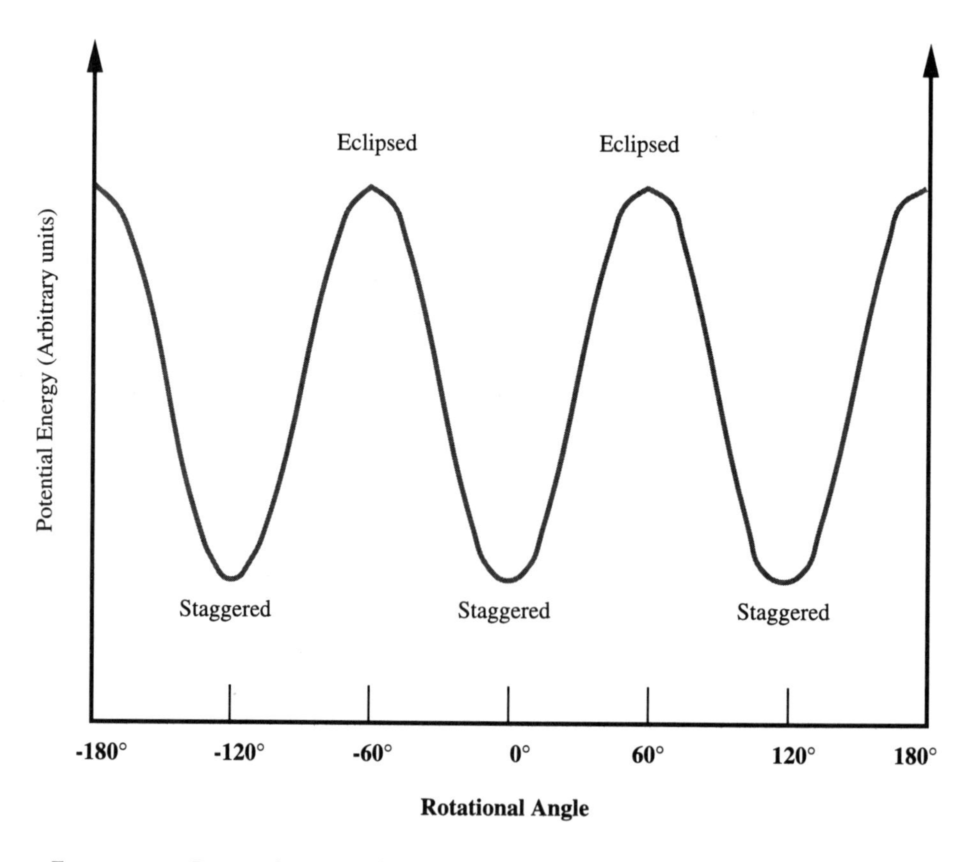

Figure 2.22 Potential energy function for hydrocarbon chains. Plot of potential energy versus angle of side chain rotation. Potential energy is minimized when the hydrocarbon side chains are staggered with respect to each other when looking down the chain from the side.

molecules and macromolecular structure in three dimensions will be discussed. Providing the sequence of every protein and nucleic acid is beyond the scope of this textbook. Instead, we will focus on the relationship between the chemistry of the repeat sequences and the resulting polymer structure. All the biological macromolecules of importance to biomaterials scientists will not be covered, but we will establish the principles so that they can be extended to the analysis of other macromolecules.

Primary and Secondary Structures of Proteins

There are four classes of protein structures that make up part or all of a biological macromolecule. Some molecules contain several different structural types connected by sequences that have other structures. The four classes that we will discuss are α helix, β sheet, collagen triple helix, and random chain structure, which are defined by the combinations of backbone angles given on the conformational plot in Figure 2.17. The one exception is the random chain structure characteristic of polypeptide sequences such as found in elastin. In this case, the conformation angles are not fixed but allow nearly free mobility of polymer chains.

α Helix

The α helix is probably the most commonly occurring helical structure in proteins and the first that was worked out. Linus Pauling determined that this helix has 3.6 amino acids per turn, an exact repeat of the helix every 5.4 Å, and an axial rise per residue of 1.5 Å (Figure 2.18). In the α helix, the amino acid side chains are directed radially away from the axis of the helix. The helix is stabilized by formation of hydrogen bonds between the carbonyl oxygen of 1 amino acid residue, which has a slight negative charge, and the hydrogen of the amino group of a residue 4 amino acids further down the chain, which has a slight positive charge. The sequence of amino acids must allow for a hydrogen bond pattern that is almost perpendicular to the axis of the molecule, which eliminates protein sequences containing either proline or hydroxyproline. α Helixes are either right- or left-handed (right-handed chains run clockwise as you look from one end to the other); right-handed forms are most commonly observed. The α helix is stable because of the numbers of hydrogen bonds formed and the linear nature of the bond.

Macromolecules of importance to the biomaterials scientist include hemoglobin, myosin, actin, fibrinogen, and keratin; some portion of these macromolecules is composed of an α-helical structure. The α helix is a rather condensed structure. The rise per residue is 1.5 Å, which is quite different from that of the collagen triple helix and the β structure of silk. The rise per residue in the two latter structures is about twice that found in the α-helix structure. For this reason, the extensibility of the α helix is greater than that of the collagen triple helix and the β structure. In keratin, tensile deformation of the α helix leads to formation of the β structure.

Although α helixes are abundant in proteins, the average length is fairly short (17 Å, containing about 11 amino acids or 3 turns). Therefore, α helixes are found predominantly in short domains within proteins and not as continuous stretches. Keratins, which make up hair and the most superficial layer of skin, contain a central domain with an α-helical component of 310 amino acids or about 46.5 nm in length. Keratin is an example of a protein with a fairly long α helix. The amino acid composition that favors α helix formation is fairly broad, with the exception of proline and serine.

β Sheet

Like α helixes, extended structures can be held together by hydrogen bonds with the hydrogen bonds running perpendicular to the chain axis. Silk is an example of a protein that is found in the β structure. The amino acid composition of silk is rich in glycine (44.5%), alanine (29.3%), and serine (12.1%), with small hydrocarbon side chains that form sets of antiparallel hydrogen bonds between molecular chains. Models of polypeptide chains with sequences of poly(Gly-Ala) and poly(Ala-Gly-Ala-Gly-Ser-Gly) show that the most probable structure contains all the Gly residues on one side of the chain and all the Ala on the other side of the chain. Therefore, by packing the chains in antiparallel fashion, the side chains fit neatly into the empty spaces between the chains (see Figure 2.19). The rise per residue in the β structure is about 3.5 Å.

Collagen Triple Helix

The other extended protein structure that we will discuss is the collagen triple helix. Our knowledge about the structure of collagen came from early studies on its amino acid composition, the structure of peptides derived from collagen, and the analysis of the X-ray diffraction pattern of collagen fibers. Early compositional studies were important in establishing that collagen was characterized by a high content of glycine, proline, and hydroxyproline. This did not fit in with the established amino acid profile for proteins that form an α helix or a β sheet structure. Therefore, a new structure needed to be proposed for this sequence. However, it was clear that the presence of proline and hydroxyproline would result in an extended structure based on the conformational plot. After cleavage of collagen with acid, it was determined that glycine accounted for about 33% of the amino acid residues and proline and hydroxyproline together accounted for another 25% of the amino acid residues. It was demonstrated by the early 1950s that the dipeptides that made up collagen were mostly Gly-Pro and Hyp-Gly. This observation led to the hypothesis that every third residue in the collagen structure was probably glycine; therefore, the proposed structure must accommodate the rigid proline and hydroxyproline residues.

The other evidence that proved important in unraveling the structure of collagen came from understanding the X-ray diffraction pattern. Before 1940, biophysicists recognized that when an X-ray beam passed through a tendon or tissue containing oriented collagen fibers, spots appeared on a photographic plate positioned behind the fiber. Near the meridian (the vertical axis) of the exposed photographic plate, arcs appeared at a position that was equivalent to a spacing of 2.86 Å. Theses arcs were found in oriented samples of different types of connective tissue, ranging from mammoth tusk to sheep intestine. The repeat of 2.86 Å was thought to be the displacement per amino acid along the axis of the molecule. In 1954, Ramachandran and Kartha proposed a model consisting of three parallel chains linked to form a cylindrical rod, with the rods being packed into a hexagonal array. A year later, Ramachandran and Kartha modified this model by adding an additional right-hand twist of 36° every three residues in a single chain (Figure 2.23). Further refinement of the structure continued during the subsequent 40 years. We now know that hydroxyproline stabilizes the molecule by forming a hydrogen bond between chains via a water molecule or by forming a direct hydrogen bond between the carbonyl group on one chain and an amide hydrogen on another chain within groups of three amino acids (Figure 2.24). In addition, it is believed that the molecule can be broken up into rigid and partially flexible regions associated with the gap and overlap regions within the fibril structure (see Structure of Fibrous Collagen for more details).

Some proteins form structures that include chain segments that are random coils; this type of polymer chain is freely flexible and is not hindered in its rotation by hydrogen bonds. An example is elastin, which is found in elastic fibers; this molecule contains α-helical regions and disorganized regions that behave like a random coil.

Primary and Secondary Structure of Polysaccharides

Most of the polysaccharides that we discuss in this textbook are *glycosaminoglycans,* polymers that contain an amino sugar in the repeat unit. Glycosaminoglycans that are abundant in mammalian tissues include hyaluronan, chondroitin sulfate, dermatan sulfate, keratan sulfate, heparin, and heparan sulfate (Table 2.4). Most of these glycosaminoglycans, with the exception of hyaluronan, are composed of short polysaccharide chains and have limited secondary structure. However, extensive studies on hyaluronan have indicated that it likely has some limited secondary structure. In 1980, Atkins and co-workers reported that the rate

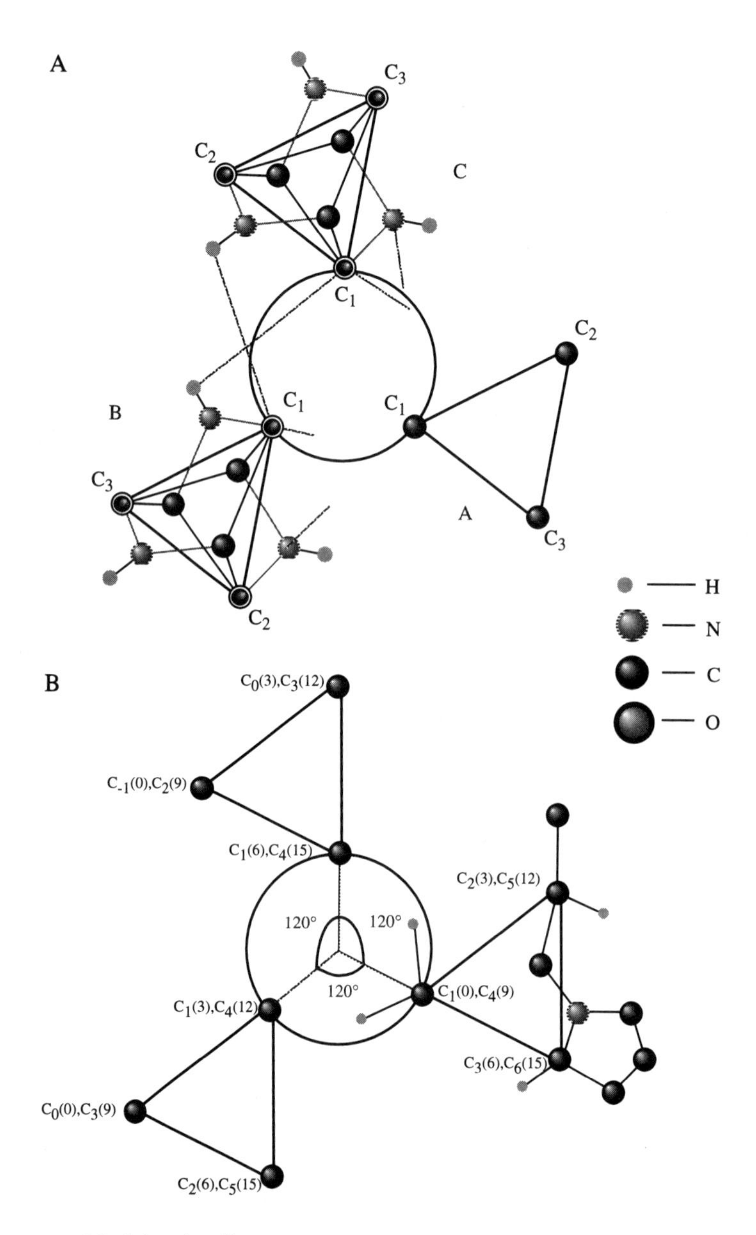

Figure 2.23 Models of collagen structure. (**A**) Model of three parallel left-handed helixes of collagen showing the location of Cα (C) for chains A, B, and C. Note all glycines are found in C-1 position because this is the only amino acid residue that can be accommodated at the center of the triple helix. Later studies by Ramachandran and co-workers indicated that the three chains are wrapped around each other (**B**) in a right-handed superhelix. The axial rise per residue is 0.29 nm, and the axial displacement of different Cα atoms is shown in parentheses in angstroms.

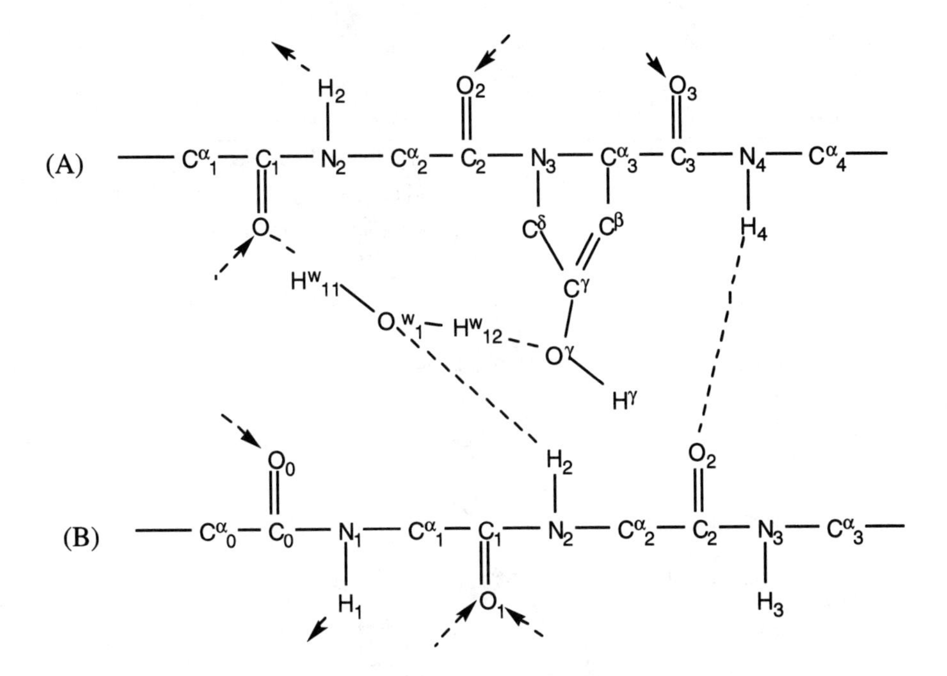

Figure 2.24 Stabilization of collagen triple helix. The diagram shows hydrogen bonding between the amide hydrogen in position 4 on chain A and the carbonyl oxygen in position 2 on chain B. A second water-mediated hydrogen bond occurs when hydroxyproline is present in position 3 on chain A between the carbonyl oxygen in position 1 on chain A and the amide hydrogen in position 2 on chain B.

of oxidation (reaction with oxygen) of hyaluronan and chondroitin sulfate showed that the C-2–C-3 glycol group in the glucuronic acid moiety was very slowly oxidized. This could be explained by steric hindrance of the glycol group (OH) in hyaluronic acid and chondroitin sulfate (which could hydrogen bond to *N*-acetyl glucosamine) but not in dermatan sulfate (which contains *N*-acetyl galactosamine and cannot hydrogen bond to the same glycol groups). The proposed hydrogen bond scheme is shown in Figure 2.5 and is consistent with a highly extended ribbonlike helix with an axial rise per disaccharide of 9.8 Å. The stiffening of hydrogen bond arrays over short segments of the molecule explains the large expanses (space-filling role) that are occupied by hyaluronan molecules.

Table 2.4 Glycosaminoglycans found in mammalian tissues

Glycosaminoglycan	Repeat disaccharide
Hyaluronic acid	β-D-Glucuronic acid + β-D-N-acetyl glucosamine
Chondroitin sulfate	β-D-Glucuronic acid + β-D-N-acetyl galactosamine
Dermatan sulfate	β-D-Glucuronic acid or β-L-Iduronic acid + α-D-N-acetyl galactosamine
Heparin, heparan sulfate	β-D-Glucuronic acid or β-L-Iduronic acid + α-D-N-acetyl glucosamine
Keratan sulfate	β-D-Galactose + β-D-glucosamine

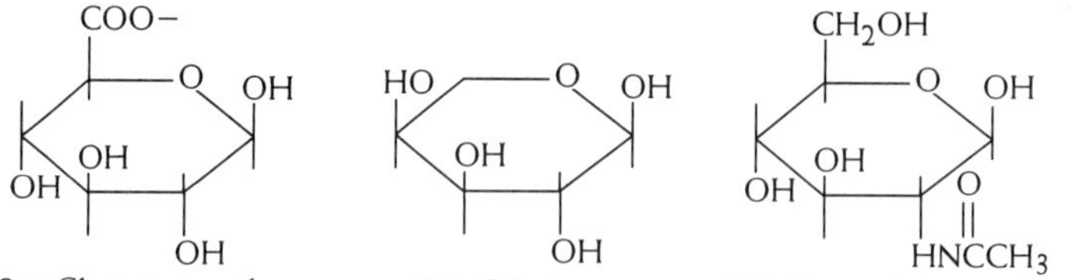

COO−, O, OH, OH, OH, OH
β-D-Glucuronic acid

HO, O, OH, OH, OH
β-D-Galactose

CH_2OH, O, OH, OH, OH, O, HNCCH3
β-D-N-acetyl glucosamine

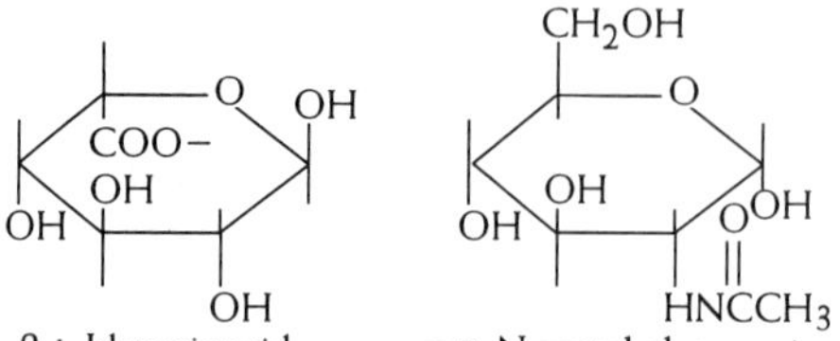

O, OH, COO−, OH, OH, OH
β-L-Iduronic acid

CH_2OH, O, OH, OH, O, OH, HNCCH3
α-D-N-acetyl glucosamine

Primary and Secondary Structure of DNA and RNA

The structure of DNA was originally thought to contain equal amounts of the four purines and pyridines. By the late 1940s, Chargaff and others found that the ratios of adenine and thymine were always close to unity, and the same was true for guanine and cytosine. This implied that for some reason, every molecule of DNA contained equal amounts of adenine and thymine and also equal amounts of guanine and cytosine. Using this chemical information together with X-ray diffraction patterns of DNA, Watson, Crick, Wilkins, and Franklin proposed a model for the structure of DNA in the early 1950s. They proposed that a molecule of DNA consists of two helical polynucleotides wound around a common axis to form a right-handed double helix. In direct contrast to the arrangements in helical polypeptides (where the amino acid side chains are directed to the outside of the helix), the purine and pyrimidine bases of each polynucleotide chain were directed toward the center of the double helix so that they faced each other. Watson and Crick further suggested, on the basis of stereochemistry principles, that there is only one possible way the nitrogen bases can be arranged within the center of the double helix that is consistent with the predicted dimensions—that in which the purine always faced the pyrimidine. Based on consideration of the possible hydrogen bond patterns between purines and pyrimidines, they concluded that adenine must be matched with thymine and guanine with cytosine.

The parameters of the double helix that is formed by DNA include a diameter of 20 Å, a rise per nucleic acid residue of 3.4 Å, and 10 residues per complete turn. The two chains that make up the molecule are antiparallel; the chains grow by adding repeat units to the 3′ or 5′ group on ribose. Therefore, one chain is joined 3′ to 5′ by phosphodiester bonds, and on the other chain the riboses are joined 5′ to 3′. The two polynucleotides are twisted around one another in such a way as to produce two helical grooves in the surface of the molecule.

The structure of RNA is different than DNA; it does not form a double helix, but folds to form different structures, such as the large and small subunit of the ribosome. The 3-D structures are discussed the next section.

Higher-Order Structures

The types of macromolecular structures found in tissues include helixes, extended structures, and random coils. Ultimately, the properties of biological polymers are dictated not only by the macromolecular structure, but also by the levels of structural hierarchy that are found in tissues. For example, elastin is easily deformed by mechanical loading. In contrast, structures containing keratin are less easily deformed—not only are they made up of α helixes, but the α helixes are packed into higher-ordered structures. In keratin and even collagen, it is the higher-ordered structures that dictate whether a tissue is soft and pliable or hard and rigid. Therefore, as we will see in Chapter 7, although helical molecules are in general more difficult to deform (they require more force per unit area) than are random coils, the presence of higher-order structure gives biological systems the ability to tailor structures and therefore physical properties.

The higher-order structures of proteins are numerous and include coiled-coil α helixes, two parallel strands of β sheets, and globular proteins that consist of one or more structural domains. Biological structures can be broken into basic structural units, as described in this section. Macromolecules such as the actin filaments of muscle should not be thought of as "thin filaments," but rather as a collection of amino acids that form domains, each of which is characterized as an α helix or some other well-defined structure that is linked to other regions of defined structure. In this manner, it is possible to relate the structure with specific biological functions. The sections that follow examine the structures of macromolecules with specific biological functions, and some of these structures will be correlated with mechanical properties in Chapter 7.

Structure of Extracellular Matrix Macromolecules

The extracellular matrix is of primary importance to the biomaterials scientist because any implant placed subcutaneously or anywhere within or on the body will immediately affect it. Any changes in the extracellular matrix and any of its components will affect the function of the tissue and may lead to device failure. In addition, any device-induced changes that lead to inflammation, immune responses, or even just physical pressure will ultimately cause activation of one or more biological systems, which is discussed in Chapter 9. Activation of these systems will lead to changes in the host's ability to respond to foreign cells. Extracellular matrix macromolecules of interest include fiber-forming collagens, elastic fibers, fibronectin, laminin, dermatan sulfate and chondroitin sulfate proteoglycans, and hyaluronan.

Structure of Fibrous Collagens

Types I, II, and III collagens form the fibrous network that prevents premature mechanical failure of most tissues. The molecular sequences of these collagens are known, and they are composed of approximately 1,000 amino acids in the form of Gly-X-Y with small nonhelical ends before and after these sequences. All these collagen types form continuous triple-helical structures that pack laterally into a quarter-stagger structure in tissues to form characteristic D-periodic fibrils, as shown in Figure 2.25. These fibrils range in diameter from about 20 nm in cornea to more than 100 nm in tendon. In tendon, collagen fibrils are packed into fibril bundles that are aligned along the tendon axis. In skin, types I and III collagen fibrils form a nonwoven network of collagen fibrils that aligns with the direction of force. In cartilage, type II collagen fibrils form oriented

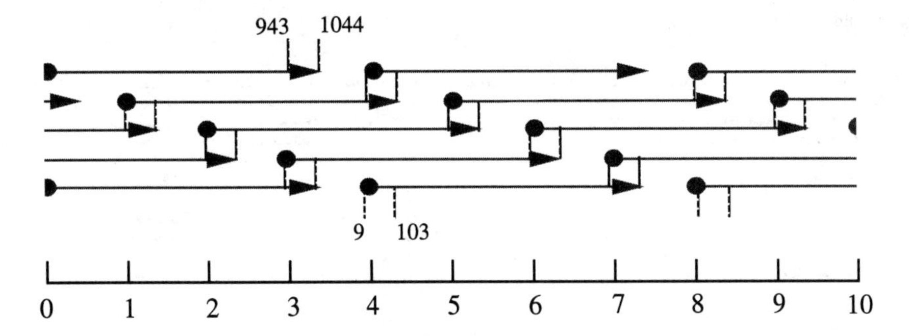

Figure 2.25 Cross-linking of collagen molecules in quarter-stagger packing pattern of collagen in fibrils. Each molecule is 4.4-D long (where D is 67 nm) and is staggered by D with respect to its nearest neighbors. A hole region of 0.6 D occurs between the head (circles) of one molecule and the tail of the preceding molecule (arrowheads).

networks that are parallel to the surface (top layer), whereas in the deeper zones, they are perpendicular to the surface.

Structure of Elastic Fibers

Elastic fibers form the network in skin and cardiovascular tissue (elastic arteries) that is associated with elastic recovery. Elastic fibers are composed of a core of elastin surrounded by microfibrils 10 to 15 nm in diameter that are composed of a family of glycoproteins recently designated *fibrillins* (Reinhardt et al. 1996). Fibrillins are a family of extracellular matrix glycoproteins (MW about 350,000) containing a large number of cysteine residues (cysteine residues form disulfide cross-links). Two members of the family have been described, fibrillin-1 and fibrillin-2. The common molecular features include N- and C-terminal ends with 47 tandemly repeated epidermal-growth-factor-like modules separated by a second repeat consisting of 8 cysteine residues and other structural elements. Several possible arrangements for fibrillin have been postulated, including unstaggered parallel arrangements and staggered parallel arrangements (Figure 2.26).

Elastin is a macromolecule synthesized as a 70,000 single peptide chain called *tropoelastin*. It is secreted into the extracellular matrix where it is rapidly cross-linked to form mature elastin. The carboxy-terminal end of elastin is highly conserved with the sequence Gly-Gly-Ala-Cys-Leu-Gly-Leu-Ala-Cys-Gly-Arg-Lys-Arg-Lys. The two cysteine residues that form disulfide cross-links are found

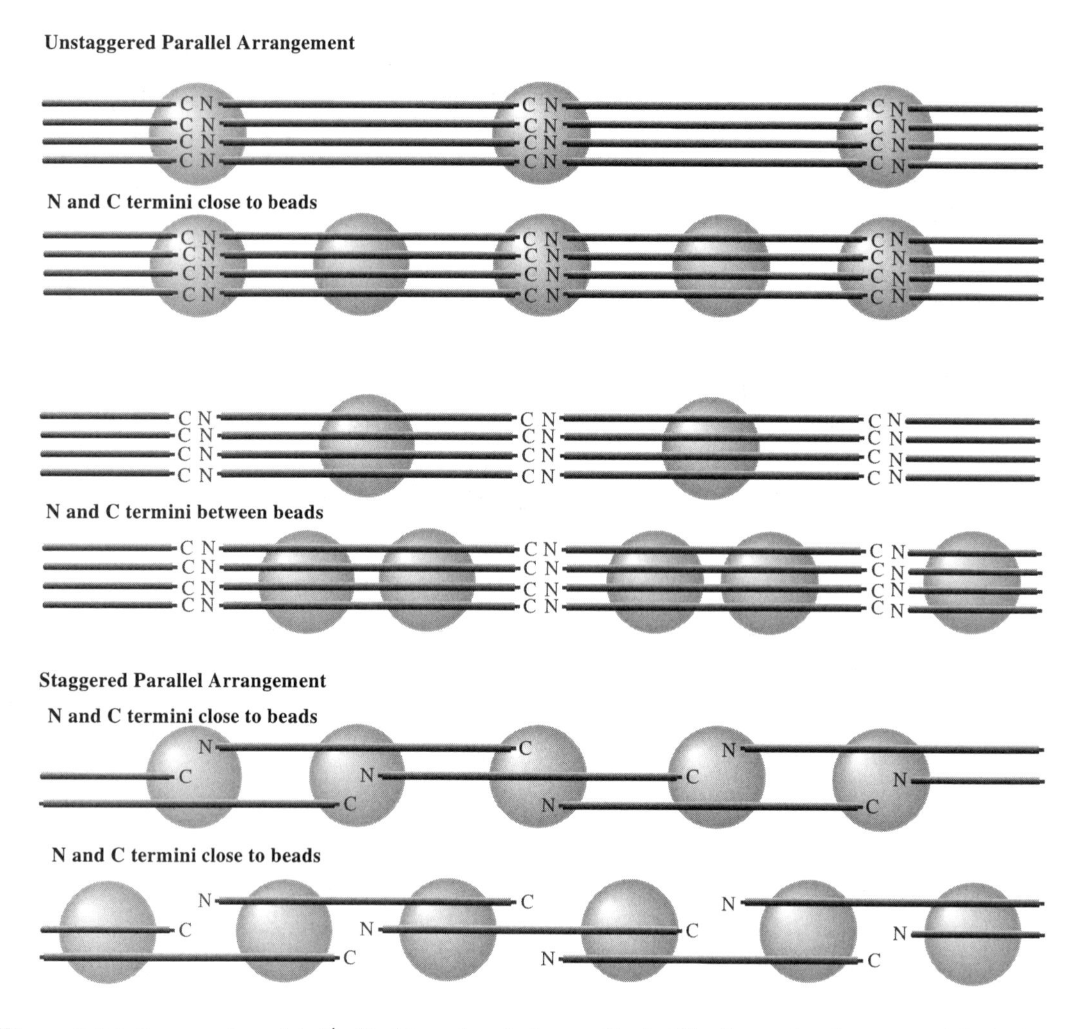

Figure 2.26 Structural models for fibrillin in beaded microfibrils. Fibrillin molecules are modeled as being arranged head to tail in parallel arrangements. In some of the models, the molecules are staggered with respect to their neighbors. The circles represent areas where the beads are observed. *Source:* Adapted from Reinhardt et al. (1996).

in this region, along with a positively charged pocket of residues that is believed to be the site of interaction with microfibrillar protein residues. Hydrophobic alanine-rich sequences near lysine residues that form cross-links between two or more chains are known to form α helixes. Alanine residues not adjacent to lysine residues found near proline and other bulky hydrophobic amino acids inhibit α helix formation. Additional evidence exists for β structures and β turns within elastin, thereby giving an overall model of the molecule that contains helical stiff segments connected by flexible segments.

Structure of Laminins

Laminins are a family of extracellular matrix proteins that are found in basement membrane and have binding sites for cell-surface integrins and other extracellular matrix components. They consist of α, β, and γ chains with molecular weights between 140,000 and 400,000. These chains associate through a large triple-helical coiled-coil domain near the C-terminal end of each chain (Figure 2.27). Eight different laminin chains have been identified: α_1, α_2, α_3, β_1, β_2, β_3, γ_1, and γ_2. The most extensively characterized of the seven forms of laminin is laminin 1 ($\alpha_1\beta_1\gamma_1$), which assembles in the presence of calcium to form higher-ordered structures in basement membranes with type IV collagen (Figure 2.28).

Structure of Fibronectins

The fibronectins are a class of high molecular weight multifunctional glycoproteins that are present in soluble form in plasma (0.3 g/L) and other bodily fluids and in fibrillar form in extracellular matrix. They bind to cell surfaces and other macromolecules. including collagen and gelatin (the unfolded form of collagen), as well as fibrinogen and DNA. Fibronectin mediates cell adhesion, embryonic cell migration, and wound healing. It is composed of two chains, α and β (MW about 260,000), that are covalently linked via two disulfide bonds near the carboxy termini. The structure may contain some β structure and contains a heparin-binding domain, a cell-binding domain, and a collagen-and-gelatin-binding domain. The molecule can adopt an extended conformation at high ionic strength or a compact one at physiologic ionic strength, suggesting that the molecule is flexible (Figure 2.29).

Dermatan Sulfate, Chondroitin Sulfate, and Heparan Sulfate Proteoglycans

Proteoglycans are a diverse family of glycosylated proteins that contain sulfated polysaccharides as a principle constituent. The diverse structures found for these

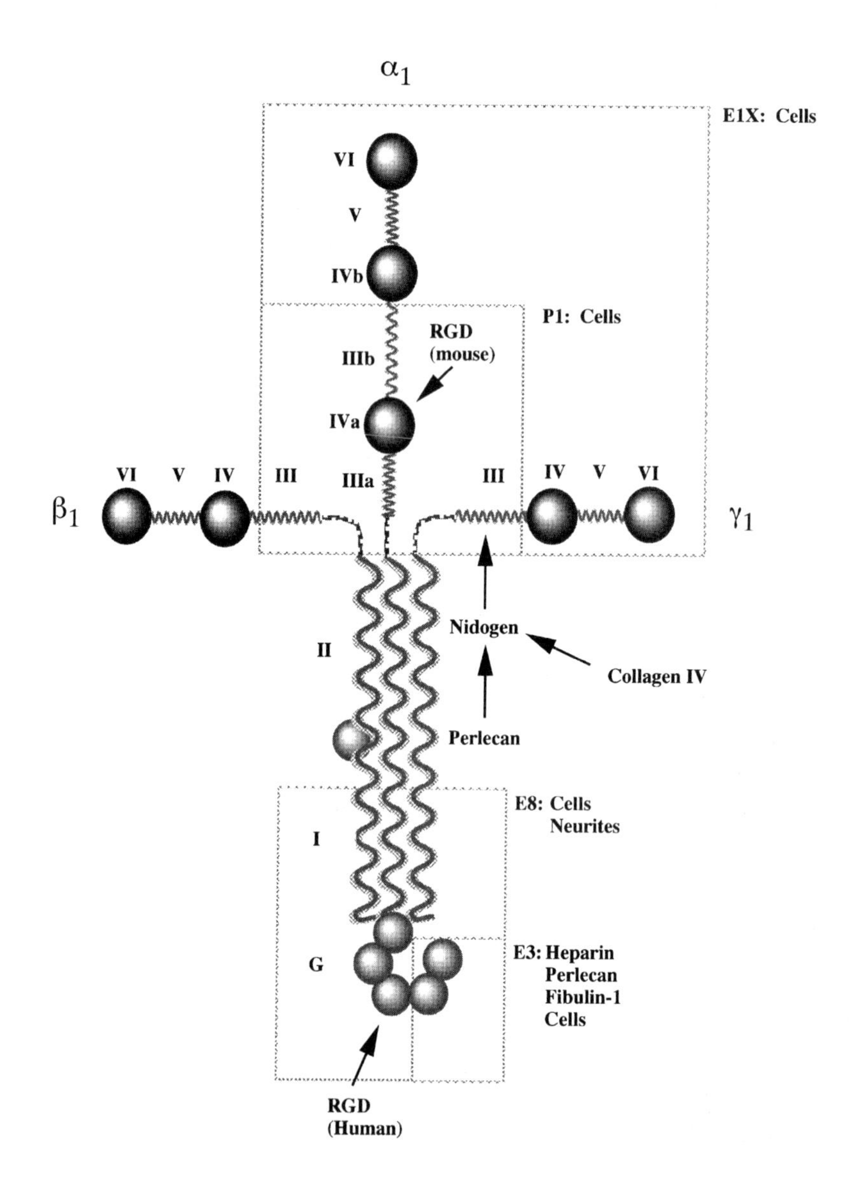

Figure 2.27 Domain structure of laminin-1. Structure of laminin-1 showing domains (Roman numerals) and binding sites for elastase (E) and pepsin (P) fragments (dotted lines). *Source:* Adapted from Timpl and Brown (1994).

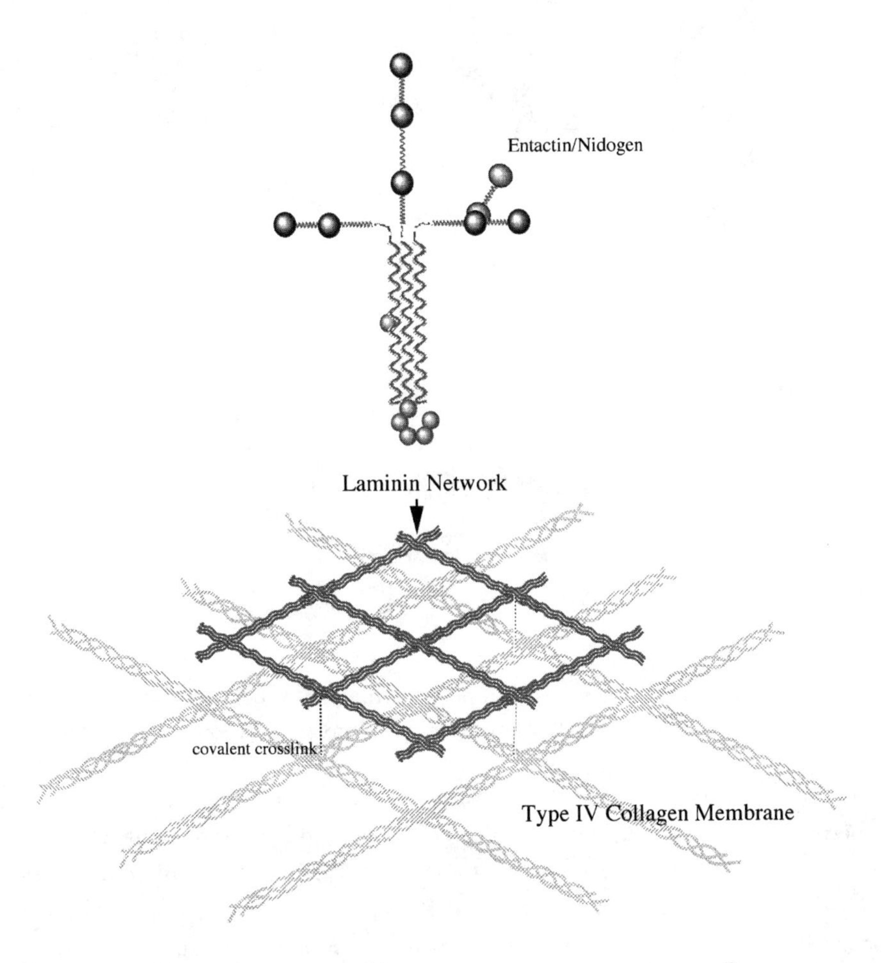

Figure 2.28 Network structure of laminin in basement membranes. Basement membranes contain a laminin network that is covalently cross-linked to a type IV collagen network. *Source:* Adapted from Yurchenco et al. (1992).

molecules depend on the type of glycosaminoglycan side chains, length, and net charge. Aggrecan, which is found in large amounts in cartilage (50 mg/g of tissue), is highly glycosylated with 200 chains containing chondroitin sulfate and keratan sulfate. The central core protein has a MW of about 220,000; the overall aggrecan weight can reach 2 to 3 million. It interacts with hyaluronan via link protein (Figure 2.30). In contrast, decorin and biglycan have relatively small core proteins (MW about 40,000) with a leucine-rich repeat and one (decorin) and two (biglycan) chondroitin sulfate and dermatan sulfate side chains (Figure 2.31). Decorin binds to collagen fibrils; biglycan does not. A unique heparan sulfate

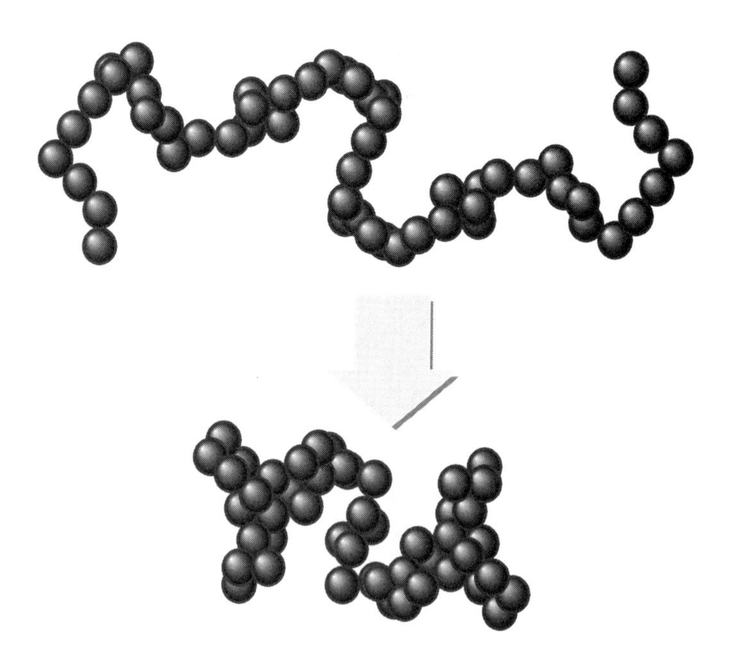

Figure 2.29 Structural changes in fibronectin. Structure of fibronectin at high (top) and low (bottom) ionic strength. Note collapse of molecule at low ionic strength.

proteoglycan, perlecan, is the major proteoglycan found in basement membranes. It has a large protein core containing about 3,500 amino acids and consists of multiple domains. Perlecan is able to self-associate or interact with several other basement membrane macromolecules, including laminin and type IV collagen. In the kidney, heparan sulfate proteoglycan contributes a negative charge to the basement membrane and is thought to exclude serum proteins from being filtered out of the blood. Although the structure of perlecan is still not completely understood, a model has been developed (Figure 2.32).

Hyaluronan

Hyaluronan is a component of every tissue or tissue fluid in higher animals. The highest concentrations are found in cartilage, vitreous humor, and umbilical cord; even blood contains some hyaluronan. At physiologic pH, the molecule can adopt a helical structure in solution; therefore, the molecule can be modeled as a series of helical segments that are connected together with flexible segments.

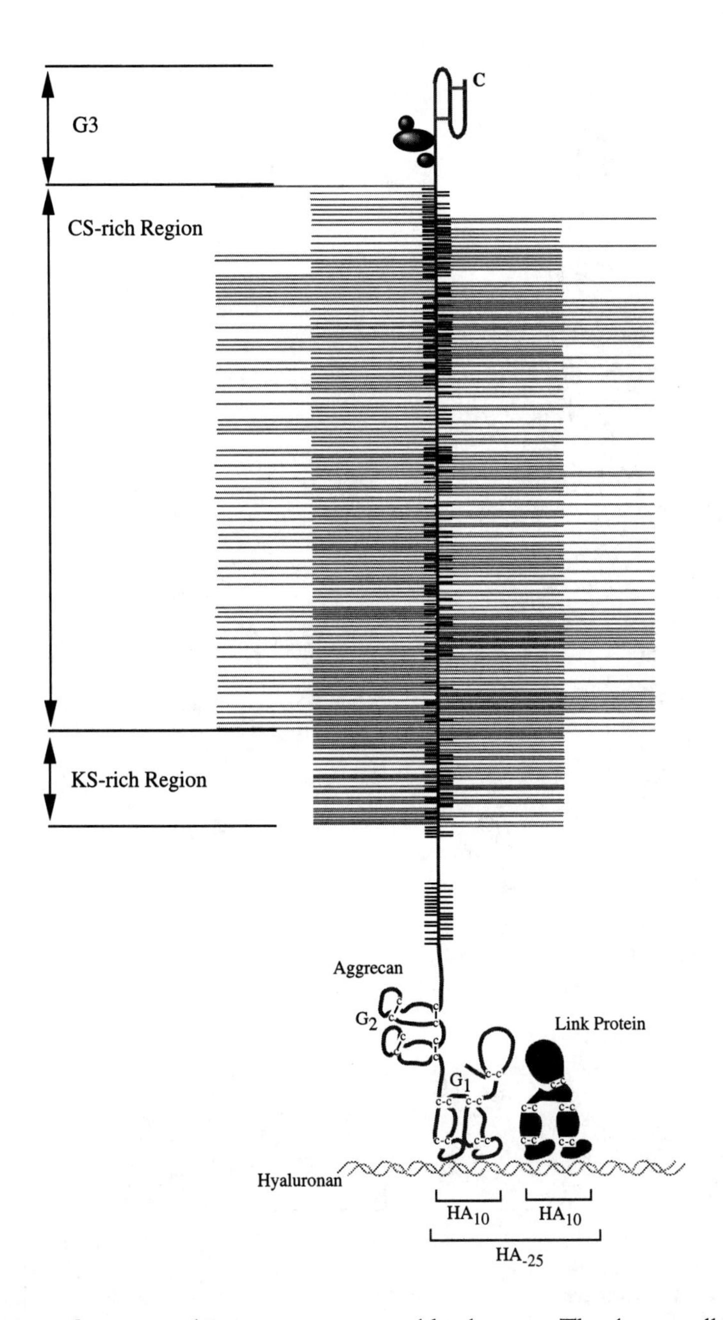

Figure 2.30 Interaction between aggrecan and hyaluronan. The diagram illustrates the interaction between large aggregating proteoglycan (aggrecan), link protein, and hyaluronan.

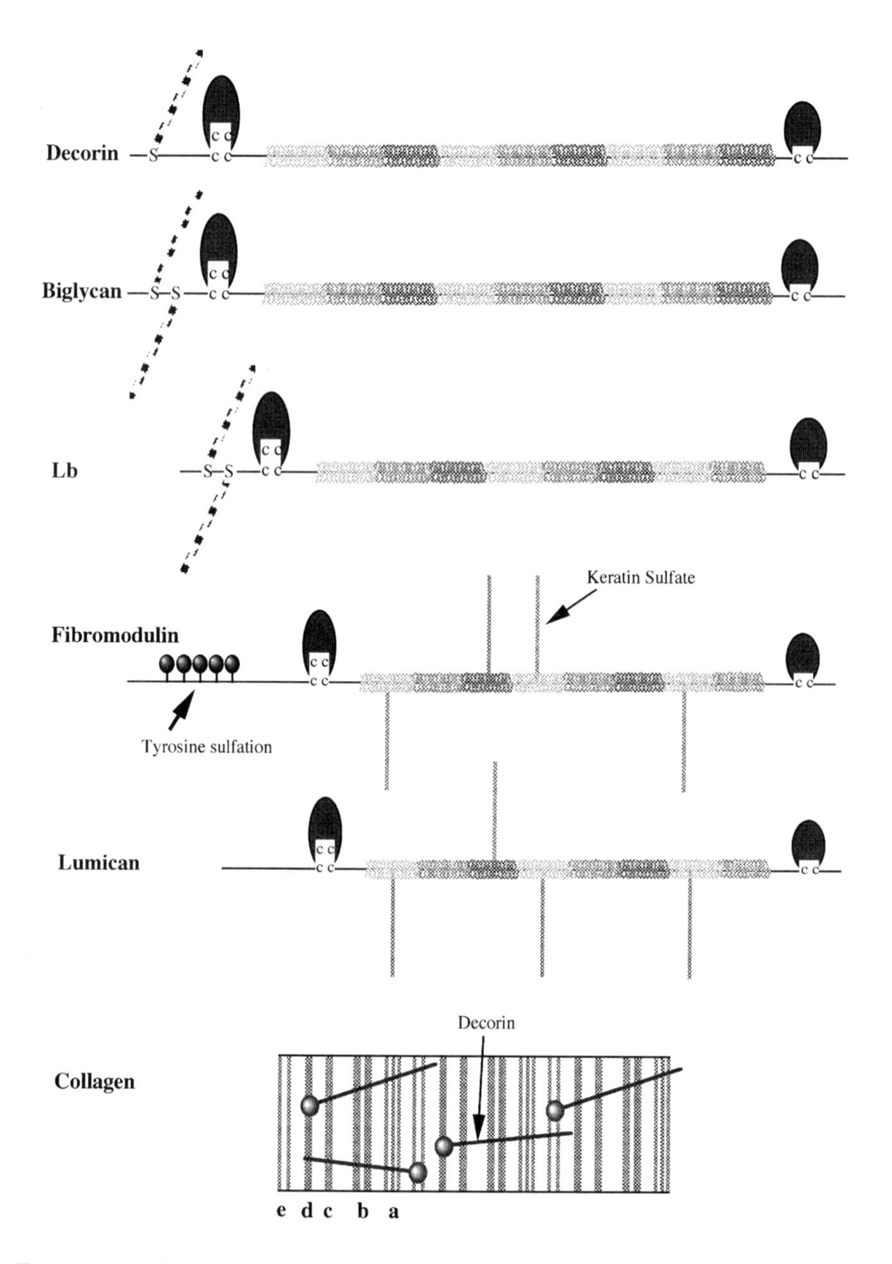

Figure 2.31 Comparison between proteoglycan structures. The diagram illustrates structures of decorin (one glycosaminoglycan side chain), biglycan (two glycosaminoglycan side chains), proteoglycan-Lb, fibromodulin, and lumican. Also shown is the specific binding of decorin to the d and e bands on collagen fibrils.

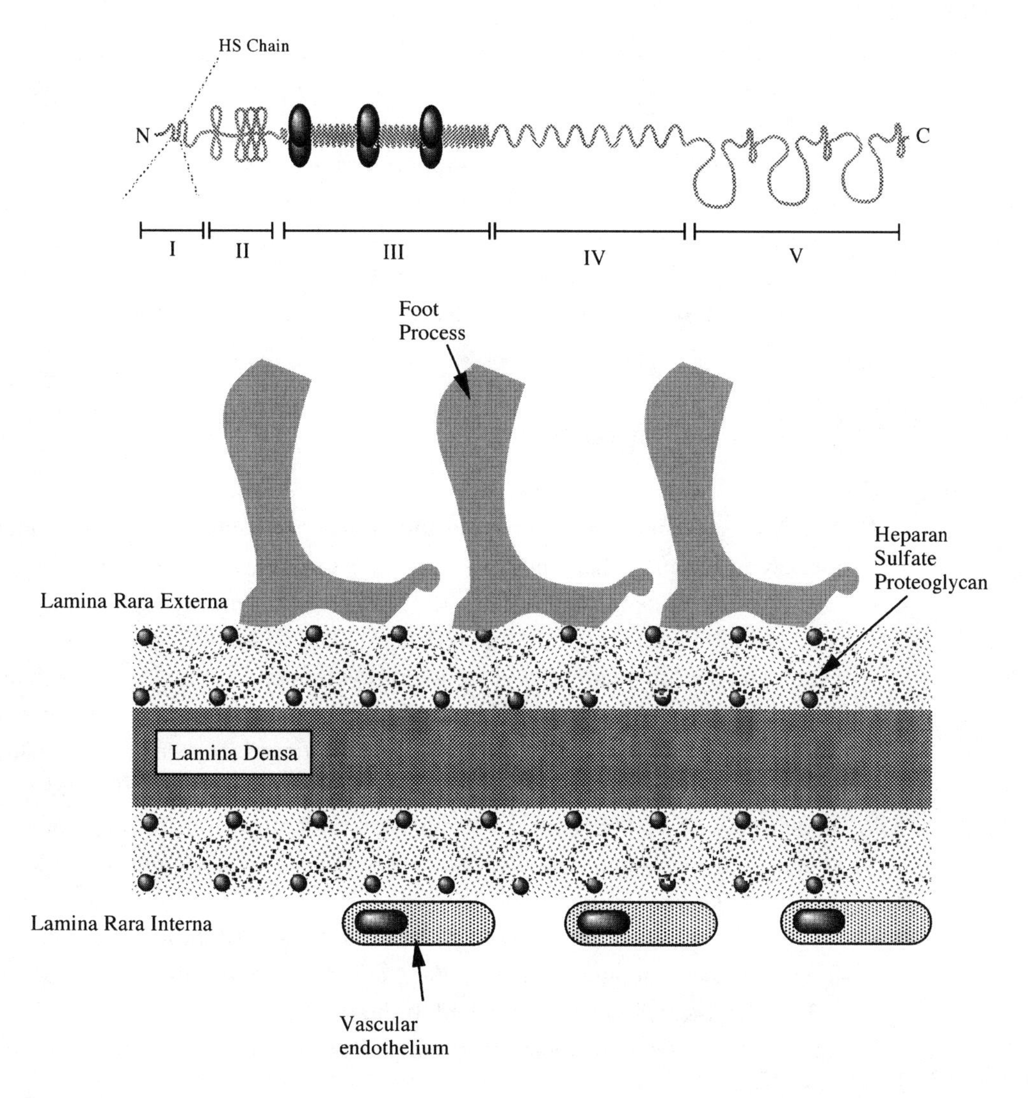

Figure 2.32 Model of perlecan in basement membranes. The illustration shows the structure of perlecan, a basement membrane proteoglycan, and localization of proteoglycans in the basement membrane.

This explains the observation that hyaluronan behaves like a wormlike coil in solution.

Cell Membrane Polymers

Several polymers associated with cell membranes are discussed in this textbook, including heparan sulfate proteoglycans, integrins, and class I and II markers of the major histocompatibility complex (MHC). These macromolecules are important to the biomaterials scientist because they provide specific recognition, attachment, and mechanical continuity between the extracellular matrix and the cell membrane as well as structural support for the cell membrane. Of importance is the fact that any changes in chemical and mechanical forces acting on the cell or changes in the fluid flow around a cell can be transduced into a change in cellular shape. These changes are transmitted through the cell membrane, which causes rearrangement of the actin filaments in the cell cytoskeleton.

Syndecan and Glypican

Nearly every mammalian cell has heparan sulfate proteoglycan as a plasma membrane component. These proteoglycans are inserted into the cell membrane through a transmembrane domain in their core protein (syndecans) or via a modified glycosaminoglycan region that is linked to the cell membrane (glypican). Heparan sulfate proteoglycans interact with numerous molecules, including growth factors, cytokines (see Chapter 8), extracellular matrix proteins, enzymes, and protease inhibitors. It is thought that they transduce signals that emanate from the interplay between components in the extracellular matrix. Syndecan is associated with epithelial cell differentiation after migration and with the intracellular actin cytoskeleton; expression of syndecan by cells appears to inhibit epithelial cell invasion into collagen.

Syndecan consists of a core protein that is inserted through the cell membrane and that contains heparan sulfate and chondroitin sulfate chains (Figure 2.33*A*). Glypican is covalently linked to the head group of membrane phospholipids in the plasma membrane (Figure 2.33*B*) and contains only heparan sulfate side chains.

Integrins

Integrins are a family of membrane glycoproteins consisting of α and β subunits. The binding site appears to contain sequences from both subunits, and their cytoplasmic domains form connections with the cytoskeleton. In this manner, integrins form a connection between the cytoskeleton and extracellular matrix. In accomplishing this, there are 11 α subunits and 6 β subunits forming at least

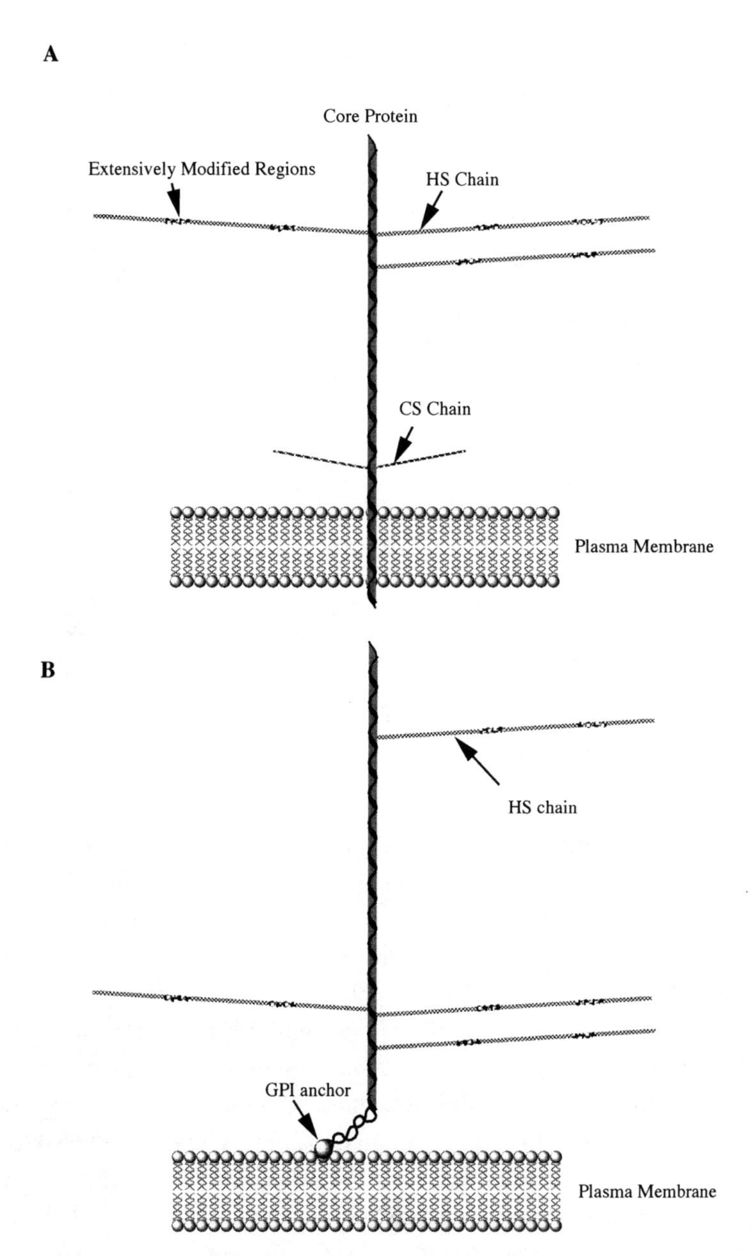

Figure 2.33 Model of cell-surface heparan sulfate proteoglycans. The diagram shows models of (**A**) syndecan and (**B**) glypican and their attachment into the plasma membrane.

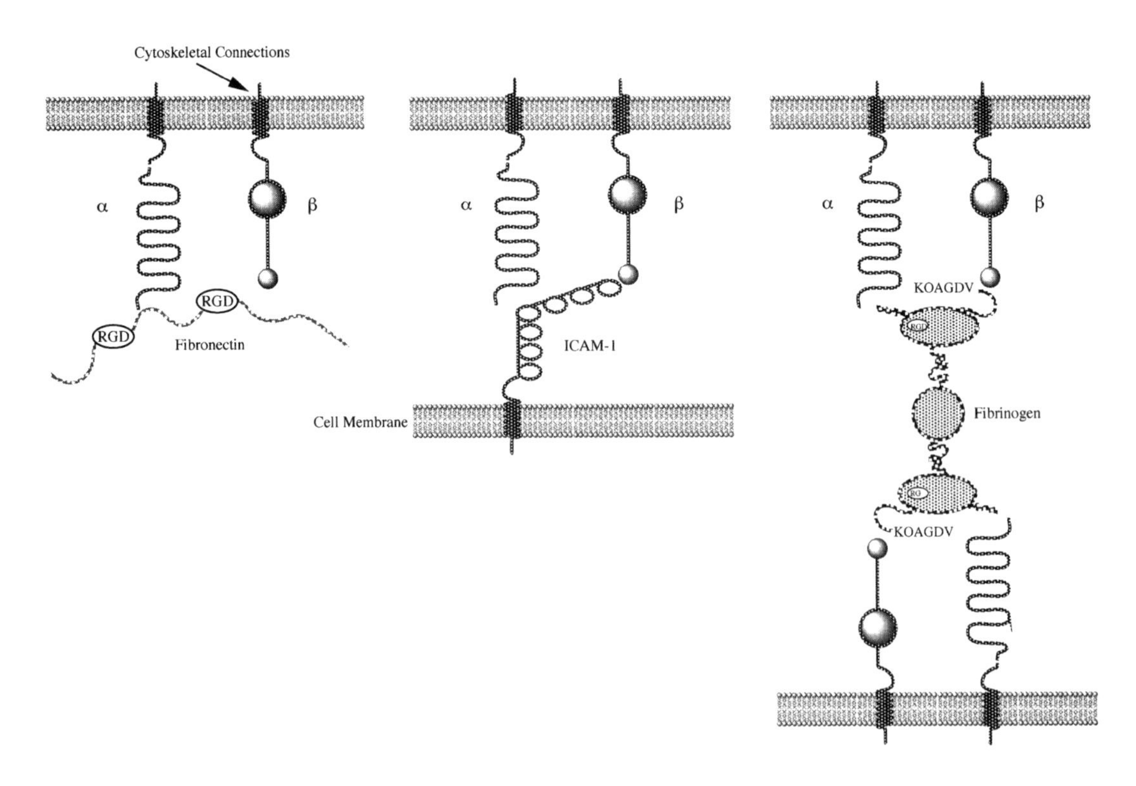

Figure 2.34 Binding of cell surface integrins to fibronectin, cell membranes, and fibrinogen. Each specific binding requires different α and β subunits. RDG = arginine-glycine-aspartic acid.

16 integrins, including glycoprotein IIb/IIa, which is expressed by megakaryocytes (cells in bone marrow from which platelets are derived) and platelets, and LFA-1, Mac-1, and p150/95, which are expressed by leukocytes. Many integrins bind to components of the extracellular matrix, such as collagen, fibrinogen, fibrin, laminin, and other proteins. Specific examples of function include the interaction of endothelial cells with core protein of basement membrane perlecan; the expression of $\alpha_v\beta_3$ integrin by blood vessels in wound granulation tissue during angiogenesis (formation of new blood vessels); adhesion of platelets to collagen involving integrin $\alpha_2\beta_1$; adhesion of natural killer cells (T cells) to fibronectin via very late activation antigens (VLA-4 and VLA-5); and adhesion of neutrophils to collagen via β_2 integrins. Other integrins bind to cell membrane proteins, mediating cell–cell adhesion. These molecules include intracellular adheşion molecules 1 and 2 (ICAM-1 and ICAM-2), leukocyte integrin (LFA-1), and vascular cell adhesion molecule (VCAM-1). The structure of $\alpha_5\beta_1$ integrin that binds fibronectin and $\alpha_L\beta_2$ that binds ICAM-1 is shown in Figure 2.34.

Class I and II Major Histocompatibility Complex Products and Antibodies

Foreign molecules and cells are recognized by the immune system by three classes of molecules: antibodies, MHC molecules, and T-cell antigen receptors (see Chapter 9). Class I and II MHC products are peptides that are synthesized in humans by each cell from the DNA on chromosome 6. These markers are exposed on the cell membrane and trigger a defense reaction when implanted into a foreign host. When recognized, these markers stimulate plasma cells to make antibodies and T cells to kill the foreign cells, which is discussed in more detail in Chapter 8. Class I molecules contain two separate polypeptide chains: an MHC-coded α or heavy chain with a MW of 44,000 and a non-MHC-encoded β chain with a MW of 12,000. Each α chain is oriented so that about 75%, including the amino-terminal end, extends into the extracellular environment; a short hydrophobic segment spans the cell membrane; and the carboxy-terminal end is found in the cytoplasm (Figure 2.35*A*). The β chain is located in the extracellular region and noncovalently interacts with the α chain. The portion of the molecule that interacts with the foreign protein consists of about 180 amino acid residues at the amino-terminal end of the class I α chain. Analysis of this region indicates that it is formed from two similar sequences, referred to as α_1 and α_2, of about 90 amino acids each. X-ray diffraction results indicate that α_1 and α_2 chains interact to form an eight-stranded β-pleated sheet supporting two α helixes. The cleft in the molecule into which the foreign polypeptide fits is formed from the two α helixes (α helixes form sides) and β-pleated sheet (sheet forms floor). The cleft is about 25 Å × 10 Å × 11 Å in its dimensions and binds fragments of a protein 10 to 20 amino acids long folded into an α helix or other conformation for presentation to T cells.

All class II MHC molecules are composed of two similar noncovalently associated chains, α and β. The α chain has a MW of 32,000 to 34,000 and is slightly larger than the β chain (29,000 to 32,000). A schematic diagram of the class II molecule is shown in Figure 2.35*B*. The binding site for the foreign molecule is similar to that for class I molecules except it is formed from both chains; in class I molecules, the binding site is formed from only the α chain. The binding groove of MHC class II molecules is normally occupied by a "dummy" peptide, a corticotropin-like intermediate lobe peptide. The dummy peptide is displaced when a vesicle containing proteolytically degraded remnants of foreign molecules causes dissociation of the dummy peptide and insertion of the foreign one. The dissociation is believed to be driven by acidic conditions associated with the vesicle.

Antibodies are proteins synthesized by B and plasma cells that react specifically with foreign proteins. All antibodies have a common core structure of two identical light chains (each light chain has a MW of about 24,000) and two identical heavy chains (the heavy chain has a MW of 55,000 or 70,000), as shown schematically in Figure 2.36. One light chain is attached to each heavy chain, and the heavy chains are attached to each other. The light and heavy chains contain a series of repeating amino acid sequences about 100 residues long that fold into a globular structure called the *immunoglobulin (Ig) domain*. All these domains contain two layers of β-pleated sheets with three or four strands

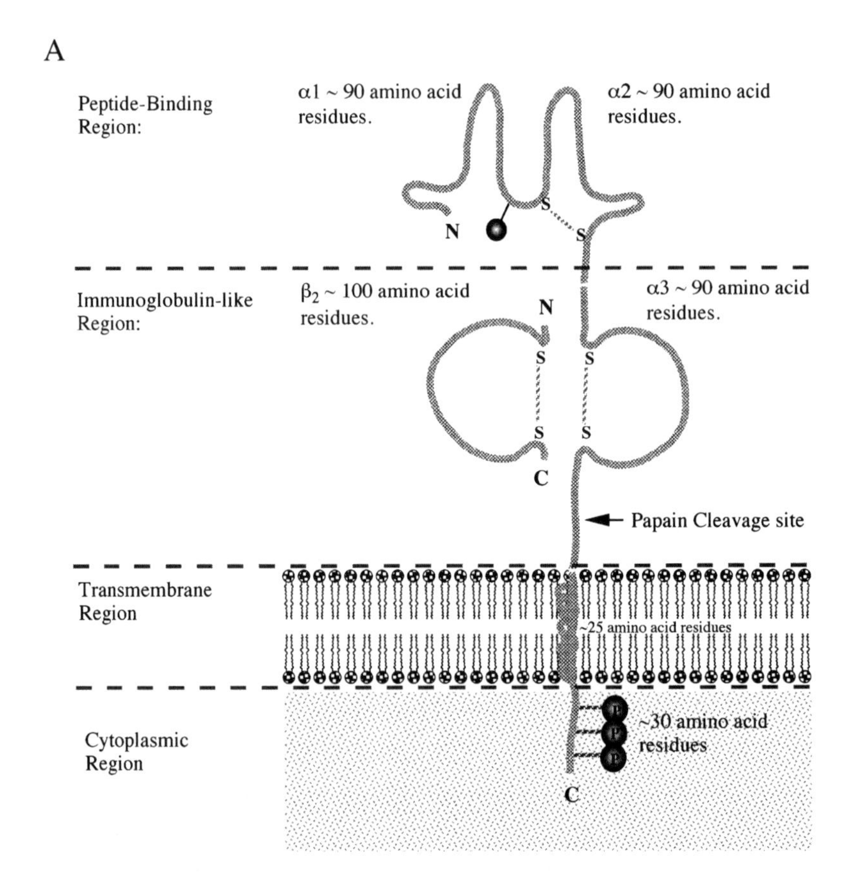

Figure 2.35 Diagram of MHC class I and II markers. The diagram shows the structure of class I (**A**) major histocompatibility markers. This marker consist of one light and one heavy chain (class I).

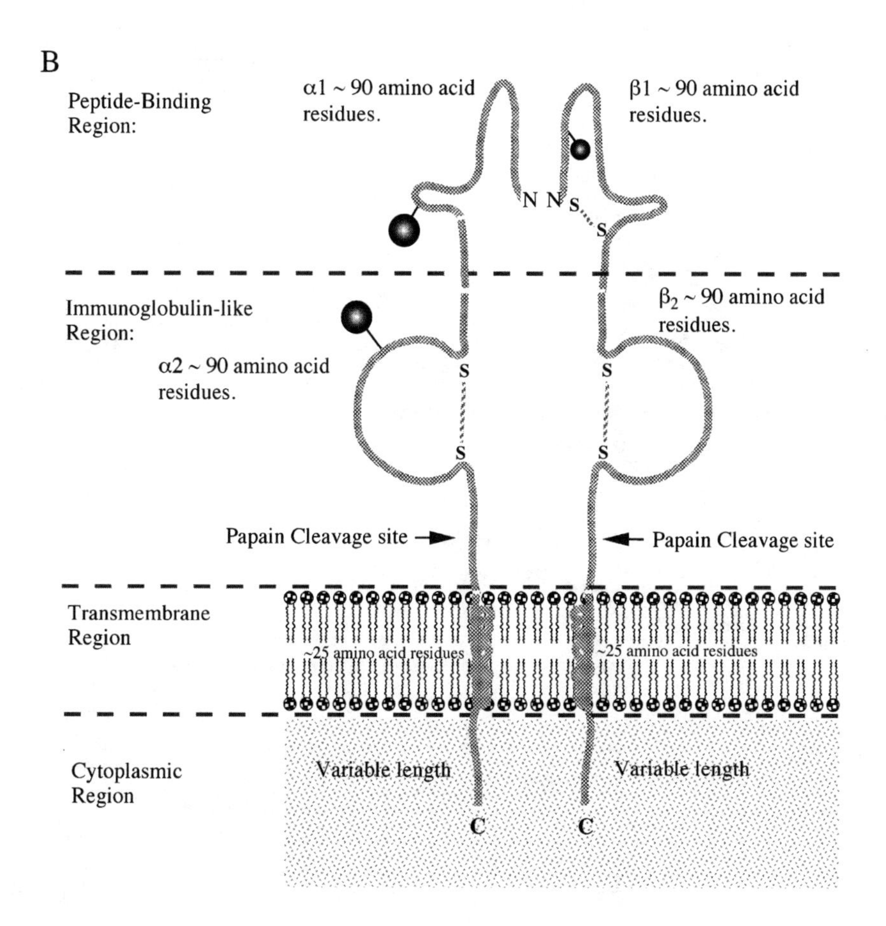

Figure 2.35 (*Continued*) Diagram of MHC class I and II markers. The diagram shows (**B**) products of the class II major histocompatibility markers. This marker consists of two heavy chains (class II).

of antiparallel polypeptide chains. Despite their similarities, antibody molecules are divided into subclasses; in humans, these are IgA, IgD, IgE, IgG, and IgM. Differences in size and antigen-binding site characterize the differences among these subclasses.

All antibodies contain light chains with two basic amino acid sequence variations or isotypes, κ and λ. All light chains of the same isotype share complete amino acid sequences of the carboxy-terminal end with other members of that isotype. Each light chain is folded into separate variable (*V*) and constant (*C*) regions (these notations identify whether the amino acid sequence varies from chain to chain or is constant) corresponding to the amino (*V*) and carboxyl

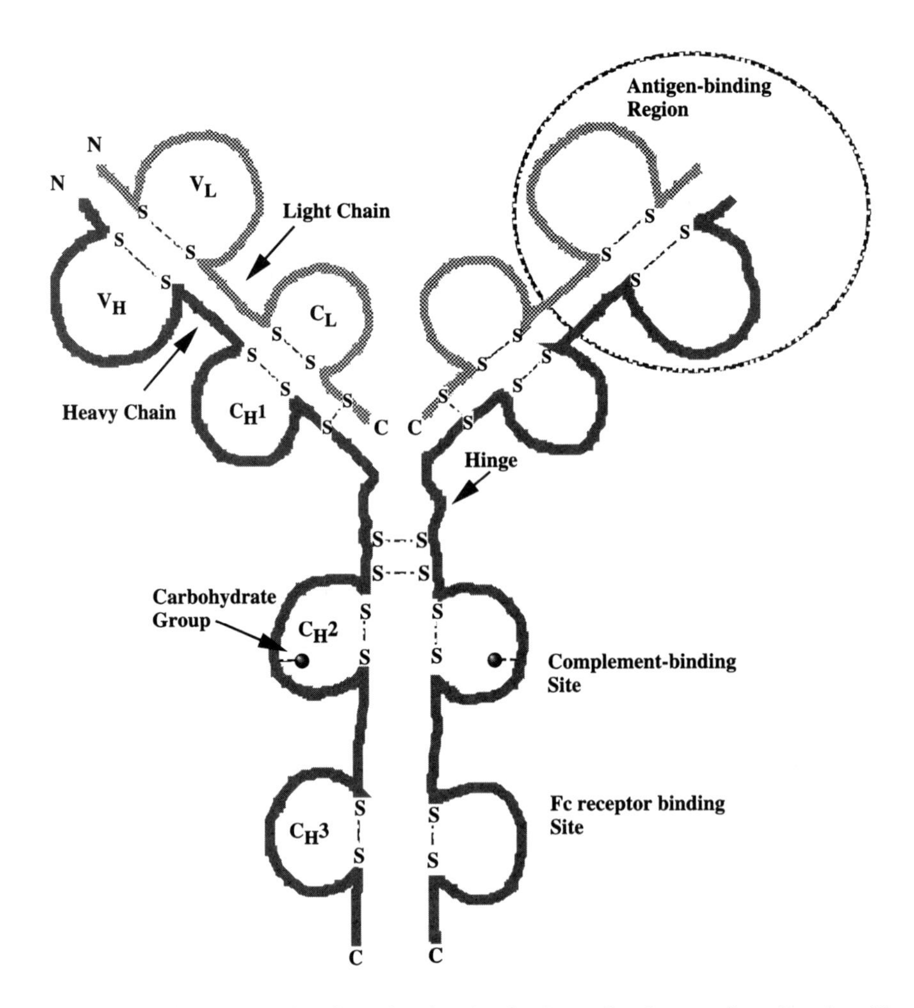

Figure 2.36 Structure of IgG molecule. Antibody molecules are found with a Y-shaped configuration composed of several regions. In the antigen-binding region, a specific amino acid sequence binds to a single antigen. The complement binding site is found near the C-terminal end of the molecule.

(C) halves of the polypeptide chain. Each domain is about 110 amino acids residues long and contains seven sequences that together assemble into a β-pleated structure.

Other Polymeric Materials

Other macromolecules of importance to biomaterials scientists include fibrinogen, actin, myosin, and keratin. By knowing the structures of these molecules, we can better understand the relationship between macromolecular structure and properties of tissues (fibrinogen and keratin) and how macromolecules contribute to force generation and structural stability of tissues, such as muscle (actin and myosin) and the cell cytoskeleton (actin).

Fibrinogen

Fibrinogen, a plasma protein formed in the liver, is the basis for the formation of blood clots. The blood concentration of this protein is about 3%. Pieces of the fibrinogen molecule are cleaved in the presence of other blood proteins, and fibrinogen is polymerized to form fibrin clot (see Chapter 9).

Fibrinogen is composed of six peptide chains, A, α, B, β, and two γ chains, giving the molecule a total MW of about 330,000. (The MWs of the A and α chains are 64,000; B and β, 56,000; and each γ, 47,000.) The Doolittle model of the molecule consists of two terminal nodules (globular regions) attached to a central nodule by two three-stranded ropes (Figure 2.37). Doolittle proposed that the structure consisted of a central nodule containing the amino-terminal regions of all six peptide chains. Two terminal or distal nodules are attached to the central nodule by two three-stranded ropes in the form of coiled coils. Each of these coiled coils contains approximately 110 amino acid residues and extends about 160 Å in space. The central nodule contains 11 of the 29 disulfide bonds in the molecule.

Keratin

Intermediate filaments are proteins 80 to 120 Å in diameter found in cytoskeletal fibers that reinforce cells. They have MWs ranging from 40,000 to 210,000 and are composed of a central domain of 310 to 350 amino acids. There are six types of intermediate filaments. Types I and II are composed of keratins, which make up the largest group of intermediate filaments. There are about 30 different protein chains that make up 20 epithelial keratins and 10 hair keratins. Epidermal keratinocytes synthesize two major pairs of keratin polypeptides: K5–K14 of the basal layer and K1–K10 of the cells in the suprabasal layer. Epithelial keratins are expressed in pairs, with one acidic chain and one neutral or basic chain in

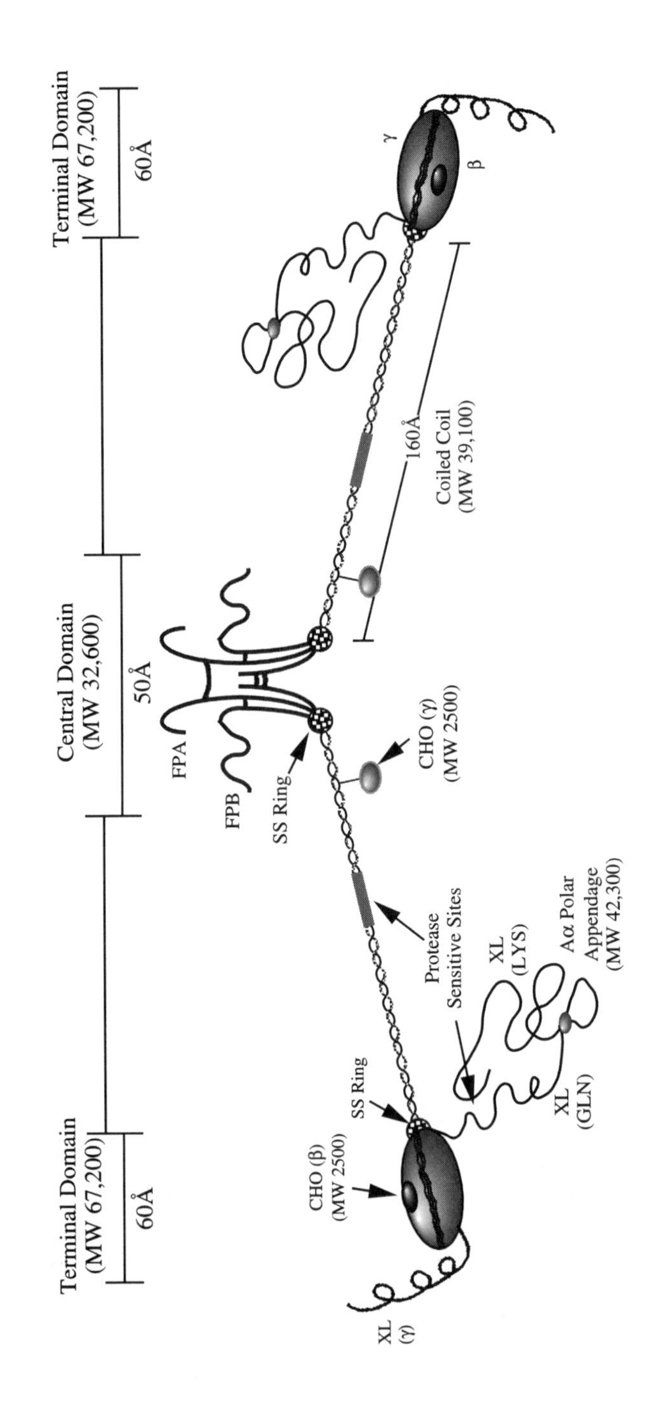

Figure 2.37 Domain structure of fibrinogen. The fibrinogen molecule is composed of globular end regions separated from a central domain by helical threads. The central domain has cross-links that hold all the polypeptide chains together. Fibrinopeptides A and B (FPA and FPB) are found in the central domain.

each molecule. Type I (K10 to 20) contains acidic groups, and type II (K1 to 9) contains neutral and basic groups.

The basic structural feature of each double-stranded (type I/type II hybrid) molecule is the presence of four α helical sequences—1A, 1B, 2A, and 2B—that are interrupted by three nonhelical sequences—L1, L1–2, and L2. In addition, there is a head and tail domain added to the interrupted α-helical domains. The head and tail regions are different for types I and II keratins. Filament assembly requires at least one type I and one type II keratin that associate parallel and in register to form a coiled coil. Two coiled coil dimers then form a ropelike structure in an antiparallel unstaggered or nearly half-staggered array, which is called a *tetramer* or *protofilament* (diameter, 2 to 3 nm). Two tetramers form a protofibril (4 to 5 nm), and four protofibrils form filaments 8 to 10 nm in diameter (Figure 2.38).

Actin and Myosin

Actin is a protein found in cells in the form of filaments in conjunction with tropomyosin and troponin. Actin exists in two states, a monomeric globular state called *G-actin* (its overall 3-D structure is spherelike), and an assembled state, called filamentous or *F-actin*. In the presence of actin-binding proteins, G-actin, Ca^{2+}, and Mg^{2+} assemble into actin filaments in the cell cytoplasm. The 3-D structure of G-actin has been determined from X-ray diffraction studies on complexes containing Ca-ATP-G-actin-DNase I containing small amounts of Mg^{2+} ions (DNase I is added to inhibit F-actin formation). The structure of G-actin at a 2.8-Å resolution consists of small and large domains, each divided into two subdomains. ATP is bound in a cleft between the two domains. The actin molecule is composed of α-helical domains connected by domains containing β-extended structures. Actin molecules polymerize into filaments that are coiled coils consisting of a left-handed helix with a pitch of 5.9 nm and a right-handed helix with a pitch of 72 nm (Figure 2.39).

Myosin is an enzyme that catalyzes hydrolysis of ATP and converts the energy released into movement through muscle contraction. During muscle contraction, an array of thick filaments containing myosin actively slides by an array of thin filaments containing actin. Myosin has a MW of about 500,000 and consists of six polypeptide chains: two heavy chains (MW 400,000) and two sets of two light chains, with each light chain having a MW of 20,000. The molecule consists of a long tail connected to two globular heads (Figure 2.40). Each glob-

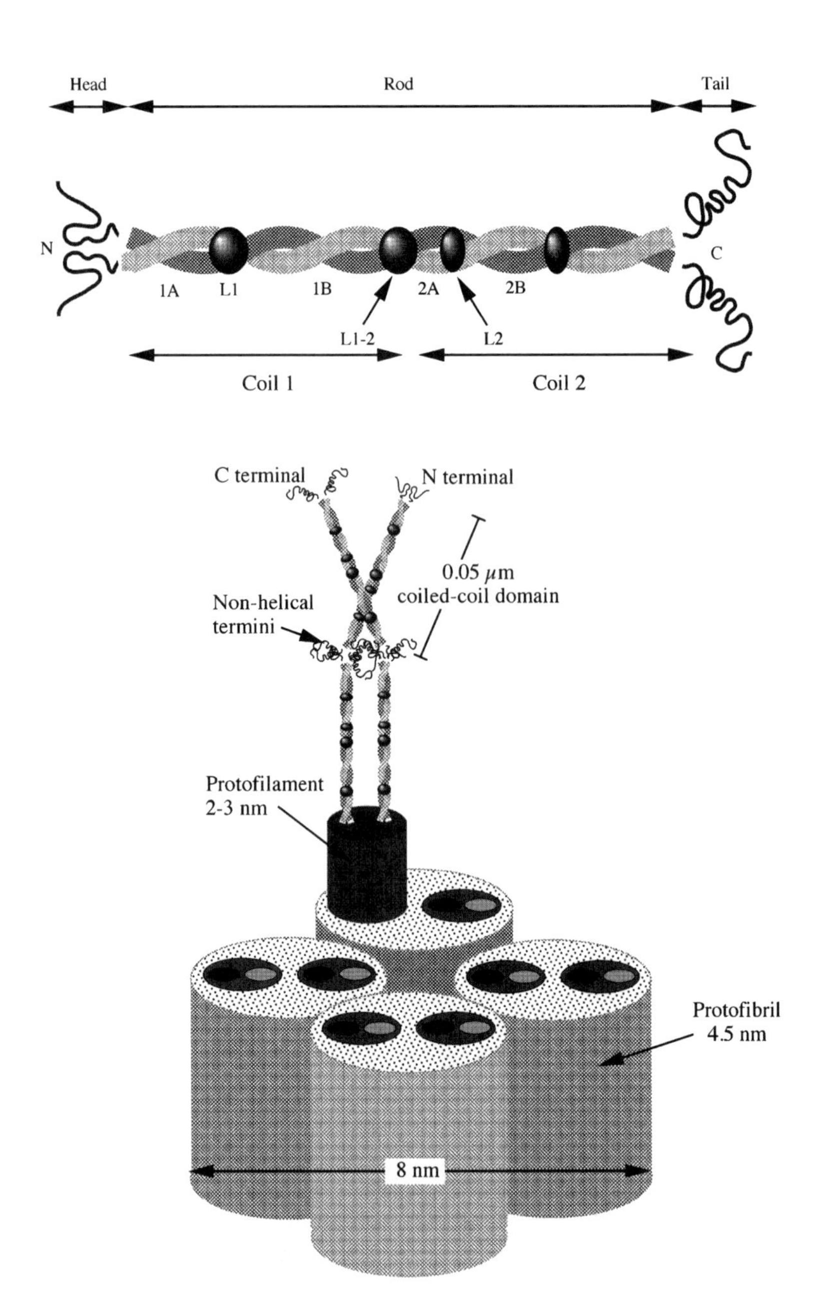

Figure 2.38 Structure of keratin protofibrils. The diagram illustrates the structure of keratin in intermediate filaments containing α helical sequences. Two coils are wound around each other and then packed into protofilaments. Eight protofilaments are packed into a filament.

Figure 2.39 Structure of F-actin. The diagram illustrates assembly of G-actin into double helical segment of F-actin.

ular head is composed of about 850 amino acids residues contributed by one of the two heavy chains and two of the light chains. The remaining portions of the heavy chains form an extended coiled coil that is about 1,500 Å in length. The globular head contains the ATP-binding sites and the actin-binding region, and the long rodlike portion of myosin forms the backbone of the thick filament. The myosin head has a length of more than 165 Å and is about 65 Å wide and 40 Å deep at its thickest end. The secondary structure is dominated by many long α helixes, and the myosin head is characterized by several prominent clefts and grooves that allow myosin to bind both ATP and actin.

Summary

Biological tissues contain many different types of macromolecules that exist in α helixes, single and double helixes, extended chain structures, β-pleated sheets and collagen coiled coils. They form supramolecular structures that maintain cell and tissue shape, resist mechanical forces, act to transmit loads, generate contractile forces, provide a mechanical link between extracellular matrix and the cell cytoskeleton, and act in recognition of foreign cells and macromolecules. The keys to understanding how these molecules form are (1) the chemistry of the repeat unit, (2) the nature of the backbone flexibility, (3) the types of hydrogen bonds that form, (4) the nature of the secondary forces other than hydrogen bonds that form between the chains, (5) the manner in which individual chains fold, (6) the way in which folded chains assemble with other chains, and (6) the manner that assembly is limited. We are only beginning to understand the intricacies of how chain folding and assembly lead to biological form and function. However, what we do know is what makes this field so exciting; nature has created clear structure–function relationships. We now know that structural materials aimed at reinforcing tissues have many levels of organization that pack molecules into a regular array. For example, materials that provide tissue integrity,

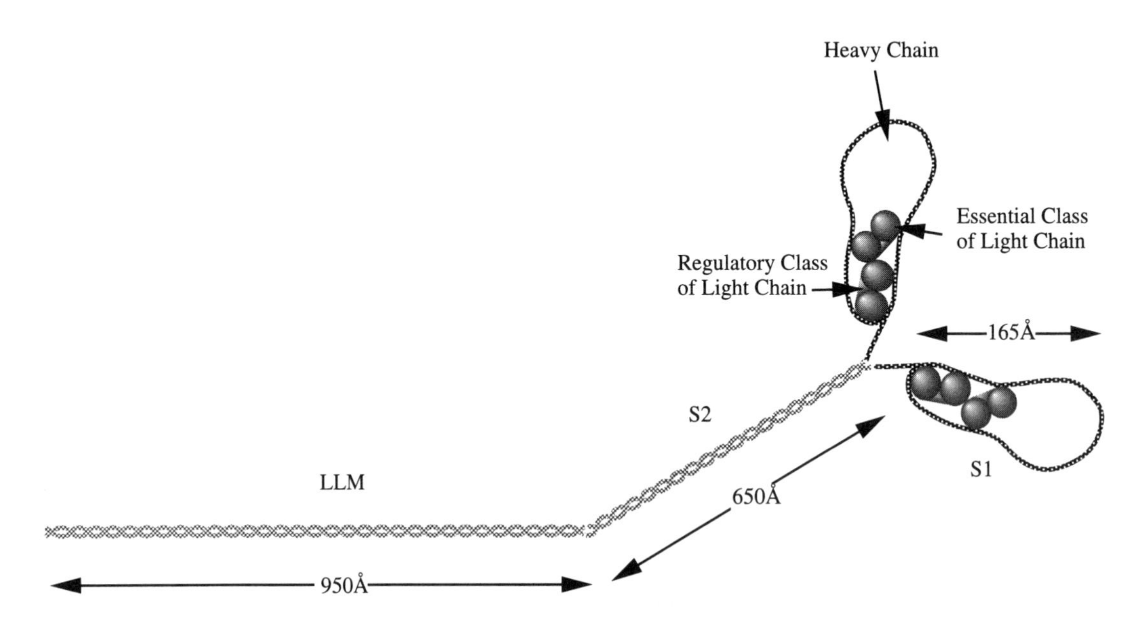

Figure 2.40 Structure of myosin. The diagram shows the structure of myosin, which is composed of a helical segment with a flexible bend that is attached to two head units.

such as keratin and collagen, contain linear regions of highly ordered structure and cross-links within the molecule to prevent extensive molecular slippage. Macromolecules that transmit force to cells, such as actin in the cell cytoskeleton and myosin in skeletal muscle, contain more globular regions that are connected by turns and bends. These molecules are often found noncovalently bonded to their nearest neighbors.

Suggested Reading

Abbas A.K., Lichtman A.H., and Pober, J.S., The Major Histocompatibility Complex, in Cellular and Molecular Immunology, W.B. Saunders Co., Philadelphia, chapter 5, pp. 104–110, 1991.

Atkins E.D.T., Meader D., and Scott J.E., Model for Hyaluronic Acid Incorporating Four Intramolecular Hydrogen Bonds, Int. J. Biol. Macromolec. 2, 318, 1980.

Carlier M.-F., Actin: Protein Structure and Filament Dynamics, J. Biol. Chem. 266, 1, 1991.

Furlan M., Structure of Fibrinogen and Fibrin, in Fibrin Stabilization and Fibrinolysis, edited by J.L. Francis, Ellis Harwood LTD, Chichester, England, chapter 1, pp. 17–64, 1988.

Guido D., Integral Membrane Heparan Sulfate Proteoglycans, FASEB J. 7, 1023, 1993.

Laurent T.C. and Fraser R.E., Hyaluronan, FASEB J. 6, 2397, 1992.

Pauling L. and Corey R.B., The Polypeptide-Chain Conformation in Hemoglobin and Other Globular Proteins, Proc. Natl. Acad. Sci. 37, 282, 1951.

Pauling L. and Corey R.B., Configurations of Polypeptide Chains with Favored Orientations Around Single Bonds: Two New Pleated Sheets, Proc. Natl. Acad. Sci. 37, 729, 1951.

Ramachandran G.N. and Kartha G., Structure of Collagen, Nature 176, 593, 1955.

Ramachandran G.N. and Sasisekharan V., Confomation of Polypeptides and Proteins, Adv. Protein Chem. 23, 283, 1968.

Rayment I. and Holden H.M., The Three-Dimensional Structure of a Molecular Motor, Trends Biochem. Sci., 19, 129, 1994.

Reinhardt D.P., Keene D.R., Corson G.M., Poschl E., Bachinger H.P., Gambee J.E., and Sakai L.Y., Fibrillin-1: Organization in Microfibrils and Structural Properties, J. Molec. Biol. 258, 104, 1996.

Rocco M., Infusini E., Daga M.E., Gogioso L., and Cuniberti C., Models of Fibronectin, EMBO J. 6, 2343, 1987.

Rosenbloom J., Abrams W., and Mecham R. Extracellular Matrix 4: The Elastic Fiber, FASEB J. 7, 1208, 1993.

Rouslahti E., Integrins, J. Clin. Invest. 87, 1, 1991.

Schulz G.E. and Shirmer R.H., Patterns of Folding and Association of Polypeptide Chains, in Principles of Protein Structure, Springer-Verlag, New York, chapter 5, pp. 66–107, 1979.

Scott J.E., Heatley F., and Hull W.E., Secondary Structure of Hyaluronate In Solution: A 1H-N.M.R. Investigation at 300 and 500 MHZ in [2H6] Dimethyl Sulphoxide Solution, Biochem. J. 220, 197, 1984.

Silver F.H., Connective Tissue Structure, in Biological Materials: Structure, Mechanical Properties, and Modeling Soft Tissues, NYU Press, New York, chapter 2, pp. 7–68, 1987.

Smack D.P., Bernhard P.K., and William D.J., Keratin and Keratinization [review], J. Acad. Derm. 30, 85, 1994.

Timpl R. and Brown J.C., The Laminins, Matrix Biol. 14, 275, 1994.

Vincent J.F.V., Proteins, in Structural Biomaterials, Halsted Press, John Wiley and Sons, New York, chapters 2 and 3, pp. 34–83, 1982.

Wertman K.F. and Drubin D.G., Actin Constitution: Guaranteeing the Right To Assemble, Science 258, 759, 1992.

Wilson I.A., Another Twist to MHC-Peptide Recognition, Science 272, 973, 1996.

Yamada K.M., Fibronectin and Other Cell Interactive Glycoproteins, in Cell Biology of Extracellular Matrix, second edition, edited by E.D. Hay, Plenum Press, New York, chapter 4, p. 111, 1991.

Yanagishita M., Function of Proteoglycans in the Extracellular Matrix, Acta Pathol. Jpn. 43, 283, 1993.

Yurchenco P.D., Cheng Y.-S., and Colognato H., Laminin Forms: An Independent Network in Basement Membranes, J. Cell Biol. 117, 1119, 1992.

3

Introduction to Structure and Properties of Polymers, Metals, and Ceramics

Introduction to Synthetic Materials

Polymers, metals, and ceramics are everyday kitchen items, from aluminum, stainless steel, and nonstick cookware to ceramic plates and mugs. These materials found their way into medicine because no other materials could simulate the properties of mineralized and nonmineralized tissues. In addition, many of these kitchen materials are stable; they do not break down into low molecular weight materials that are harmful to the body's homeostasis.

Much of our knowledge and use of polymers derived from practical needs, such as the need for rubber during World War II. Before this time, the need for rubber was met by the abundance of natural rubber, which is a product derived from the sap of the plant *Hevea brasiliensis*. This plant is indigenous to South

America, especially the Amazon valley, and eventually was cultivated in Malaysia and Indonesia to maintain the supply.

In 1493, Columbus observed natives in Haiti playing a game with a ball made from smoked dried latex of a tree (Semegen and Fah, 1978). Thomas Hancock invented the masticator in 1820, which allowed solid rubber to be softened and shaped and resulted in the production of rubber articles. In 1839, Charles Goodyear observed that heating rubber with sulfur yielded a more stable rubber compound, one that would not melt in the sun or become rigid in the cold. During World War II, the expediency of war efforts led to the rapid development of styrene-butadiene synthetic rubber.

Since then, many different types of polymers have been produced, including rubbers, thermoplastics, and semicrystalline materials. Many of these long-chained molecules were polymerized from oil derivatives because of the abundance of the raw materials. Rubbers are polymers that have glass-transition temperatures below room temperature and are usually cross-linked so that they do not flow. Thermoplastics are either amorphous or semicrystalline and melt by application of heat. Semicrystalline polymers have short- and long-range order and have melting temperatures above room temperature.

A few polymers, including poly(ethylene terephthalate), poly(methyl methacrylate), poly(dimethylsiloxane), and polyurethane, are commonly used in medical devices. These polymers are used in hip, tendon, ligament, facial, vascular, ophthalmic, orthopedic, breast, and skin implants. Their frequency of use has been established based on successful use in implantable devices during the past three or four decades.

Metals are crystalline materials made of atoms that have highly mobile electrons. The use of metals dates back to ancient times; gold, copper, and silver were used to make art because they were easy materials to work. Gold bands were used by the Phoenicians and Etruscans for construction of partial dentures (Phillips, 1991). Research studies on the use of dental amalgams (alloys of mercury, silver, and tin) date back to about 1850, but it was not until the 1920s that the first courses in dental materials were established. In the 1930s, advances in metals technology and improved surgical techniques led to the use of metallic alloys as bone and joint implants. The first artificial hip implant, made of stainless steel, was developed by Charnley in the 1950s. Since then, implants made from cobalt- and titanium-based alloys have been introduced.

Ceramic materials can be crystalline or noncrystalline, and they contain atoms that are both metallic and nonmetallic in nature. These materials are held together primarily by highly directional and stable ionic bonds; this makes these materials stiffer and inert. Ceramics tend to be more brittle than metals and polymers and cannot be used in applications where toughness and ductility are required. Natural ceramics, such as coral and hydroxyapatite found in bone, have been used as implants for some time. These ceramics are composed of calcium, carbonate, and hydroxyl groups. Dental porcelains contain silica or aluminum oxide, or both, in either crystalline or amorphous forms and are used for tooth restoration. There has been interest in making hip and knee implants out of similar materials.

In this chapter, we discuss the structure and properties of polymers, metals, and ceramics to better understand medical device design using synthetic materials. Most medical devices are made of synthetic materials because of their availability, ability to be terminally sterilized, and long history of use in implants.

Polymer Structures

Polymers are widely used in medicine because they have physical properties that are most similar to that of soft tissues. They are used in a number of devices, including wound dressings, tendon and ligament replacements, vascular prostheses, intraocular lenses, catheters, and as the lining of total joint replacements. Many of the applications listed involve use of several polymers, including poly(methyl methacrylate) (bone cement and intraocular lenses), polyethylene (hip implants, tendons and ligaments, and facial implants), polypropylene (sutures), and polyurethanes (wound dressings, breast implants, and vascular implants). Although all these materials are composed of long chains, they are manufactured by two different processes.

Polymerization Processes

Polymers are synthesized by one of two methods, condensation or addition polymerization. In condensation polymerization, one or more bifunctional small molecules react, releasing a byproduct with the formation of a polymer chain (Figure 3.1). An example of condensation polymerization is the reaction of dimethyl terephthalate and ethylene glycol in the presence of heat to form poly(ethylene terephthalate), PET. This polymer is made into fiber with the trade name Dacron. PET is used in vascular grafts, sutures, sewing rings around bioprosthetic heart valves, and facial implants. Condensation polymers are characterized by a broad range of polymer chain molecular weights because the condensation reaction is just as likely to occur with another monomer as with a growing polymer chain.

In contrast, addition polymerization occurs by propagation of a growing chain that is initiated by the reaction of a free radical with a double bond (see Figure 3.1). Such a reaction is catalyzed by the presence of transition metals, including vanadium, triethylaluminum, and titanium tetrachloride. Polyethylene is produced when ethylene is exposed to a free radical. Formation of free radicals is achieved by the decomposition of a molecule such as dibenzylperoxide into two peroxide free radicals. The free radicals attack the double bond in an ethylene molecule and create a chain that grows by attacking other ethylene mol-

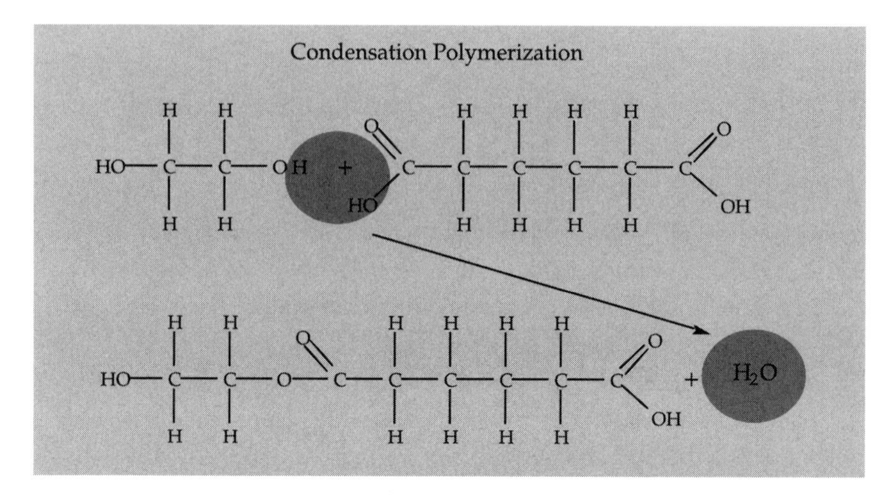

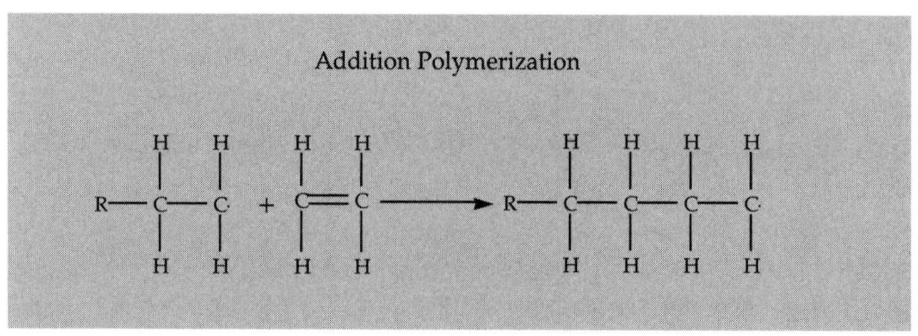

Figure 3.1 Types of polymerization processes. Condensation polymers (top) are formed from the reaction between the ends of two difunctional reagents, generating a byproduct such as water. Addition polymerization (bottom) occurs when the free radial ends of the growing chains propagate by attacking the double bond in a monomer.

ecules. The process is terminated by adding a free radical scavenger such as vitamin C or when two free radicals combine.

A number of different physical forms of free radical polymers exist when the monomer that is polymerized has one atom that is not hydrogen. For example, when vinyl chloride is polymerized, the chlorine group can react so that adjacent chlorines are either on the same side (*cis*) or opposite sides (*trans*) of the polymer backbone (Figure 3.2). If the chlorine group is on the same side of the backbone as one goes along the chain, the polymer is referred to as *isotactic*

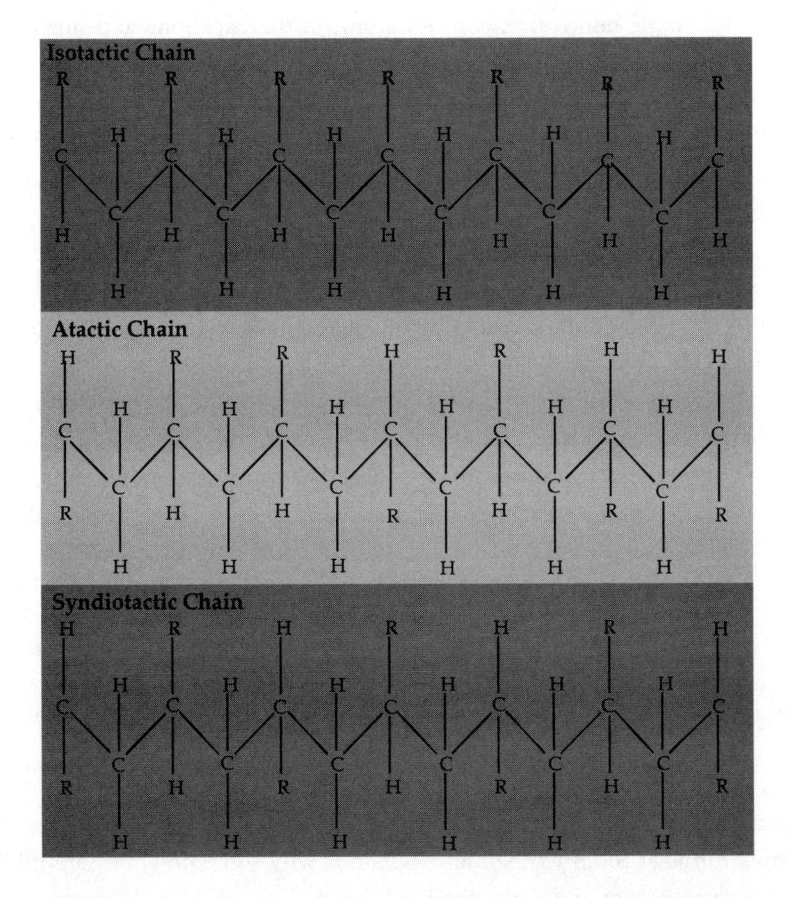

Figure 3.2 Structural differences in vinyl polymers. Polymers formed from monomers with an asymmetrical carbon (a carbon with four different groups attached) can form three different arrangements of the side chains. Isotactic polymers are recognized because they have all the R groups on the same side of the backbone; syndiotactic polymers have the R groups on alternate sides; and atactic polymers have a random arrangement of these groups.

poly(vinyl chloride). If the chlorine group alternates on either side along the backbone, the product is called *syndiotactic*. Finally, randomly oriented side groups are called *atactic*. Vinyl chloride polymers have been used as blood bags because of their flexibility when plasticized by adding low molecular weight liquids during fabrication. However, the loss or leaching of plasticizers out of vinyl polymers and into the body has limited their applications.

The behavior of polymer molecules is related to the size of the repeat unit, the type of bonds that are present in the backbone, and the size of the side chains. When heated above the crystalline-melting temperature, unbranched chains with single bonds that connect atoms in the backbone and small atoms in the side chains, such as polyethylene, are flexible. These chains are highly mobile and pack into crystalline arrays. When one of the hydrogen atoms on polyethylene is replaced with a styrene group containing a ring, the chain is no longer flexible enough to pack into a crystal. Large side chains prevent crystal formation.

In noncrystalline polymers, the chains exist in a state of random or amorphous structure. The flexibility of the polymer chains is directly related to the length of the repeat unit, the number of single bonds in the backbone, and the size of the side chains. Long repeat units with double bonds or rings in the backbone are not as flexible as chains with single bonds in the backbone. Large side chains also reduce the chain flexibility, as does chain branching. The presence of any of these conditions reduces the glass-transition temperature, which is the temperature at which the chain backbone stops rotating around its long axis. Table 3.1 lists several repeat units of polymers of interest in medicine. Table 3.2 lists the transition (T_g) and glass-melting temperatures (T_m) of these polymers.

Physical Structure of Polymer Chains

As discussed briefly in the previous section, polymer chains exist in several macroscopic forms at room temperature, which is why some polymers are rubbery at room temperature and some are hard. Examples include poly(methyl methacrylate), a rigid clear material at room temperature; polyethylene, a rigid opaque material; and poly(hydroxyethyl methacrylate), a water-swollen gel. This difference in physical properties at room temperature is a result of the hydrophilicity (whether they swell in an aqueous environment) of the chains and the chain flexibility.

Table 3.1 Repeat structure of common medical polymers

Polymer	Repeat unit
Polyethylene	$\left[-CH_2- \right]_n$
Poly(tetrafluoroethylene)	$\left[-CF_2- \right]_n$
Poly(ethylene terephthalate) (Dacron)	$\left[-C(=O)-C_6H_4-C(=O)OCH_2CH_2O- \right]_n$
Poly(vinyl chloride)	$\left[-CH_2-CHCl- \right]_n$
Poly(vinyl fluoride)	$\left[-CH_2-CF_2- \right]_n$
Poly(methyl methacrylate) (Plexiglas)	$\left[-CH_2-C(CH_3)(C(=O)-O-CH_3)- \right]_n$
Poly(2-hydroxyethyl methacrylate) (HEMA-hydrogel)	$\left[-CH_2-C(CH_3)(C(=O)-CH_2CH_2OH)- \right]_n$
Poly(dimethylsiloxane)	$\left[-Si(CH_3)_2-O- \right]_n$
Polysulfone	$\left[-C_6H_4-O-C_6H_4-S(=O)_2- \right]_n$
Polyamide	$\left[-C(=O)-NH- \right]_n$
Polyurethane	$\left[-O-C(=O)-NH- \right]_n$
Polyester	$\left[-C(=O)-O- \right]_n$

Table 3.1 (*Continued*) Repeat structure of common medical polymers

Polyurethanes
$\left[(CH_2)_4 - O - \overset{O}{\overset{\|}{C}} - \underset{}{\overset{H}{\overset{\|}{N}}} - C_6H_4 - CH_2 - C_6H_4 - \overset{H}{\overset{\|}{N}} - \overset{O}{\overset{\|}{C}} - O - \left[(CH_2)_4 - O \right]_n - \overset{O}{\overset{\|}{C}} - \overset{H}{\overset{\|}{N}} - C_6H_4 - CH_2 - C_6H_4 - \overset{H}{\overset{\|}{N}} - \overset{O}{\overset{\|}{C}} - O - (CH_2)_4 \right]$
Pellethane (Upjohn)
$\left[(CH_2)_4 - \overset{H}{\overset{\|}{N}} - \overset{O}{\overset{\|}{C}} - \overset{H}{\overset{\|}{N}} - C_6H_4 - CH_2 - C_6H_4 - \overset{H}{\overset{\|}{N}} - \overset{O}{\overset{\|}{C}} - O - \left[(CH_2)_4 - O \right]_n - \overset{O}{\overset{\|}{C}} - \overset{H}{\overset{\|}{N}} - C_6H_4 - CH_2 - C_6H_4 - \overset{H}{\overset{\|}{N}} - \overset{O}{\overset{\|}{C}} - \overset{H}{\overset{\|}{N}} - (CH_2)_4 \right]$
Biomer (Ethicon)
$\left[(CH_2)_4 - O - \overset{O}{\overset{\|}{C}} - \overset{H}{\overset{\|}{N}} - C_6H_{10} - CH_2 - C_6H_{10} - \overset{H}{\overset{\|}{N}} - \overset{O}{\overset{\|}{C}} - O - \left[(CH_2)_4 - O \right]_n - \overset{O}{\overset{\|}{C}} - \overset{H}{\overset{\|}{N}} - C_6H_{10} - CH_2 - C_6H_{10} - \overset{H}{\overset{\|}{N}} - \overset{O}{\overset{\|}{C}} - O - (CH_2)_4 \right]$
Tecoflex (Thermedics)

Table 3.2 Effects of different polymeric side chains on the glass-transition and melting temperatures

Polymer	T_m (°C)	T_g (transition temperature at 1 Hz [°C])
Polypropylene	—	0
Polystyrene	137	116
Poly(vinyl chloride)	212	87
Poly(vinyl acetate)	—	29
Poly(methyl methacrylate)	—	105
Poly(ethylene terephthalate)	267	69
Poly(dimethylsiloxane)	−54	−123
Polyethylene	137	−120
cis-Polyisoprene	28	−70

In the absence of water molecules, polymer chains can exist in either amorphous or crystalline states. Large chains with bulky side chains and rings or double bonds in the backbone do not crystallize and are randomly folded. Smaller chains with no branching and small side chains can easily fold into crystals. The physical properties of polymer chains to a first approximation are related to the glass-transition and glass-melting temperatures. The relationship between physical structure and properties is discussed in the next section. In the absence of crystallinity, polymer chains form random chain structures, and the chains are highly folded (Figure 3.3). In semicrystalline polymers, the chains associate in some regions to form crystals, but in other regions, the random chain structure is prevalent (see Figure 3.3). Finally, in crystalline polymers, the chains form planar zigzags and then associate laterally to make crystals. The nature of the interactions between chains dictates the types of crystals that form and the mechanical properties of the resultant polymer.

Polymer Mechanical Properties

The mechanical properties of polymers are related to the physical state of the chains. Mechanical properties refer to the relationship between applied load and resulting strain. In the simplest test, the polymer specimen is loaded in tension and then pulled at a constant rate of loading until failure occurs (Figure 3.4).

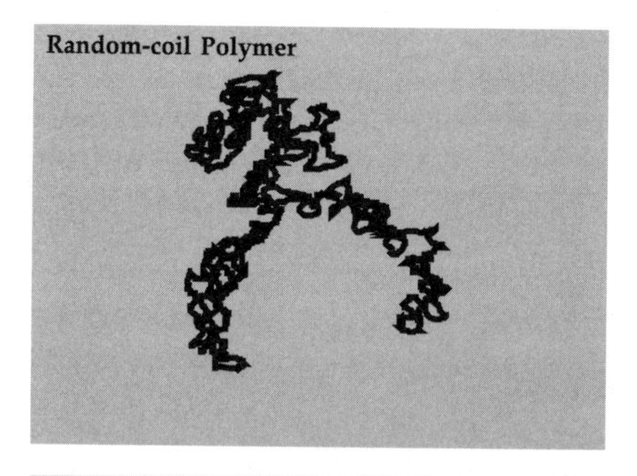

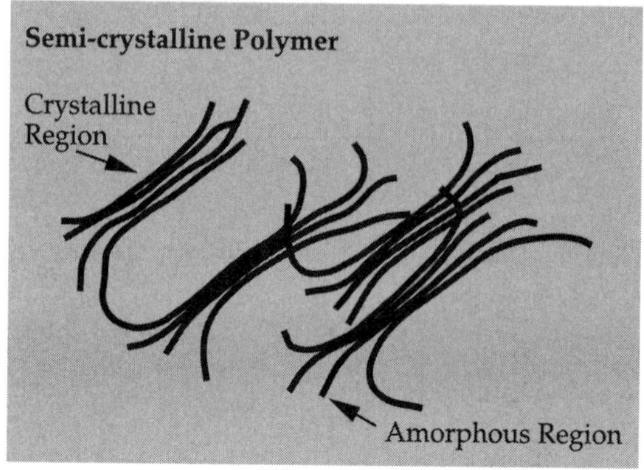

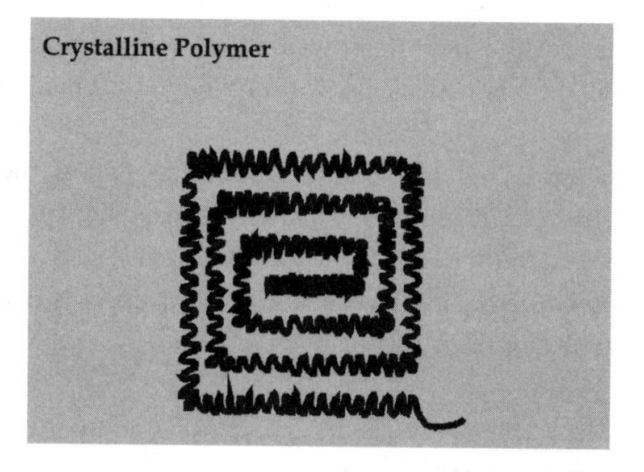

Figure 3.3 Models of polymer structures. Ideal types of polymer structures include random coil (top), composed of random chains; semicrystalline (middle), containing regions where polymer chains align, form crystalline regions, and are surrounded by amorphous regions; and crystalline polymers (bottom), which form folded chain crystals.

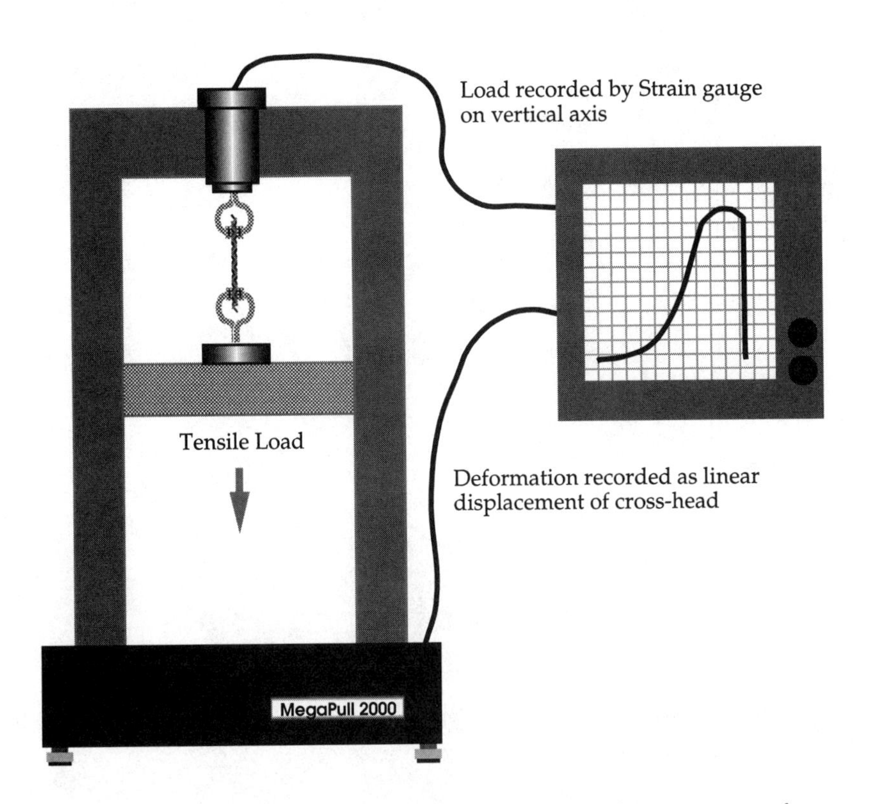

Figure 3.4 Mechanical testing device. This tensile testing device consists of a cross-head in which a specimen is moved at a constant rate of extension. Jaws fix the specimen to the cross-head at one end and to the load cell at the other end. The load is recorded by the load cell, and the deformation is recorded by the displacement of the cross-head.

The force required to maintain a constant rate of strain is divided by the original cross-section area of the sample; plotted against the strain, which typically is the change in length that results; and divided by the original length. The shape of this plot is almost linear for glassy polymers, concave upward for rubbery materials, and linear bending toward the strain axis for crystalline polymers (Figure 3.5). The slope of this curve is called the *modulus* or *stiffness* and reflects the resistance to deformation.

At temperatures above the melting-transition and glass-transition temperatures, all polymers exist as rubbers. As such, they are characterized by stress–strain behaviors that reflect those of rubbery materials, as discussed later. At low temperatures, polymers are either amorphous glasses or crystalline polymers, and their stress–strain behaviors are almost linear. Amorphous chains at temperatures

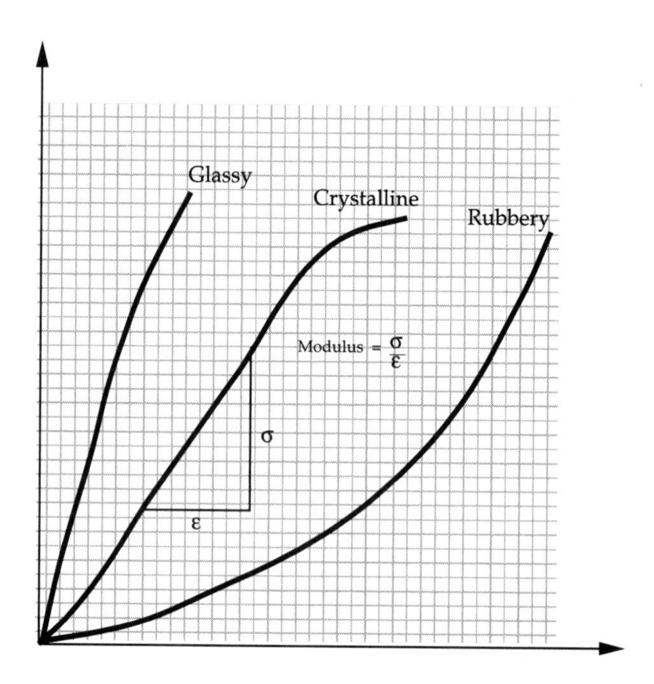

Figure 3.5 Stress–strain curves for polymers. Stress–strain curves for polymers, including glassy, crystalline, and rubbery polymers.

below the glass transition behave as glasses and have moduli of about 10^9 MPa (Figure 3.6). The modulus falls to 10^5 MPa above the glass-transition temperature and remains there as the temperature is increased if there are chemical cross-links to prevent chain sliding. If the chains are not chemically cross-linked, then eventually they will slide by each other, and the polymer will fall apart. Crystalline polymers have moduli above that of amorphous polymers below the crystalline-melting temperature. Above the crystalline-melting temperature, the modulus becomes identical to that of amorphous polymers.

Williams and co-workers (1955) developed an empirical relationship for the modulus of all amorphous polymers between T_g and T_g + 100 °C. At some temperature (T) between T_g and T_g + 100 °C, the shear modulus (G) is related to the shear modulus at T_g (G_g) via equation 3.1:

$$\log (G/G_g) = -17.44dT/(51.66 + dT) \tag{3.1}$$

where $G_g = 10_9$ MPa and $dT = T - T_g$.

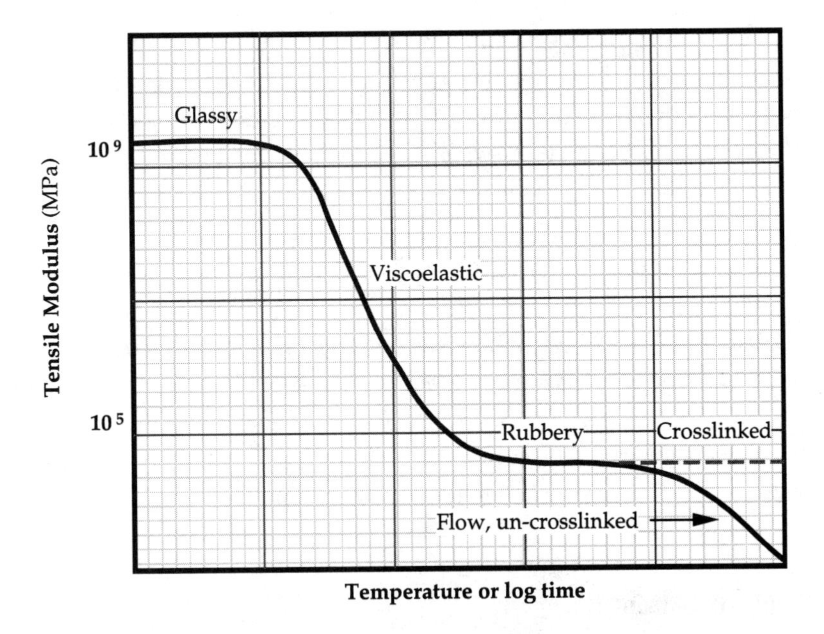

Figure 3.6 Modulus profiles for amorphous polymers. Amorphous polymers that have no crystalline structure exhibit decreased tensile modulus with increased temperature or time. Glassy polymers have a tensile modulus of about 10^9 MPa. Upon heating or loading, they become rubbery and have a modulus of 10^7 MPa. Rubbery polymers that are cross-linked retain a modulus of about 10^5 up to the point that they pyrolyze. Uncross-linked polymers flow when further heated.

Rubber Elasticity Theory

The behavior of rubbers has been derived from an analysis of the thermodynamics of the equilibrium force required to stretch this type of material. The reader is referred to a standard polymer science textbook such as Rodriguez (1982) for the details of the derivation that relates the stress required at equilibrium to stretch a rubbery material to an extension ratio (λ), where λ is defined as L/L_0 (where L and L_0 are the length at any time and at time zero). The relationship is given by equation 3.2, where R is the gas constant and N is the number of moles of polymer chains per unit volume:

$$\sigma = RTN(\lambda - 1/\lambda^2) \tag{3.2}$$

The Mooney-Rivlin equation (3.3) is a modified form of equation 3.2 and applies to polymer chains swollen in solvent. The latter equation contains a constant (C_2), the polymer density (ρ), number average molecular weight (M_n), and the volume fraction of polymer (V_2). The stress and extension ratio are based on the nonswollen cross-sectional area.

$$\sigma = [(N - (2\rho/M_n)RT + 2C_2/\lambda][\lambda - (1/\lambda^2)][1/V_2]^{1/3} \tag{3.3}$$

Equation 3.3 is the stress-extension ratio for rubbery materials that are swollen in solvent. By examining this equation, it should be clear that σ is linear with $\lambda - (1/\lambda^2)$ at equilibrium and that the slope of that plot is inversely related to the molecular weight of the average chain. The molecular weight of the average chain is inversely related to the density of cross-links that link the chains.

Polymer Viscoelasticity

Equation 3.3 gives the stress-extension ratio for rubbers at equilibrium; unfortunately, the stress required to stretch a polymer to a fixed extension varies with time. Initially, the force or stress required to stretch a polymer is high, and then it eventually falls as time increases to an equilibrium value. Polymers and biological tissues exhibit a time-dependent behavior during tensile loading. The behavior is characterized by changes in polymer chain conformation or sample geometry. The time dependence continues over a period until cross-links prevent any further changes in specimen structure. At that point, the structure remains stable, as do the resultant mechanical properties. The time dependence of the stress at constant specimen external dimensions is referred to as a creep test (Figure 3.7). The mechanisms associated with creep include geometrical alignment of the sample with the loading direction and extension of rubber chains with interchain slippage. To apply equation 3.3, one must wait until equilibrium occurs and then determine the force and the stress-extension ratio plot.

Polymers and biological tissues exhibit a time-dependent change in specimen dimensions when a fixed load is applied. The behavior is characterized by either an increased specimen length if the load is tensile or a decreased thickness if the sample is compressed. This manifests itself in a time- and load-dependent modulus or stiffness of these materials. This is the reason the modulus is temperature dependent; the same relaxation phenomena that cause the modulus to relax with increased time also cause it to relax with increased temperatures. If

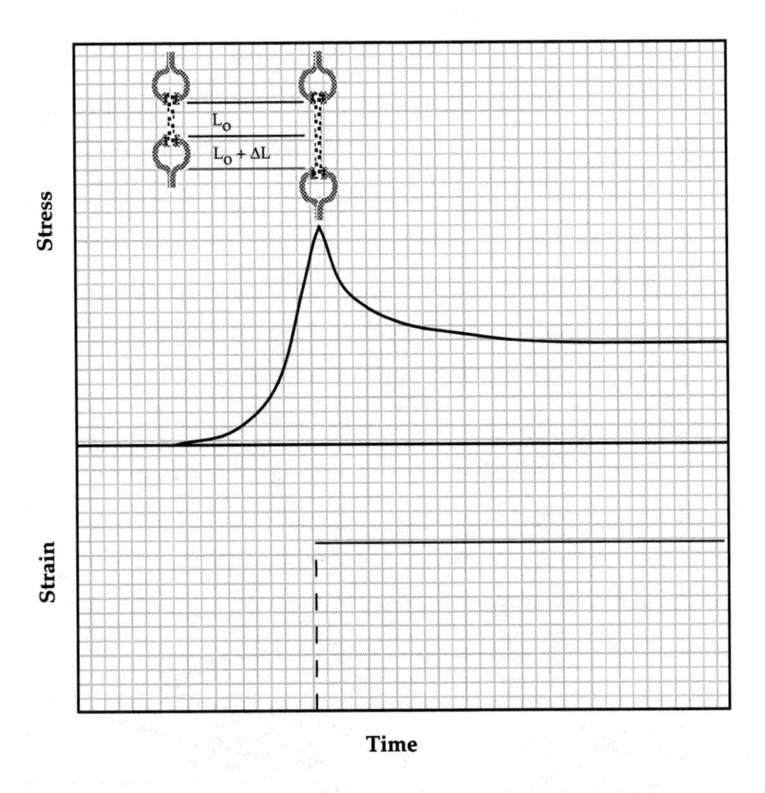

Figure 3.7 Stress relaxation of polymers. Stress is a function of the time a polymer is stretched to a fixed length. The stress is initially high but then decays by viscous processes until it reaches an equilibrium.

the log of the relaxation stiffness at a fixed time period is obtained at different temperatures from a relaxation curve like the one in Figure 3.6, then it is possible using equation 3.1 to shift all the data with respect to the glass-transition temperature to superimpose the data to form a single curve, as shown in Figure 3.6.

Figure 3.6 can be used to interpret the effects of molecular weight and cross-link density on the mechanical properties of amorphous polymers at some specific time or temperature. According to equation 3.2, at high extensions where the term $1/\lambda^2$ vanishes, the ratio of stress to strain, which is the stiffness, is proportional to the temperature and the number of moles of polymer chain per unit volume. Therefore, as diagrammed in Figure 3.8, the modulus increases with increased molecular weight of the polymer chains, increased cross-link density, and increased temperature.

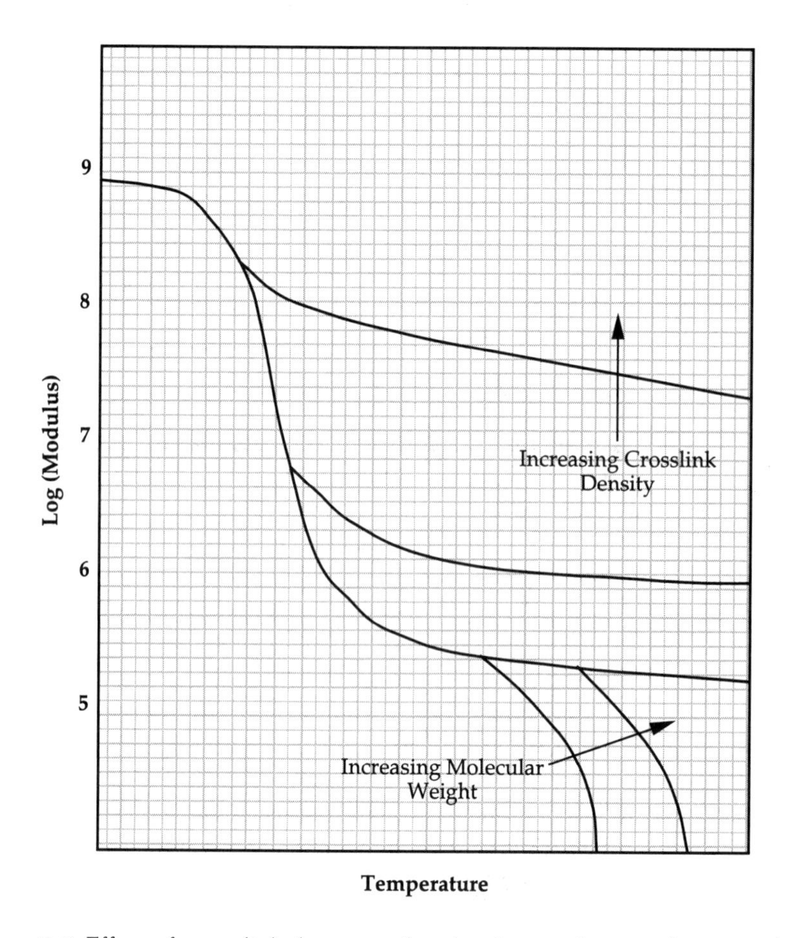

Figure 3.8 Effect of cross-link density and molecular weight on polymer modulus. The polymer modulus increases as a function of temperature, number of cross-links, and polymer molecular weight (chain length).

Types of Mechanical Tests

Several mechanical tests are used to characterize polymers, including constant strain-rate, breaking energy, creep, dynamic, fatigue, and impact tests. The constant strain-rate experiments are described in Figure 3.4 for tensile testing. In this case, a constant rate of strain is applied to the specimen in tension or compression, and the resultant stress is measured. The result is a plot of stress or load versus strain for a solid piece of polymer, usually in the form of a dumbbell. If the material is a fiber, then the force at failure is divided by the diameter because the cross-sectional area is difficult to measure. From the area under the stress–

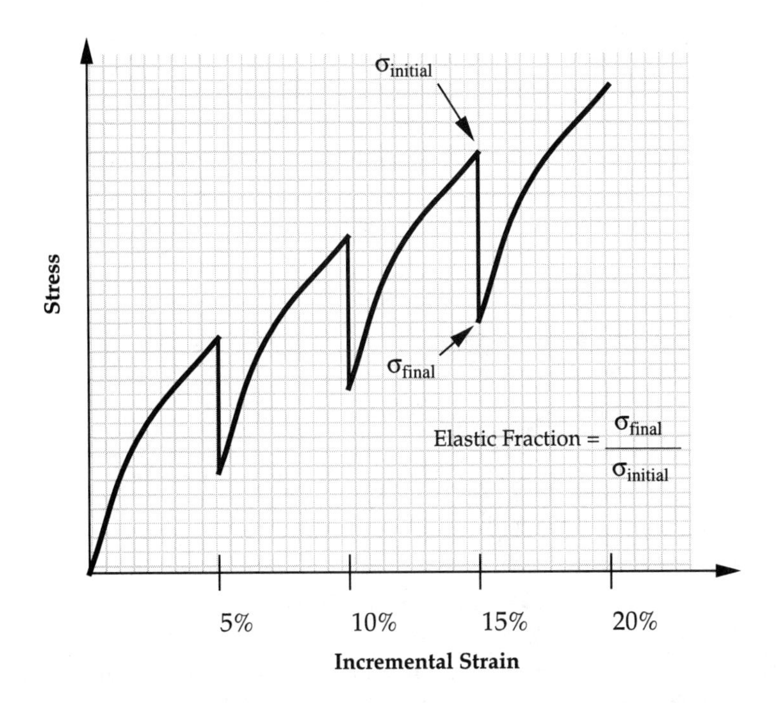

Figure 3.9 Incremental stress–strain curves. Stress relaxation can be assessed by stretching a sample, allowing it to relax to equilibrium, and then repeating this process until the sample fails. After each equilibrium point is obtained, the sample is stretched again. The elastic fraction is defined as the final stress divided by the original stress.

strain curve, the energy required for failure is calculated. Creep testing is demonstrated in Figure 3.7 and determines the time required for equilibrium to be reached and the ratio of the final to the initial force. Near the glass-transition temperature and the melting temperature, the creep rate increases rapidly. The elastic fraction is the fraction of stress that remains at equilibrium and is diagrammed in Figure 3.9.

Dynamic testing involves placing an oscillating load on a material and then measuring the lag between the stress and the resultant strain. The stress in phase with the strain is used to calculate the elastic modulus, and the stress out of phase with the strain is used to calculate the storage modulus. Dynamic testing is also used to establish the fatigue endurance limit of a material. Fatigue tests are conducted by cycling a polymeric material repeatedly to a predetermined stress and then counting the number of cycles until failure occurs. Some polymers, such as poly(ethylene terephthalate) and poly(methyl methacrylate), ex-

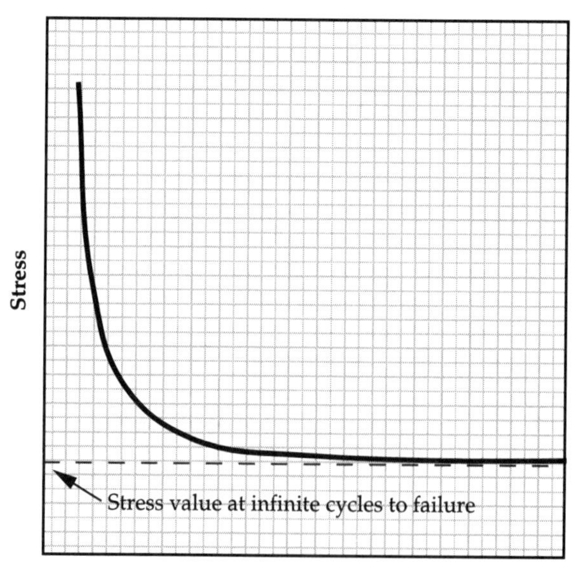

Figure 3.10 Endurance limit testing. The endurance limit of a material is determined from a plot of the stress versus number of cycles required to cause failure. The endurance limit is the stress at which the sample will undergo at least 1 million cycles before failure occurs. Each cycle consists of application of a load followed by removal of the load.

hibit an endurance limit below which the polymer will cycle infinitely without failing.

The endurance limit is the stress at which a material will undergo an infinite number of cycles before failure. Figure 3.10 is a hypothetical diagram illustrating endurance testing of synthetic polymers.

Impact testing is used to assess the toughness of a material at different operating temperatures. Older impact testers measure the difference in height that a swinging pendulum reaches after causing a notched specimen to fail. The impact strength is the energy required per unit thickness to break a notched sample (Figure 3.11). It is calculated by taking the difference in potential energy per unit sample thickness. The impact strength of a polymer drops discontinuously below the glass-transition temperature and above the crystalline-melting temperature; therefore, polymers are selected for applications where they will be dynamically loaded, based on values of T_g and T_m.

Most synthetic polymers are not formed into parts or devices in a pure form. To raw polymer is added a number of compounding agents, including

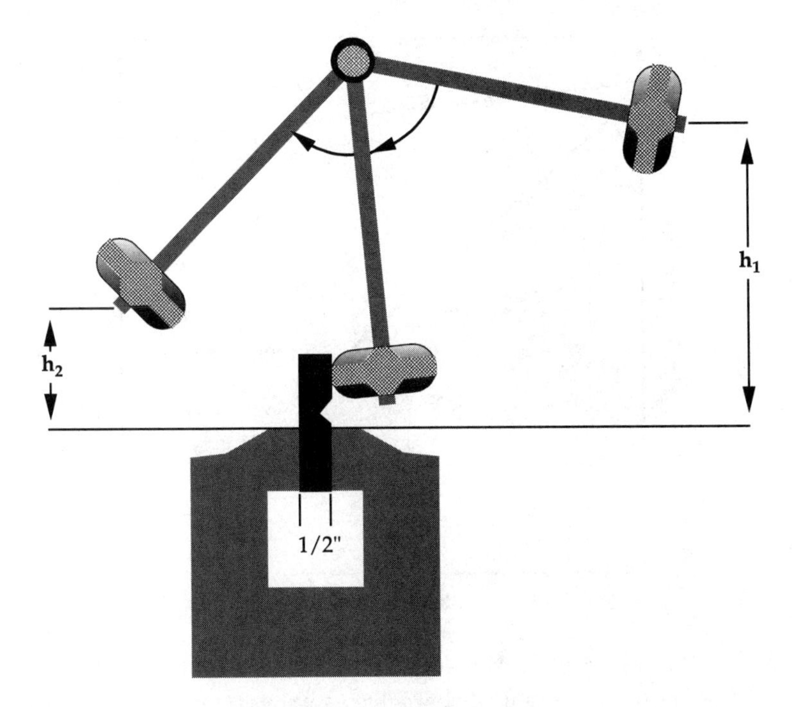

Figure 3.11 Fracture energy determination. The fracture energy is the amount of energy required to cause a notched specimen to fail. It is determined from the amount of potential energy lost after a weight causes failure of the specimen. The difference in heights ($h_1 - h_2$) is directly proportional to the energy required to cause failure. The fracture energy decreases as the temperature is lowered below the glass-transition temperature.

particulates in the form of silica, plasticizers in the form of oils and other liquid hydrocarbons, antioxidants, and cross-linking agents. These components are added to improve the ability to process the polymer into different shapes, improve the mechanical properties, and prevent premature component failure. The final properties of polymers are affected by the type of compounds added. For example, the flex temperature is the temperature at which a part cracks when flexed or bent. The flex temperature is affected in turn by the type of plasticizer added (Figure 3.12). It should be noted that the biological response is also affected by the type of plasticizer added; therefore, the final composition is a balance of mechanical properties and biological response.

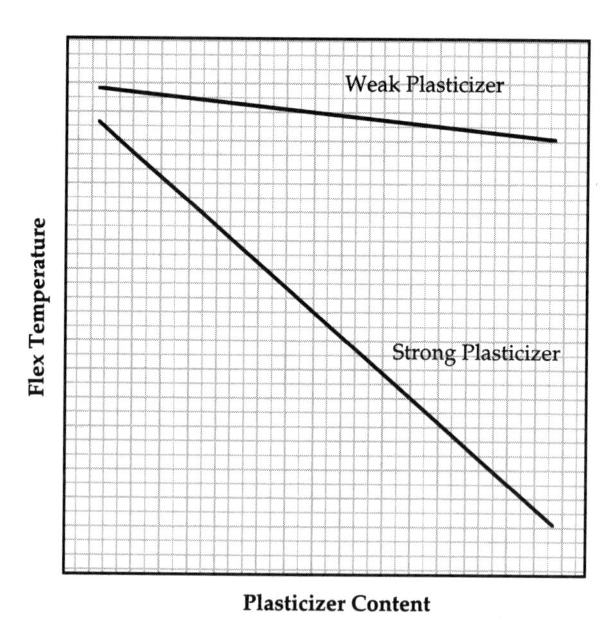

Figure 3.12 Role of plasticizers in lowering the flex temperature. The flex temperature is defined as the temperature below which a polymer sample undergoes brittle failure and is not able to be folded on itself. Polymer molecules remain flexible below the glass-transition temperature when low molecular weight molecules that are soluble in the polymer are added. The closer the match between the solubility parameter of the plasticizer and that of the polymer, the greater the strength of the plasticizer. The flex temperature is related to the strength of the plasticizer and the amount added.

Structure of Metals

Metals are crystalline materials that form from atoms that have highly mobile electrons. They are built up of repeating unit cells containing a specific number of atoms in specific positions. Several types of unit cells that maximize the numbers of nearest-neighbor atomic contacts are frequently observed, including face-centered cubic, hexagonal-close packed, and body-centered cubic. Other unit cells are modifications of these unit cells and include rhombohedral, orthorhombic, monoclinic, triclinic, tetragonal, simple hexagonal, close-packed hexagonal, and rhombic (Figure 3.13). These structures are characterized by the distance between nearest neighbors and between unit cells.

Most of the alloys of interest in medicine and dentistry form face-centered cubic structures, including iron, gold, platinum, silver, and copper. Other atomic

Crystal System	Axial Relationship	Interaxial Angles	Unit Cell Geometry
Cubic	$a = b = c$	$\alpha = \beta = \gamma = 90°$	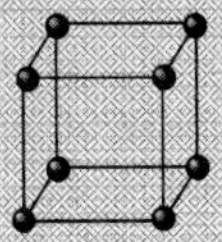
Hexagonal	$a = b \neq c$	$\alpha = \beta = 90°, \gamma = 120°$	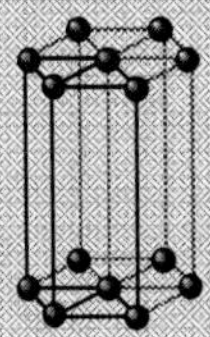
Tetragonal	$a = b \neq c$	$\alpha = \beta = \gamma = 90°$	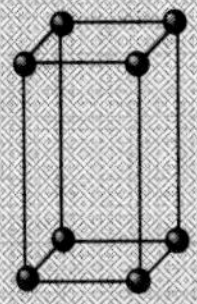
Rhombohedral	$a = b = c$	$\alpha = \beta = \gamma \neq 90°$	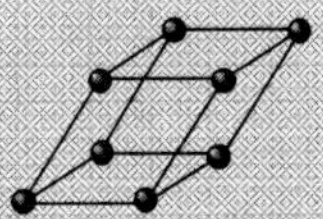
Orthorhombic	$a \neq b \neq c$	$\alpha = \beta = \gamma = 90°$	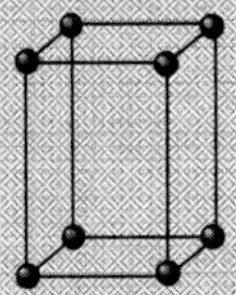
Monoclinic	$a \neq b \neq c$	$\alpha = \gamma = 90° \neq \beta$	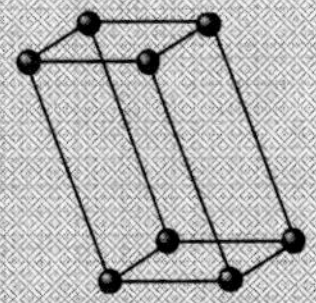
Triclinic	$a \neq b \neq c$	$\alpha \neq \beta \neq \gamma \neq 90°$	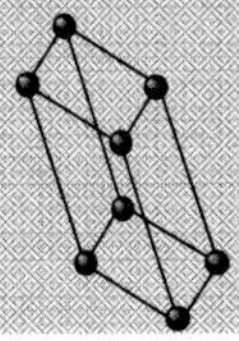

Figure 3.13 Unit cells found in crystalline lattice structures. A number of unit cells are observed repeatedly in analyzing crystalline materials. Unit cells are defined by the type of geometry, such as cubic, hexagonal, tetragonal, or other shape as well as by the dimensions of the unit cells, where *a*, *b*, and *c* are the sizes of the height, length, and width. The last parameter is the angles between the sides, defined as α, β, and γ; in the case of a cube, they are all 90°.

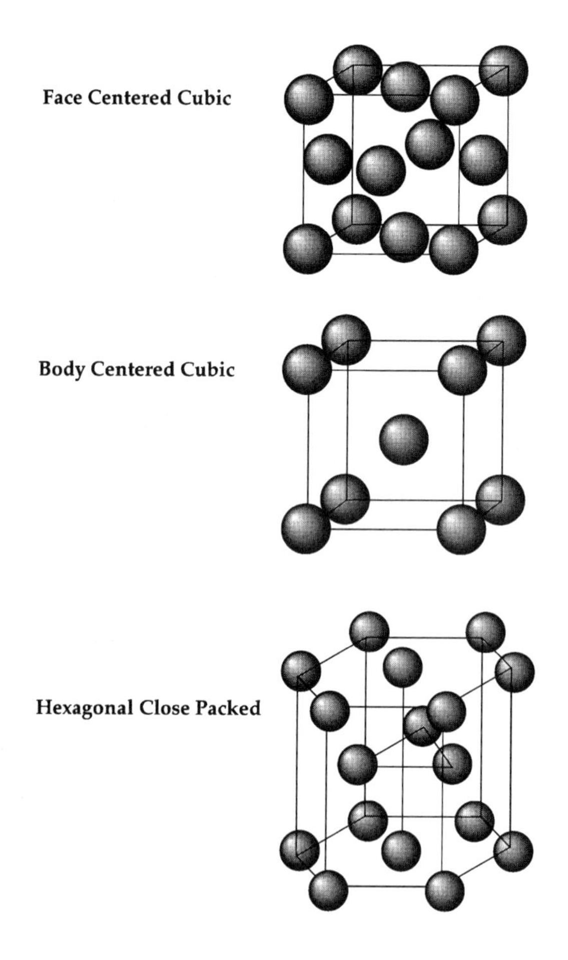

Figure 3.14 Common crystalline lattice structures. Common lattice structures seen in metals include face-centered cubic, where an atom is found in the center of each face of the cubic unit cell as well as at each corner; body-centered cubic, where an atom is found in the center of the cube as well as each corner; and hexagonal close-packed, where atoms are found in the center of the upper and lower faces as well as the center of the unit cell and at the corners.

structures include body-centered tetragonal and close-packed hexagonal (Figure 3.14). These structures are held together by metallic bonds that share outer electrons between all atoms of the unit cell. The free valence or outer electrons are mobile and are able to move around in the metal space lattice and give rise to the ability of metals to conduct heat and electricity.

The use of pure metals is limited because of their softness and tendency to corrode rapidly. To overcome these limitations, most metals are commonly used as mixtures, called *alloys*. In other cases, such as steel, a nonmetal is added (carbon). These mixtures are commonly produced by heating the elements above their melting point. Upon cooling, liquid metallic solutions solidify at temperatures below the fusion point. The transition that occurs is from a solution phase where all the atoms diffuse freely, occasionally forming small unstable atomic clusters called nuclei to that of a solid. As the solidification temperature is approached, the nuclei are present in higher numbers, and they grow to a point where further growth is spontaneous. If the formation of the nuclei is random, then the process is *homogeneous*; however, if the solid solution of metal atoms is seeded with nuclei in the form of fine metal particles, then the process is *heterogeneous*.

The process of crystallization can proceed to form a treelike morphology called *dendrite*. This occurs as nuclei form and grow in three dimensions, because long thin aggregates have more surface area for atoms to add onto than do spheres. Figure 3.15 illustrates the formation of an alloy composed of dendrites as well as the formation of alloy grains during solidification. Growth of the crystal starts after nuclei are formed in separate areas of the solution, and then they grow toward each other by adding metal atoms (Figure 3.15A). Each crystal continues to grow until it reaches the boundary with another crystal, at which time it stops (Figure 3.15B). Finally, the entire space is filled with crystals; each crystal remains a separate unit known as a *grain*, and the boundary between the crystals is the *grain boundary*. The crystals form dendrites (Figure 3.15C), which make up the solid material.

The grain size of metals is an important parameter because many physical properties depend on it. The deformation of metals occurs by sliding rows of atoms past each other in a crystal. This only works when there are atomic defects, called *dislocations*, in the crystals because there is no other space within the crystal for deformation of atoms to occur. The dislocations are eventually moved to grain boundaries where they stop because neighboring crystals have different orientations. Smaller grains limit the deformation and strain that an alloy will undergo before failure; however, the stiffness and strength are inversely related to the square root of the grain size obtained during casting.

Because grains crystallize from nuclei during solidification, the number of nuclei formed during crystallization dictates the number of grains that will form. The number of nuclei formed can be dictated by the temperature to which the alloy solution is cooled below the fusion temperature, that is, the degree of supercoiling and the rate of cooling.

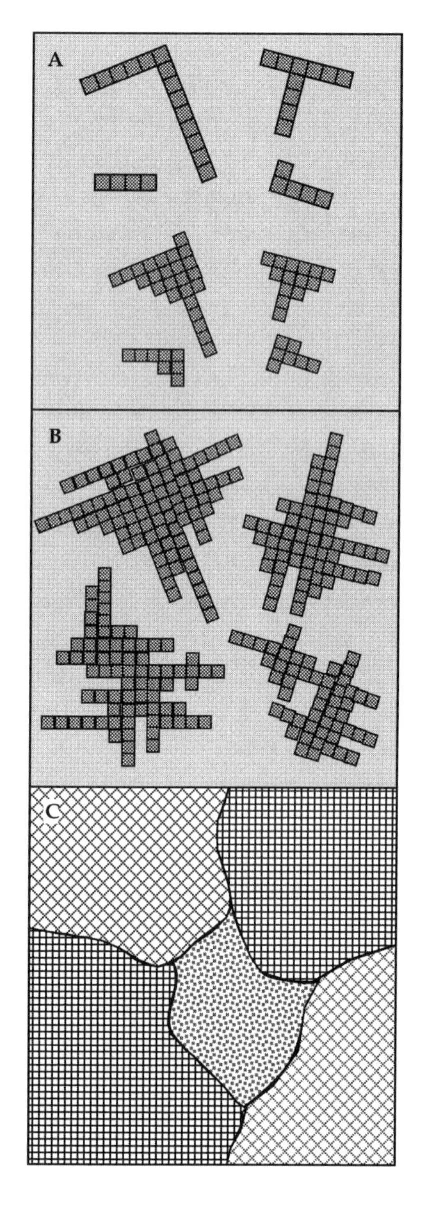

Figure 3.15 Formation of metallic grains. Stages in the formation of a metal include nucleation of crystals (A), which later grow by addition of atoms (B), which finally grow until the grain boundaries prevent further crystal growth (C).

Processing of Alloys

Cast alloys can be further processed to form medical devices via a number of processes. A cast metal that is plastically deformed into wire by drawing through a die or other process is called *wrought metal*. When an alloy is deformed beyond the limit of Hookean behavior into the plastic region, it undergoes a permanent deformation. During this permanent deformation, the material becomes strain hardened, taking on new mechanical properties that in some cases are more suitable to performing mechanical functions. Drawing, rolling, machining, extruding, forging, and other mechanical processes are used to harden alloys. The hardening occurs as a result of mechanical deformation to the material.

The effects of work hardening a metal can be reversed by simply heating the metal, a process called annealing. During annealing, the metal undergoes recrystallization and grain growth. During this process, the final grain size can be chosen by carefully selecting the conditions for annealing.

Types of Alloys Used in Medical Devices

Several alloys are used in medical devices, including (1) silver–tin–mercury–copper alloys that are used for dental amalgam; (2) cobalt–chromium alloys in dental appliances, heart valves, joint components, screws, and fracture plates; (3) titanium alloys used in conductive leads, screws, joint components, and nails; and (4) stainless steel found in fracture plates and vascular stents, and guide wires (Table 3.3).

Elements used in the formation of metallic implants include aluminum, cobalt, chromium, copper, gold, iridium, iron, manganese, mercury, molybdenum, nickel, niobium, palladium, platinum, tantalum, tin, titanium, vanadium, silver, tungsten, and zirconium. The base elements include silver (dental amalgams), cobalt (joint implants), iron (stainless steel implants for orthopedic and vascular surgery), gold (dental fillings), platinum (dental fillings), and titanium (joint implants). The remainder are alloying elements.

Stainless steels are composed of iron and carbon atoms in the presence of other alloying elements, such as chromium, nickel, manganese, molybdenum, and vanadium. These elements are needed to provide corrosion resistance to steel, because without this resistance, implants such as fracture plates would fail by corrosion. Cobalt-based alloys and titanium-based alloys are used extensively in joint replacements because of their corrosion resistance. Silver–tin–copper

Table 3.3 Medical applications of metals

Application	Metal
Conductive leads	Titanium alloys
Dental amalgams	Silver–tin–copper alloys; gold and platinum fillings
Dental appliances	Cobalt–chromium alloys
Fracture plates	Stainless steel, cobalt–chromium alloys
Guide wires	Stainless steel
Heart valves	Cobalt–chromium alloys
Joint components	Cobalt–chromium alloys, titanium alloys
Nails	Cobalt–chromium alloys, titanium alloys
Pacemaker cases	Titanium alloys
Screws	Cobalt–chromium alloys, titanium alloys
Vascular stents	Stainless steel, nitinol

Table 3.4 Mechanical properties of stainless steels

Type and condition	UTS (MPa)	Yield at 2%[a]	US (%)
F55, F138, annealed	480–515	170–205	40
F55, F138, cold worked	655–860	310–690	12–28
F745, annealed	480 min	205 min	30 min

Source: Adapted from ASTM, vol. 13.01.

Note: UTS = ultimate tensile strength; US = ultimate strain; min = minimum.

[a]Yield strength at 2% offset in MPa.

alloys are used as dental filling materials in the form of amalgams made using mercury. Gold and platinum materials are used alone or as layers that are bonded together (foils) as filling materials.

Mechanical Properties of Metals

Mechanical properties of metals used in medical applications are given in Tables 3.4 through 3.8. Stainless steel has ultimate tensile strength values that range from 480 to 860 MPa and ultimate tensile strain values from 12 to 40%. In

Table 3.5 Mechanical properties of cobalt-based alloys

Type and condition	UTS (MPa)	Yield at 2%[a]	US (%)
F75, cast	655	450	8
F90, annealed	896	379	30–45 min
F562, solution annealed	793–1,000	241–448	50
F562, cold worked	1,793 min	1,586 min	8
F563, annealed	600	276	50
F563, cold worked and aged	1,000–1,586	827–1,310	12–18

Source: Adapted from ASTM, vol. 13.01.
Note: UTS = ultimate tensile strength; US = ultimate strain; min = minimum.
[a]Yield strength at 2% offset in MPa.

Table 3.6 Mechanical properties of titanium-based alloys

Type and condition	UTS (MPa)	Yield at 2%[a]	US (%)
F67	240–550	170–485	15–24 min
F136	860–896	795–827	10 min

Source: Adapted from ASTM, vol. 13.01.
Note: UTS = ultimate tensile strength; US = ultimate strain; min = minimum.
[a]Yield strength at 2% offset in MPa.

Table 3.7 Comparison of compressive strength and creep of low-copper and high-copper silver–tin amalgams

	Compressive strength (MPa)			Tensile strength (MPa)
Amalgam	**1 h**	**7 day**	**Creep (%)**	**24 h**
Low copper	145	343	2.0	60
High copper	262	510	0.13	64

Table 3.8 Physical properties of gold foil used for fillings

Processing	Transverse strength (MPa)
Hand	296
Mechanical	265
Combined	273

comparison, cobalt-based alloys have ultimate tensile strength values that can exceed 1,500 MPa and ultimate tensile strain values between 8 and 50%. Titanium alloys have ultimate tensile strength values up to about 900 MPa and ultimate tensile strain values between 10 and 24%. Low copper and high copper amalgams have ultimate tensile strength values up to 343 and 500 MPa, respectively, and gold foil has a ultimate tensile strength value of up to 300 MPa. In comparison to polymers, metals have higher ultimate tensile strength and moduli but lower strains at failure. However, in comparison to ceramics, metals have lower strengths and moduli with higher strains to failure.

Structure of Ceramics

Materials classified as ceramics include inorganic materials such as silicon dioxide (SiO_2) or other metal oxides in crystal and glassy phases that are similar to polymers. Ceramics have been traditionally thought of as stiff, hard, brittle materials that are insoluble in water. Their primary advantages include high strength and stiffness; however, their disadvantages include low ductility, which reflects the ionic bonding between atoms that compose these materials. Ionic bonds are directional and require very high energies to dissociate. For this reason, they tend to be resistant to chemical and mechanical dissolution. Ceramics include forms of the mineral phase of bone that are used as implants for orthopedic and dental materials. They are used for repair or replacement of bone and dentin.

Although pyrolytic carbons are not held together in the same manner as are other ceramics, they are classified in this group of materials. Carbons typically are made by heating carbon-containing polymers to high temperatures in the absence of oxygen, making pyrolytic carbons. These materials were considered for replacement of heart valves, ligaments, and dental implants; however, some problems with migration of carbon fibers into local lymph nodes was observed in ligament replacements.

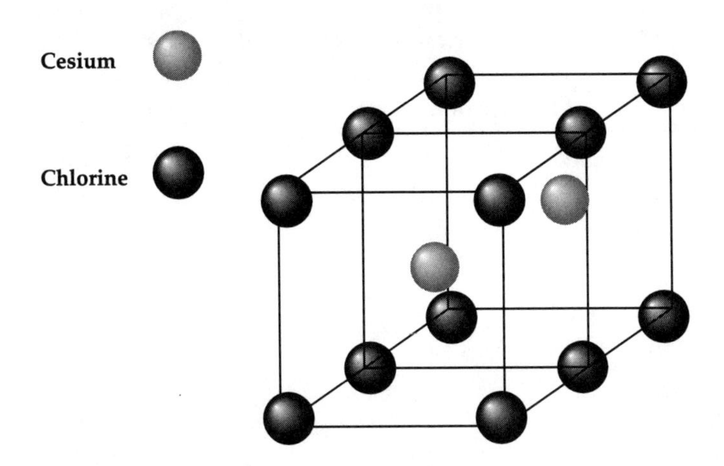

Figure 3.16 Crystal structure of cesium chloride. This diagram shows the unit cell in cesium chloride, a ceramic material held together by ionic bonds between the positively charged cesium atoms and the negatively charged chlorine atoms.

Ceramics are represented structurally as ionic compounds, such as sodium chloride. Atoms in sodium chloride are ionized by electron transfer, making an ionic compound as a result of coulombic attractions. Negatively charged ions have increased atomic radii because they gain an electron, and positively charged atoms have smaller radii because they lose an electron. The electronic attraction of the counter ions in the crystal structure gives rise to its stability. An example is that of the face-centered cubic structure of cesium chloride ions (Figure 3.16).

Ceramics can be classified by their components, because they contain metal (M) and nonmetal (N) atoms, using the general formula M_xN_y, where x and y represent the number of atoms of each species in a unit crystal cell. In the simplest case, the unit cell has equal numbers of atoms of each species and is represented by the formula MN. If the atoms have similar radii, for example, if the ratio of the radius of metal to nonmetal is more than 0.732, then the unit cell becomes a simple cubic structure, as shown for cesium chloride in Figure 3.16. If the radii ratio is less than 0.732, the unit cell becomes face-centered cubic. Examples of face-centered cubic structure include sodium chloride and zinc sulfate. If unequal numbers of metal and nonmetal atoms are found in the unit cell, then the formula is more complicated, such is the case for Al_2O_3. In this ceramic, the O^{2-} ions are in hexagonal close-packed structure, and aluminum ions fill two-thirds of the spaces between the negatively charged ions.

Dental porcelains are ceramics used to make denture teeth, single-unit crowns, and partial dentures. Dental porcelains are based on silica that can exist

Table 3.9 Mechanical properties of ceramics and carbons

Ceramic	Modulus of elasticity (GPa)	Compressive strength (MPa)
Al_2O_3 (crystals)	379	345–1,034
Fused alumina	—	519
Sintered Al_2O_3	365	207–345
Silica glass	72	107
Dental porcelains	—	75–140
Pyrex glass	69	69
Carbons		
Graphite	24	138
Glassy	24	172
Pyrolytic	28	517

Note: Compiled from Black (1988) and Phillips (1991).

in crystalline form, such as quartz or as an amorphous glass called fused silica. Dental porcelains contain other metallic ions, including boric oxide (B_2O_3), alumina (Al_2O_3), potassium oxide (K_2O), sodium oxide (Na_2O), calcium oxide (CaO), zinc oxide (ZnO), and zirconium oxide (ZrO_2). Dental ceramics are made by using naturally occurring feldspar, which is composed of potash (K_2O), alumina (Al_2O_3), and silica (SiO_2). Feldspar is used for preparation of almost all dental ceramics because it can be mixed with other metal oxides by firing, allowing the particles to fuse together.

Mechanical Properties of Ceramics

Ceramics are known for their high hardness and high values of the modulus. On the Moh's hardness scale, diamond, which is a carbon-based ceramic, has a hardness of 10; alumina, 9; quartz, 8; and hydroxyapatite, 5. Values of moduli and compressive strengths of ceramics are given in Table 3.9.

The mechanical properties of carbons such as graphite depend in part on the organization of the atoms and aggregates of atoms. Graphite is composed of aggregates of crystallites containing parallel layers of carbon atoms in a hexagonal close-packed structure. The layers of crystals are held together by carbon-to-carbon cross-links. Large numbers of cross-links are present in pyrolytic carbons, giving these materials their high compressive strengths.

Glassy or vitreous carbons are made by heating cross-linked carbon-based polymers such as phenol–formaldehyde in the absence of oxygen. Although short range crystallinity does exist, a well-defined crystal structure is absent. The absence of a well-defined crystal structure explains the lower value of the compressive strength of glassy carbons.

Unlike metals, ceramics tend to deform minimally during loading, a result that is a consequence of the electronic charge that prevents atoms from moving large distances during deformation. This phenomenon makes ceramics very brittle, and they tend to fracture at lower strengths in tension than in compression. The low value of the tensile strength is a consequence of stress concentration that occurs around cracks or flaws in these materials.

Structure of Composites

Composite materials consist of mixtures of polymers, metals, and ceramics to form materials with a balance of properties. For example, the most commonly used composite is fiberglass, a mixture of glass fibers coated with a polymeric matrix. Its utility is that it is strong and lightweight. Composite materials are not used extensively in medicine, partially because of the limited amount of research that has been conducted evaluating these materials. The most widely studied system consists of high-modulus carbon fibers embedded in a polymeric matrix such as poly(lactic acid) and carbon-fiber-reinforced high molecular weight polyethylene. These materials have been used for tendon and ligament replacements and for joint and facial implants. One problem associated with use of these materials is that breakdown of the implant leads to migration of the carbon fibers into the neighboring lymph nodes. This led to dissatisfaction with carbon-fiber tendon and ligament replacement in the United States and difficulty using these composites for other applications.

Biological composite materials are much more prevalent. Connective tissue is an example of a composite material because it contains collagen fibers embedded in a matrix containing proteoglycans and water. The collagen fibers are arranged either parallel to the load axis in tendon or biaxially in skin. The collagen fibers are typically aligned along the direction(s) of primary loading. The diameter and density of the fibers also play a role in defining the mechanical properties.

Mechanical Properties of Composites

The mechanical properties of composite materials depend on the direction of loading and the volume fraction of fibers. At low volume fraction of fibers, the

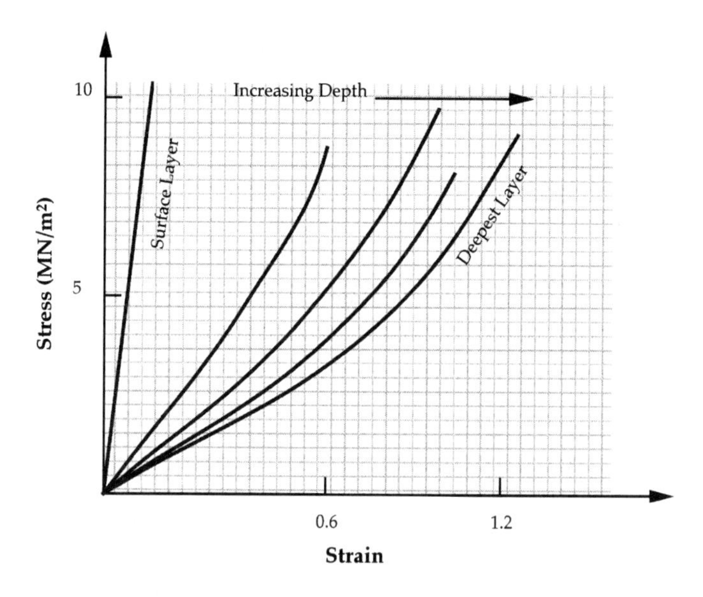

Figure 3.17 Stress–strain curves for articular cartilage. These stress–strain curves were obtained from tensile testing of specimens obtained from different layers of cartilage cut parallel to the surface. Note the slope of the curve—the modulus—increases from the deepest layer to the surface layer, which is believed to coincide with the degree of orientation of the collagen fibers with the direction of the specimen. Another way of stating this is that the collagen fibers in the superficial zone are found parallel to the surface; therefore, this layer has the highest modulus.

modulus of the composite is somewhat less than that of the contribution of the fibers; however, above a critical volume fraction, the modulus of the composite approaches that of the fibers. The strain at failure is usually somewhat higher than that of the fibers but less than that of the matrix, which is usually more extensible.

Examples of the mechanical properties of composites include the modulus of cartilage, a composite of collagen fibers in a matrix of glycosaminoglycans, and the tensile strength of skin, a composite of thick collagen fibers and glycosaminoglycans. Figure 3.17 shows a plot of stress versus strain for articular cartilage, demonstrating that the slope and ultimate tensile strength is a function of the depth of layer tested. The collagen fibers in the surface layer of cartilage are oriented parallel to the surface of the joint; they are stiffer and stronger than the deeper layers, where the fibers are oriented in different directions. Figure 3.18 illustrates that the tensile strength for wound tissue is proportional to the fiber

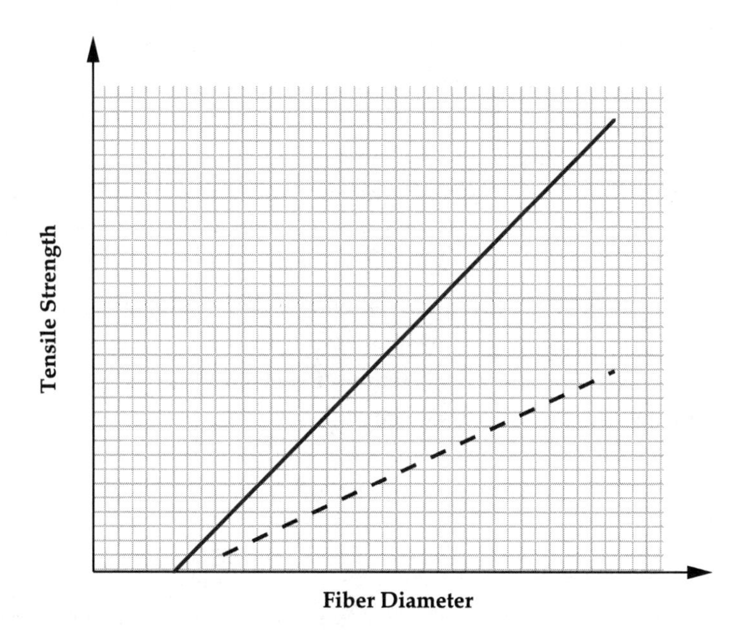

Figure 3.18 Dependence of tensile strength on fiber diameter for wound tissue. The drawing shows the relationship between tensile strength and strength of wound tissue. Larger collagen fibers are believed to have higher tensile strengths because they are more mature and contain more cross-links.

diameter for dermal wound tissue. Because the collagen fiber content increases as the fiber diameter increases, this illustrates how fiber volume fraction is related to composite mechanical properties.

Materials Degradation

Polymers, metals, ceramics, and composites may be intended as biodegradable or permanent implants. Polymers that are condensation products tend to degrade in water as a result of hydrolysis or enzyme-induced degradation. Addition polymers are somewhat more stable, but they also will degrade as a result of attack by free radicals, such as the superoxide anion that is present in biological tissues. Although somewhat more stable than polymers, metals also degrade by releasing ions as a result of corrosion. Although corrosion has been reduced as a result of limiting the use of steel and stainless steel to short-term applications such as guide wires, metallic particles that are generated by implant wear and tear lead not only to implant failure but also to hypersensitive responses. Ceramics tend to be the most stable types of materials and do not degrade unless mechanical trauma causes brittle fracture.

Degradation products of any implant are important to consider, as is leaching of any material used to make the implant. The release of low molecular weight materials from degradation or corrosion or any component of the implant will lead to inflammation and immunological complications that may require implant removal. Even the release of amino acids from a protein used in an implant over a prolonged period of time may cause autoimmune responses that may lead to implant failure. For this reason, the stable makeup of the implant is important to the function of any device.

Summary

Medical devices are made of polymers, metals, ceramics, or mixtures of these materials. These materials are used to try to mimic the mechanical properties of the tissues that are replaced or augmented with implants. Although the properties of tissues are often different than those of synthetic materials, it is possible to use these materials to prolong the functioning of tissues and organs by the careful selection of materials. Prolonged biodegradation of implant materials ultimately risks setting into motion a series of biological cascades that can lead to implant failure and even autoimmunity.

Suggested Reading

Black J., Orthopaedic Biomaterials in Research and Practice, Churchill Livingstone, New York, chapters 6, 7, and 8, 1988.

Phillips R.W., Science of Dental Materials, W.B. Saunders Company, Philadelphia, pp. 1–4, 1991.

Rodriguez F., Principles of Polymer Systems, 2nd edition, Hemisphere Publishing Corporation, pp. 201–206, 1982.

Semegen S.T. and Fah C.S., Natural Rubber, in The Vanderbilt Rubber Handbook, edited by R.O. Babbit, R.T. Vanderbilt Co., Inc., Norwalk, CT, pp. 18–19, 1978.

Silver F.H., Biomaterials, Medical Devices and Tissue Engineering: An Integrated Approach, Chapman &Hall, London, pp. 4–25, 1994.

Williams M.L., Landel R.R., and Ferry J.D., J. Am. Chem. Soc. 74, 3701, 1955.

4

Microscopic and Macroscopic Structure of Tissue

Introduction to Methods for Cellular and Tissue Analysis

Our understanding of tissue structure comes from an accumulation of knowledge that includes gross and microscopic observations. Gross observations give the biomaterials scientist information about the size, shape, surface, and mechanical properties of a tissue. Light and electron microscopes provide images at low magnification (light microscope 4× to 1000×) and high magnification (electron microscope up to 300,000×). At low magnification, the biomaterials scientist can determine what are the structural units that make up a tissue or organ (Figure 4.1). The structural units usually contain connective tissue and extracellular matrix in some 3-D arrangement that is characteristic to that class of tissue. The connective tissue contains specific cell types; the distribution and

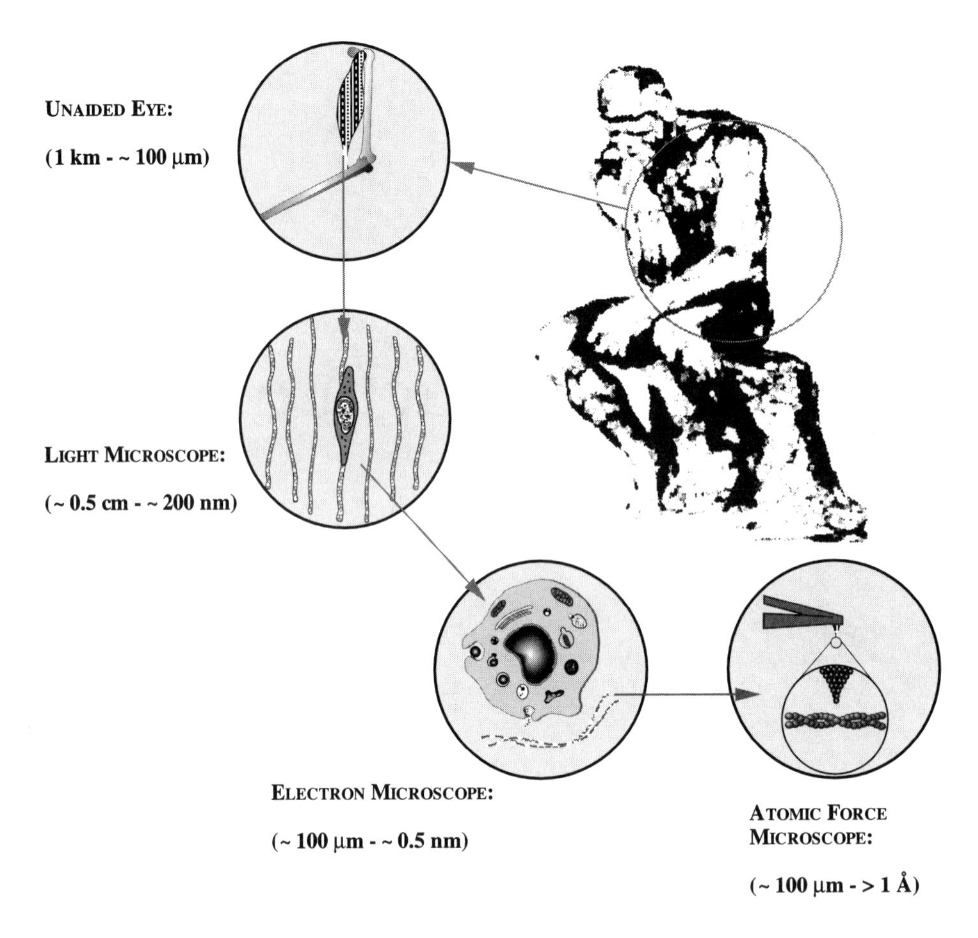

Figure 4.1 Levels of structural analysis of tissues. The levels of structural hierarchy of tissues include levels viewed by eye, light microscope, electron microscope, and atomic force microscope.

type of connective tissue components and cell types are unique for each specific tissue type. At higher magnification in the electron microscope, cell organelles and specific macromolecules can be identified (see Figure 4.1). Because it is the behavior of the macromolecules and their packing arrangement that ultimately dictate how a tissue will behave, it is important to understand the arrangement of specific macromolecules inside cells and in the extracellular environment as well as how structural units fit together to make up gross tissue structure (see Chapter 2).

Several techniques are used to elucidate the structure of tissues at the light and electron microscopic levels. At the light level, a number of different techniques are used to process tissue for making standard paraffin sections, which are the gold standard of light microscopic sample preparation. The steps used to prepare paraffin sections include fixation, decalcification (for mineralized samples), dehydration, embedding, section preparation, and staining (Figure 4.2). Fixation of tissue involves chemical cross-linking to stabilize the tissue to prevent postmortem changes, preserve various cell constituents, and harden tissue for easier processing. Fixatives include formaldehyde (also known as formalin), glu-

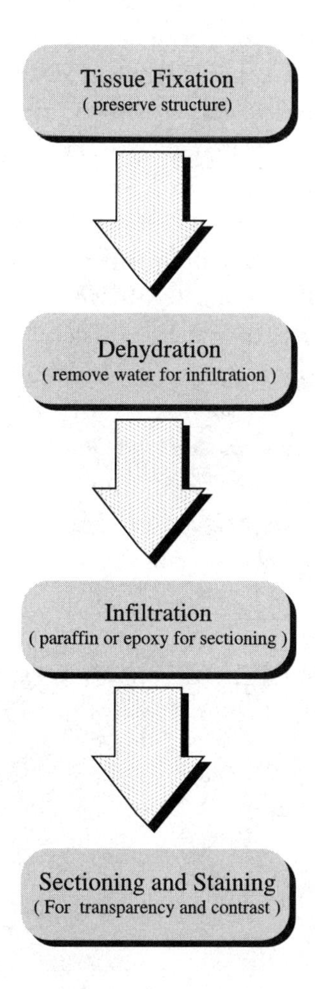

Figure 4.2 Processing of tissues for microscopy. When processing tissues for light and electron microscopy, tissues are isolated and fixed with solutions containing aldehydes and then dehydrated through alcohol solutions to remove all water. Dry tissue samples are infiltrated with either paraffin wax or epoxy to stiffen the sample. Samples are then cut with a knife into sections several millimeters thick, through which light or electron beams can pass.

taraldehyde, paraformaldehyde, and buffered solutions containing these cross-linking materials. Decalcification can be accomplished using nitric acid or formic acid–sodium citrate or by immersion in a solution containing a chelating agent (an agent that binds calcium) such as ethylenediaminetetraacetic acid (EDTA). Dehydration of the tissue is then accomplished by passing the tissue through a graded series of alcohol solutions with increasing concentrations of alcohol. During this process, the tissue dimensions shrink, which must be accounted for when using quantitative light microscopy. The dehydrant is replaced with a clearing agent such as xylene, which is miscible in the dehydrant and the embedding material. The clearing agent transforms the translucent tissue into a transparent material. The section is next infiltrated with liquid paraffin to provide support for the tissue block that is then cut into thin sections during sectioning. Sections about 3 to 6 μm are cut using a microtome. The cut sections are floated without wrinkles onto slides and attached to the section using an adhesive such as gelatin. Sections are deparaffinized and stained with dyes such as hematoxylin and eosin (H&E), van Gieson's, Masson's trichrome, and alcian blue.

Hematoxylin is a natural dye substance that combines with a metal salt such as an aluminum salt. After staining with H&E, the nuclear structures are stained dark purple or blue, and most of the cytoplasmic and intracellular structures are stained various shades of pink. Collagen fibers in the extracellular matrix are stained light pink with H&E (Figure 4.3). Masson's trichrome is used to stain collagen in tissues; collagen fibers stain blue in contrast to cells that stain purple

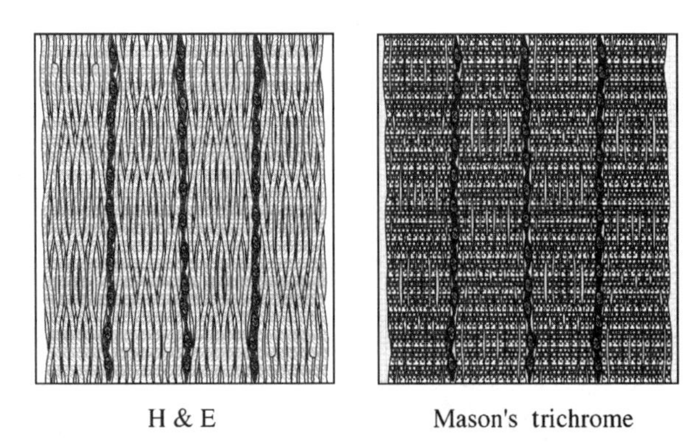

Figure 4.3 Staining of connective tissue with Masson's trichrome and H & E. Staining of dense regular connective tissue with H & E stains collagen fibers pink, and the cellular nuclear material stains blue-purple. Masson's trichrome stains collagen fibers blue and the nuclear material dark.

using this stain (see Figure 4.3). Elastic tissue stains black with van Gieson's solution, and the rest of the extracellular matrix stains off-white as a background. Glycosaminoglycans stain blue with alcian blue, and the surrounding extracellular matrix stains pink with H&E as a counterstain.

Sometimes tissues are frozen and sectioned frozen so that they can be cut directly without the need for embedding. In other cases, tissues are embedded in a polymer such as methacrylate instead of paraffin after fixation. Several variations can be used to make tissue sections to view under the light microscope. In addition to staining with dyes, fluorescent molecules can be specifically attached to macromolecules using antibodies.

At the electron microscopic level, there are two approaches to revealing tissue structure: scanning electron microscopy and transmission electron microscopy. Scanning microscopy involves the analysis of surface structure by bouncing electrons off the surface of a tissue fragment and then forming an image. Secondary electrons from the surface of the object are collected by scanning the surface. The power of this approach is that the composition of the surface can be determined by analysis of the energy spectrum of the electrons, which is characteristic of the atoms on the surface. Images obtained by scanning electron microscopy give good resolution up to a magnification of about 40,000× (Figure 4.4). In transmission electron microscopy, the electrons of the imaging beam go through the specimen (therefore, the specimen must be sectioned) and impinge on photographic paper to make an image. The resolution obtained using this technique is much greater than that obtained using scanning microscopy and can be as high as several hundred thousand. Using these two techniques, the structure of individual macromolecules as well their assemblies can be elucidated. However, in each technique the macromolecule must be first fixed and then either coated with metallic ions by vapor deposition (scanning microscopy) or stained with salts of heavy metals (transmission electron microscopy) to scatter the electrons of the imaging beam. The next section discusses how light and electron microscopic techniques help us elucidate the structure of biological tissues.

Surface and Internal Linings

As discussed in Chapter 1, surface and internal lining structures keep what is outside from entering and mixing with host internal organ structures. This translates into a number of specific functions, including (1) selective adsorption of

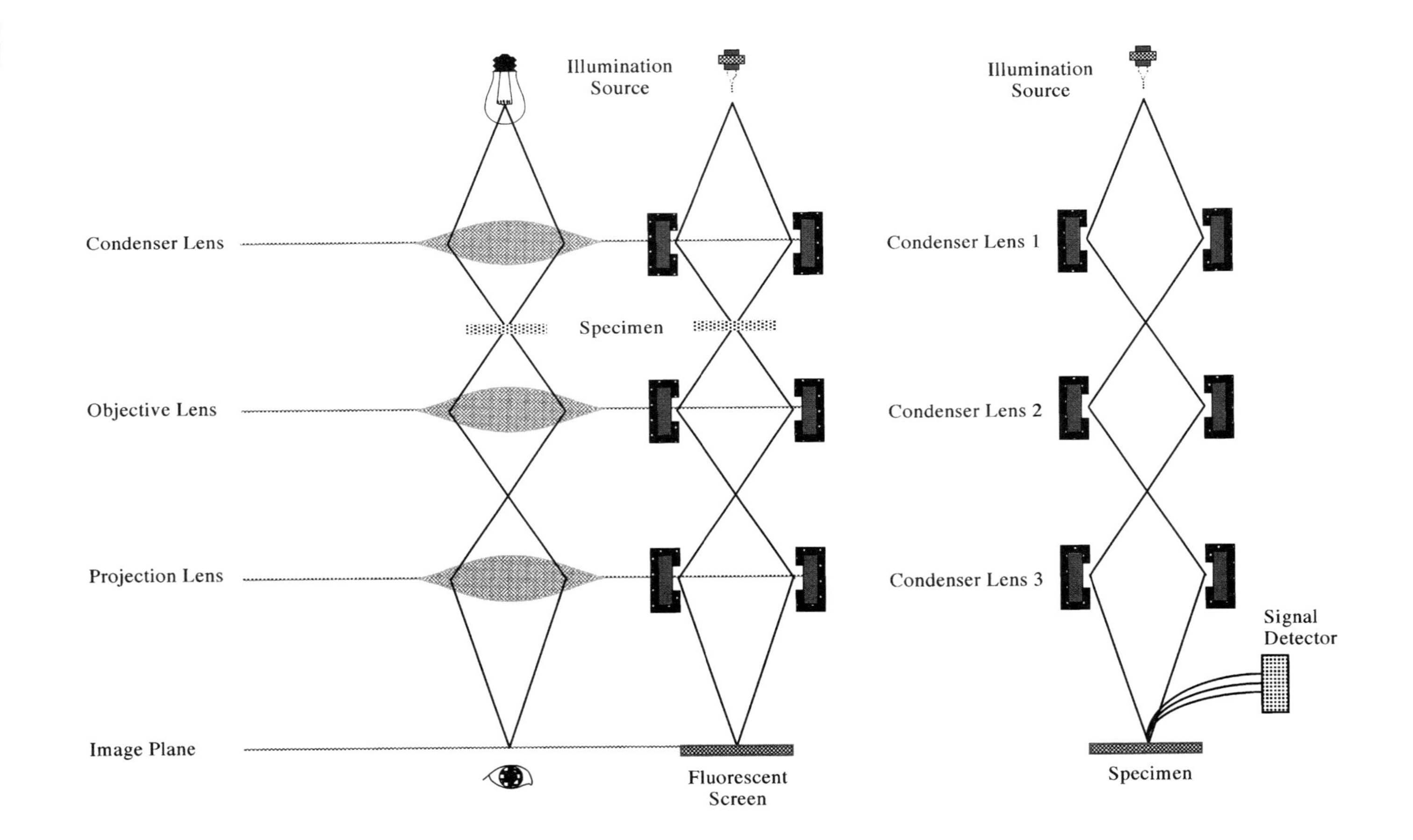

Figure 4.4 Schematic diagram of light and electron optics for the light microscope and transmission and scanning electron microscopes. The optical systems for the light microscope and transmission electron microscope are comparable with either the incident light or electron beam transmitted through the sample and focused by the objective and projection lens on the image plane. In scanning electron microscopy, electrons from the source are focused onto the sample using three condenser lenses. Electrons emitted from the surface of the sample are collected by a signal detector.

gases (alveoli); (2) prevention of external mechanical and chemical injury to the eye (cornea), oral tissues (mucosal lining), skin (cornified epithelium), and vagina (mucosa); and (3) prevention against internal mechanical injury to stomach and chest organs (peritoneum and pleura).

All these structures have an epithelial lining that lies at the interface as well as an extracellular matrix, including basement membranes and loose connective tissue, as indicated in Table 4.1. These tissues are similar in their general structure: they all have an inner cellular layer, supportive connective tissue, and an outer cellular layer. The biomaterials scientist must be familiar with these structures because it is not possible to evaluate the response of tissue to implants or implantation procedures without being able to recognize the details of normal tissue structure.

Histology of Alveoli and Bronchi

The alveolus is the gas exchange unit of the respiratory system. It is in this unit that environmental air comes in contact with the surface lining of functional lung tissue. Alveoli consist of a maze of air spaces surrounded by a thin layer of tissue, and they are lined with squamous epithelium (see Table 4.1 and Figure 4.5). Adjacent alveoli have a common wall (interalveolar septum) containing capillaries supported by small amounts of connective-tissue-containing fibroblasts. The capillaries are close to the squamous epithelium so that oxygen and carbon dioxide are exchanged by diffusion through the vessels and alveolar walls.

The bronchi are lined with pseudostratified columnar epithelium, surrounded by a thin lamina propria containing fine collagen and elastic fibers. A thin layer of smooth muscle surrounds the lamina propria. Glands are found in the submucosa, and hyaline cartilage and the pulmonary arteries are found in the outer layer (adventitia).

Histology of Cornea

The outer surface of the cornea is covered with a smooth layer of stratified corneal epithelium (Figure 4.6). A scratched cornea causes excruciating pain when the epithelia are lost. The lower layer of cells is columnar in shape and rests on a basement membrane that sits on top of Bowman's membrane, a thick

Table 4.1 Cellular and noncellular composition of lining structures

Structure	Cellular layers	Noncellular layers	Function
Alveolus	Squamous epithelium	Basement membrane	Lines inside of alveolus
	Capillaries and fibroblasts	Connective tissue	O_2 absorption
Cornea	Stratified squamous		Protects eye
	Columnar epithelium	Bowman's membrane	Supports epithelium
	Stromal fibroblasts	Collagen lamellae	Provides refraction
	Posterior epithelium	Descemet's membrane	Controls fluid transport
Mouth	Stratified squamous		
Mucosa			
	Columnar cells	Basement membrane	Protects mouth
	Capillaries	Connective tissue	Supports epithelium
Submucosa	Blood vessels	Connective tissue	Provides nutrition
	Nerves and fat		
Peritoneum	Squamous epithelium	Basement membrane	Protects organs
Pleura	(Mesothelium)		
	Fibroblasts	Loose connective tissue	Supports epithelium
Skin			
Epidermis	Stratified squamous epithelium		Provides resistance to shear injury
	Stratum corneum		
	Stratum spinosum	Basement membrane	
	Stratum basale		
Dermis	Fibroblasts, histiocytes immune cells	Loose connective tissue	Supports epidermis
	Epithelium of accessory structures	Dense connective tissue	Contains accessory structures such as sweat glands and hair follicles
	Endothelium	Blood vessel walls	
	Adiposites		
	Smooth muscle cells		
Uterus			
Endometrium	Columnar epithelium	Basement membrane	Secretes substances
	Fibroblasts and glandular cells		
	Basal epithelium	Loose connective tissue	
	Endothelium	Blood vessel walls	
Myometrium	Smooth muscle cells	Loose connective tissue	Supports endometrium
	Fibroblasts	Smooth muscle fibers	
	Endothelium	Blood vessel walls	

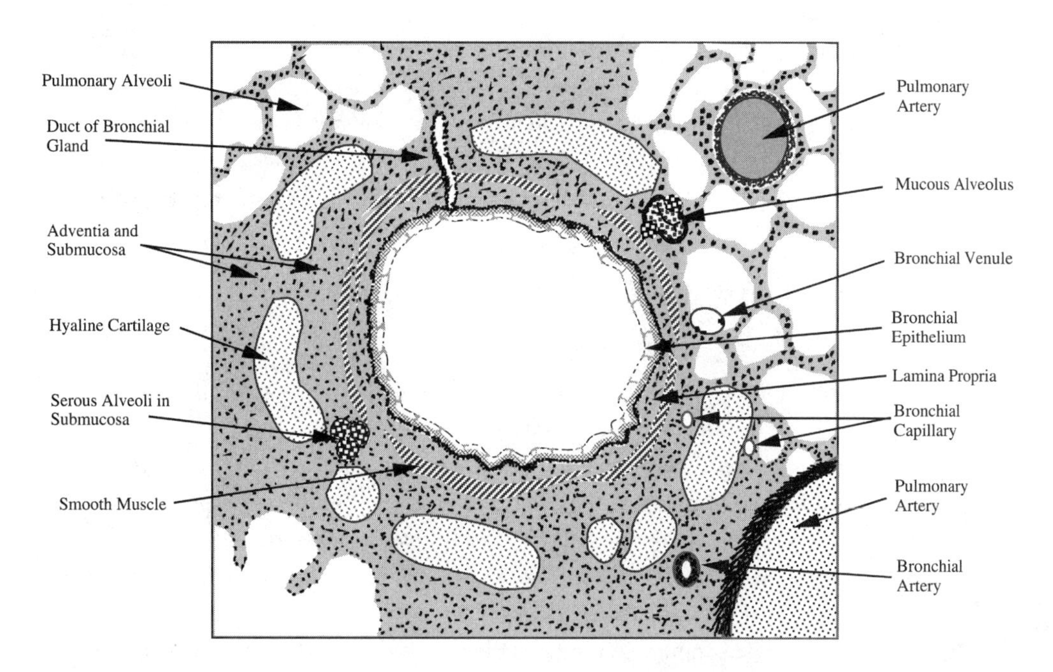

Figure 4.5 Structure of alveoli and bronchi. The alveoli absorb oxygen from the air and transport it into the capillaries. The bronchi are tubular structures through which air flows to the alveoli.

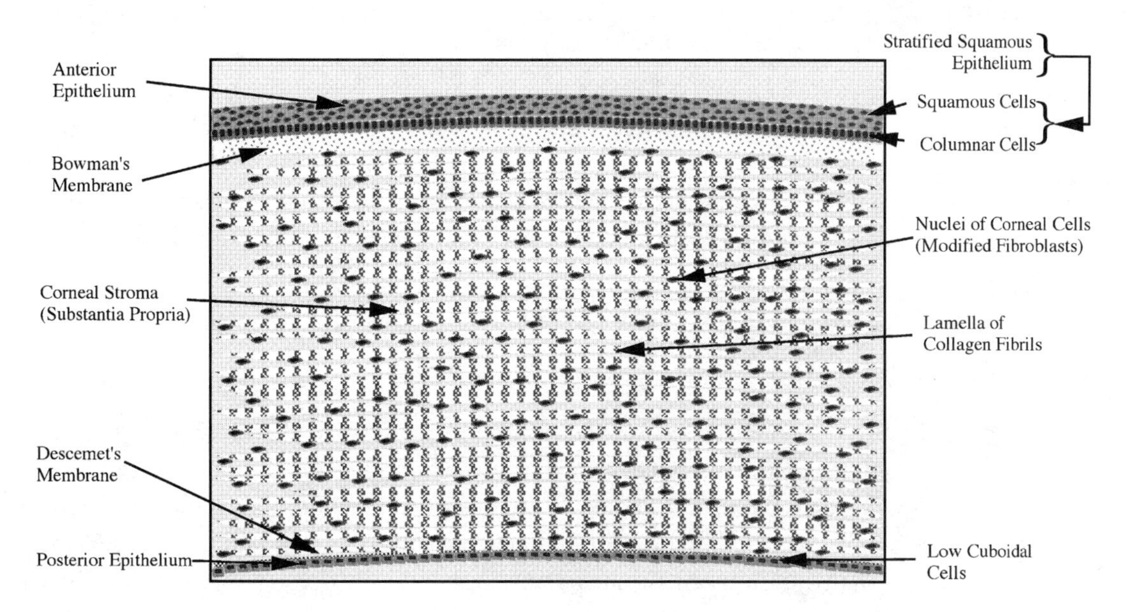

Figure 4.6 Structure of cornea. The structural components of cornea include epithelium, corneal stroma, collagen lamellae, Bowman's and Descemet's membranes.

limiting structure derived from the corneal stroma below. The corneal stroma is composed of lamellae, parallel bundles of collagen fibrils, and keratocytes, rows or layers of branching corneal fibroblasts. The posterior of the cornea is covered with a low cuboidal epithelium that has a wide basement membrane (Descemet's membrane) and rests on the posterior portion of the corneal stroma.

Oral Histology

The mouth and esophagus are composed of two layers: the mucosa and submucosa (Figure 4.7). The mucosa is lined on its outer surface by a stratified squamous epithelium that has layers of polyhedral cells in the intermediate layers and low columnar cells in the basal layer. Below the cellular layer is the lamina propria, which contains loose connective tissue with blood vessels and small aggregates of lymphocytes. Smooth muscle within the mucosa (muscularis mucosal layer) is seen as small bundles.

The submucosa contains blood vessels, nerves, adipose tissue, and skeletal muscle, similar to the subcutaneous tissue seen in skin.

Histology of Peritoneum and Pleura

The pleura and peritoneum are composed of a thin mesothelial cell lining that rests on a basement membrane and a thin layer of connective tissue that contains endothelium and smooth muscle cells (peritoneum) and fibroblasts (Figure 4.8). These cell linings are continuous with internal tissues that lie beneath.

The Histology of Skin

Skin is composed of two layers: the epidermis and dermis (Figure 4.9). The epidermis is a stratified squamous epithelium composed of several cell layers, including the keratinized outer cell layer (stratum corneum), polygonal cells of the stratum spinosum, and the inner layer of basal cells (stratum basale). Basal cells sit on a basement membrane, below which is composed the fine-fibered dense connective tissue of the papillary dermis. The next layer, the reticular layer, contains thick collagen fibers. Reticular dermis is above the subcutaneous layer and contains fat (adipose tissue), blood vessels, and skeletal muscle. The acces-

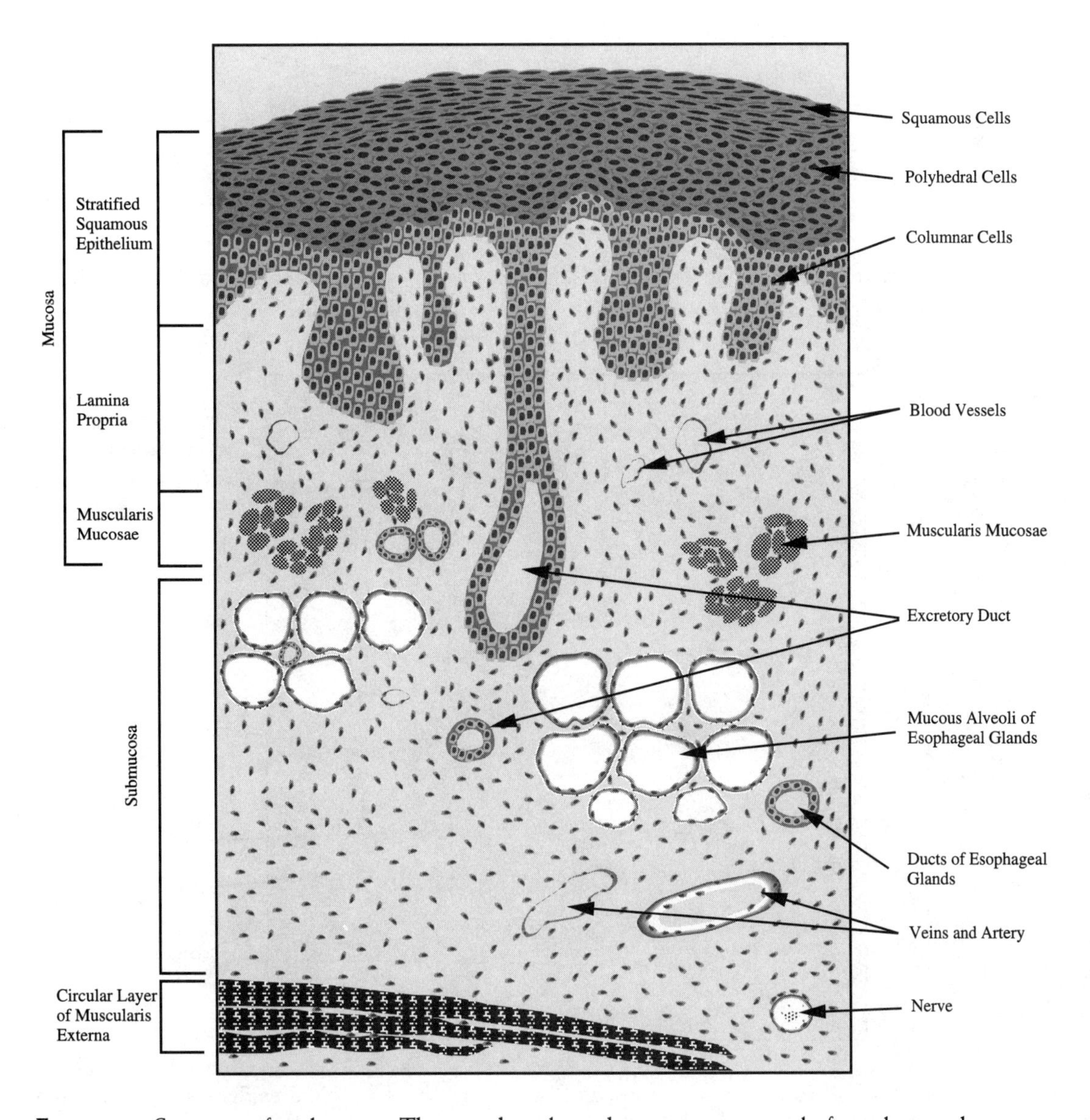

Figure 4.7 Structure of oral tissues. The mouth and esophagus are composed of two layers: the mucosa and submucosa. The mucosa is composed of layers of squamous epithelium, lamina propria, and muscularis. The submucosa contains blood vessels, nerves, adipose tissue, and skeletal muscle.

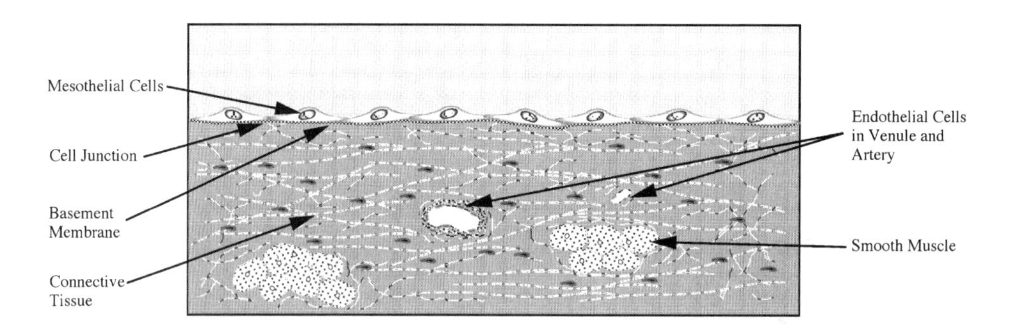

Figure 4.8 Structure of peritoneum and pleura. The peritoneum and pleura contain a thin mesothelial cell lining, a basement membrane, connective tissue, and smooth muscle cells.

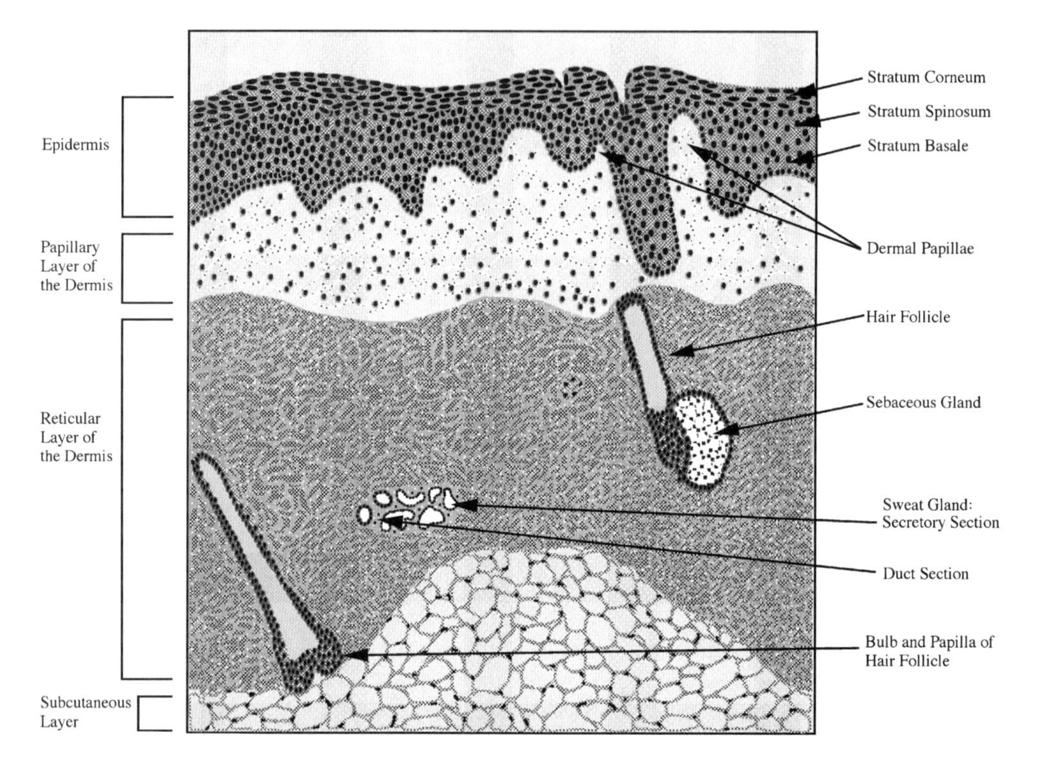

Figure 4.9 Structure of skin. Skin is composed of two layers: the dermis and epidermis. Epidermis is a stratified squamous epithelium containing stratum corneum, stratum spinosum, and stratum basale (basal cell layer). The dermis contains papillary and reticular layers and a subcutaneous layer containing fat, blood vessels, and skeletal muscle.

sory structures in the skin are found in the dermis and include hair follicles, sweat glands, and glands that secrete a waxy substance (sebaceous glands).

Histology of Uterus

Uterus is composed of two layers: the endometrium and myometrium (Figure 4.10). On the outside is perimetrium, a mesothelial lining layer similar to peritoneum. The endometrium is composed of two layers: functionalis and basalis. The functionalis is composed of an outer layer of columnar secretory epithelium overlying laminar propria, a connective tissue layer. The long tubular uterine glands are present in the laminar propria. The glands are usually straight in the superficial portion and coiled in deeper portions.

The myometrium contains compactly arranged smooth muscle fiber bundles separated by thin partitions of connective tissue as well as blood vessels that provide nutrition to the organ.

Conduit and Holding Structures

Conduit and holding structures are another category of structures important to the biomaterials scientist. These structures include conduits for distributing blood (blood vessels), food (stomach, small intestine, and large intestine), and waste products (bladder and ureter). Most of these structures contain three layers, similar to blood vessels. They are composed of an inner cellular lining (intima), a muscular layer (media), and an outer layer (adventitia) containing connective tissue that anchors the tissue to the surrounding structures. Each of these subclassifications is discussed in this section, including the names of the arrangements of the cells and the extracellular matrix (Table 4.2).

Structure of Blood Vessels and Lymphatics

Blood vessels include capillaries, arterioles, arteries, and veins. Each of these structures is composed of three layers: the intima, media, and adventitia (Figure 4.11). For example, elastic arteries consist of an intimal layer containing endothelial cells and connective tissue; a media containing smooth muscle cells, collagen, and elastic fibers; and an adventitia containing collagen fibers, nerves, and

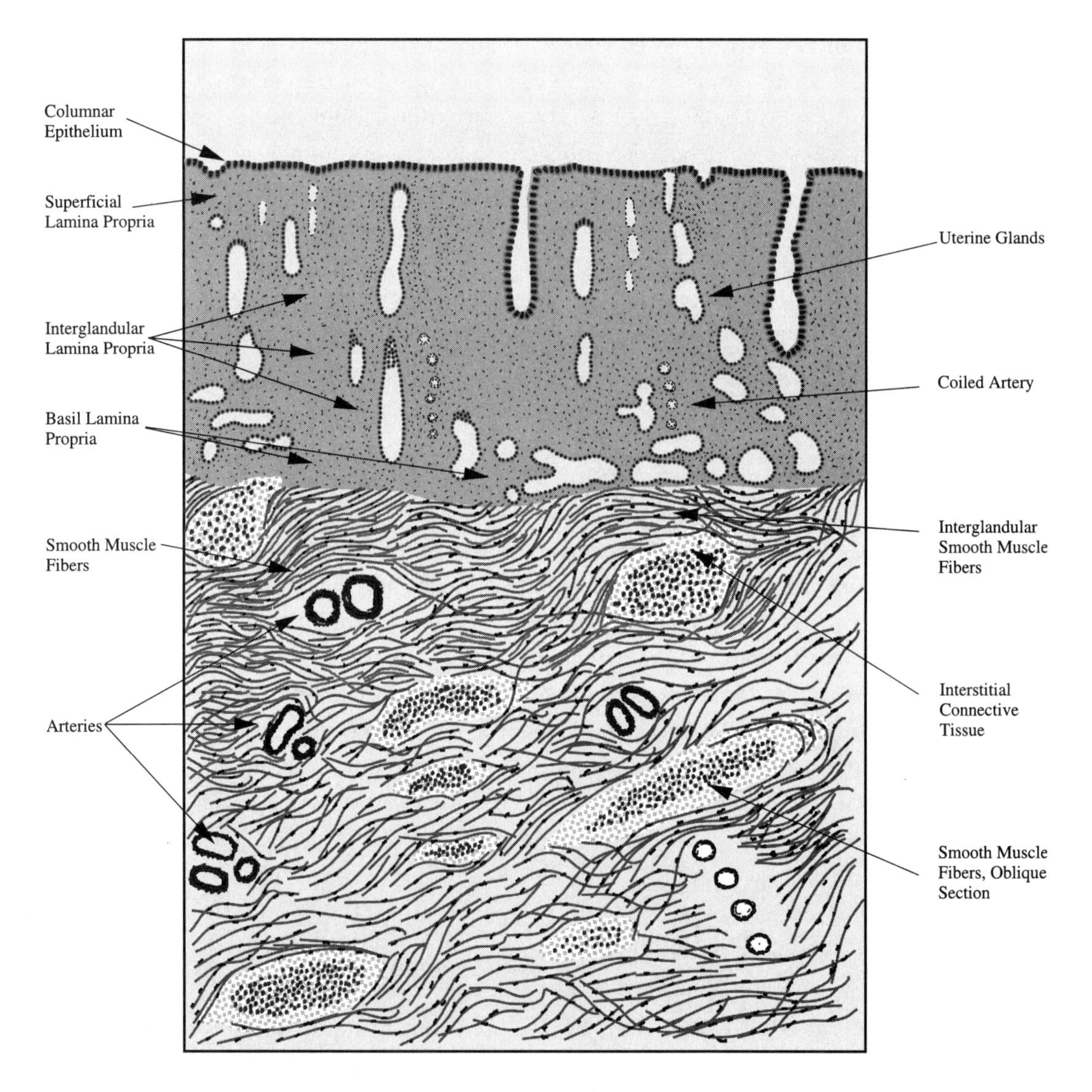

Figure 4.10 Structure of uterus. The uterus contains two layers: the endometrium and the myometrium. Endometrium contains columnar epithelium, lamina propria, uterine glands, and arteries. Myometrium contains smooth muscle fibers, interstitial collagen, and arteries.

Table 4.2 Cellular and noncellular composition of conduit and holding structures

Structure	Cellular layers	Noncellular layers	Function
Blood vessels			
Intima	Endothelium	Basement membrane	Maintain blood flow
Media	Smooth muscle	Connective tissue	Prevents blowout
	Fibroblasts	Elastic fibers	Auxillary pump
Adventitia	Fibroblasts	Connective tissue	Anchors vessel
	Nerve	Connective tissue	
	Blood vessels	Connective tissue	
Stomach and intestines			
Mucosa	Columnar epithelium	Basement membrane	Food transport
	Glandular cells or pyloric cells	Connective tissue	Release of material
	Fibroblasts	Connective tissue	
Submucosa	Fibroblasts	Connective tissue	
	Glands (intestine)		
Muscularis	Smooth muscle	Connective tissue	Contractile motion
Serosa	Fibroblasts	Connective tissue	Part of peritoneum

blood vessels. In cross-section, arteries and arterioles are circular; the only difference is that arteries are larger in diameter and have a more prominent media. Capillaries are also circular and usually contain red blood cells but are much narrower. In contrast, veins and venules are oval and contain a less prominent media. Lymphatic vessels are very thin walled and typically are recognized by the valves within the lumen.

Structure of Stomach and Intestines

The stomach and intestines contain several layers, including the mucosa, submucosa, muscularis, and serosa. In the stomach, the inner lining is made up of a mucous layer with columnar epithelium that extend into the gastric pits about one fourth of the thickness of the wall (Figure 4.12). Below the epithelium is the lamina propria, which includes the bases of the gastric pits where the gastric glands release their products. At the base of the mucosa is the muscularis mucosa, a band made up of smooth muscle, below which is found the submucosa.

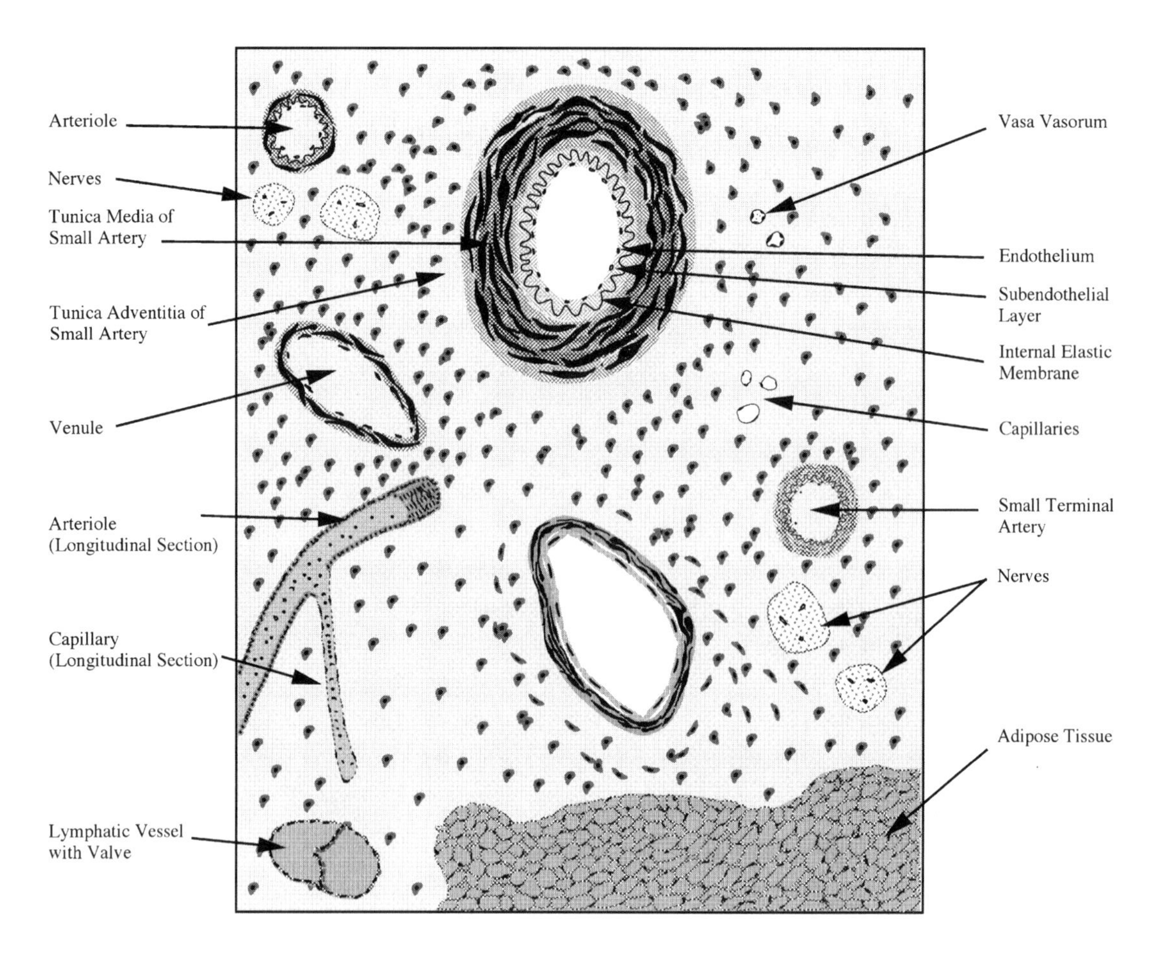

Figure 4.11 Structure of vessels. All vessels contain three layers: the intima, media, and adventitia. In large elastic arteries, the intima is found beneath the internal elastic membrane and interfaces with the lumen. The media is found between the internal and external elastic membranes, and the adventitia is found outside the external elastic membrane. The media is less prominent in the other types of vessels.

The submucosa is made up of connective-tissue-containing small blood vessels.

The next layer is the muscularis externa, which contains circular muscle fibers on the top layers and longitudinal muscle fibers on the outer layers. The serosa is continuous with the peritoneum.

In contrast, in the small intestine (Figure 4.13), the mucosa is composed of surface projections or villi and has a core of lamina propria and muscle fibers. In the lamina propria proper are found intestinal glands, fine connective tissues, reticular cells, and lymphatic tissue. The submucosa is filled with glands in the

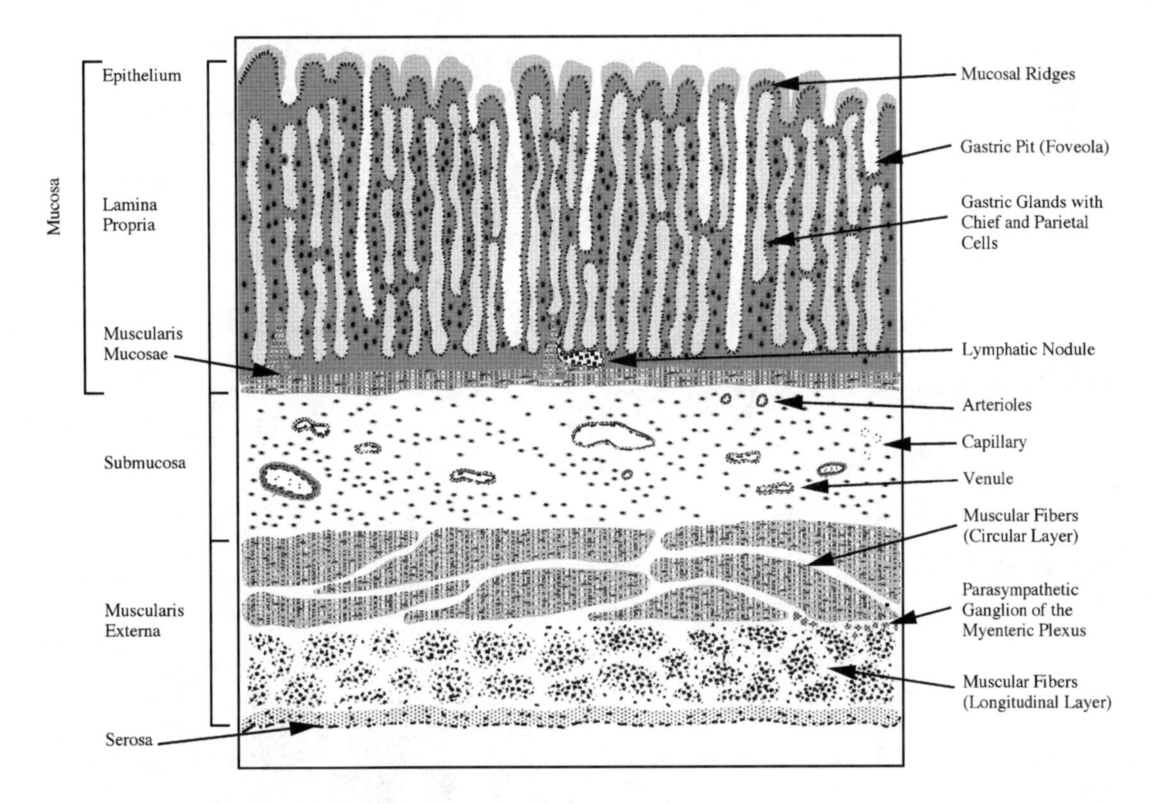

Figure 4.12 Structure of the stomach. Stomach contains four layers, including the mucosa, submucosa, muscularis, and serosa. The inner infolded mucous layer is made up of columnar epithelium, and gastric pits are found within the infoldings. Submucosa contains connective tissue and vessels, and the muscularis contains muscle fibers. The serosa is on the outside of the stomach and is continuous with the peritoneum.

duodenum, and in the muscularis externa, an inner circular layer and an outer longitudinal layer of muscle are present. Parasympathetic ganglion cells are seen in the thin layer of connective tissue between the two muscle layers. The serosa forms the outermost layer and is continuous with the peritoneum.

Structure of Bladder and Ureter

The structure of the bladder, ureter, and urethra are similar in that they contain three layers: the mucosa, muscularis, and serosa. In the bladder (Figure 4.14), the inner layer (mucosa) when empty is infolded, and it is made up of transitional epithelium. The lamina propria below contains collagen and elastic fibers in the

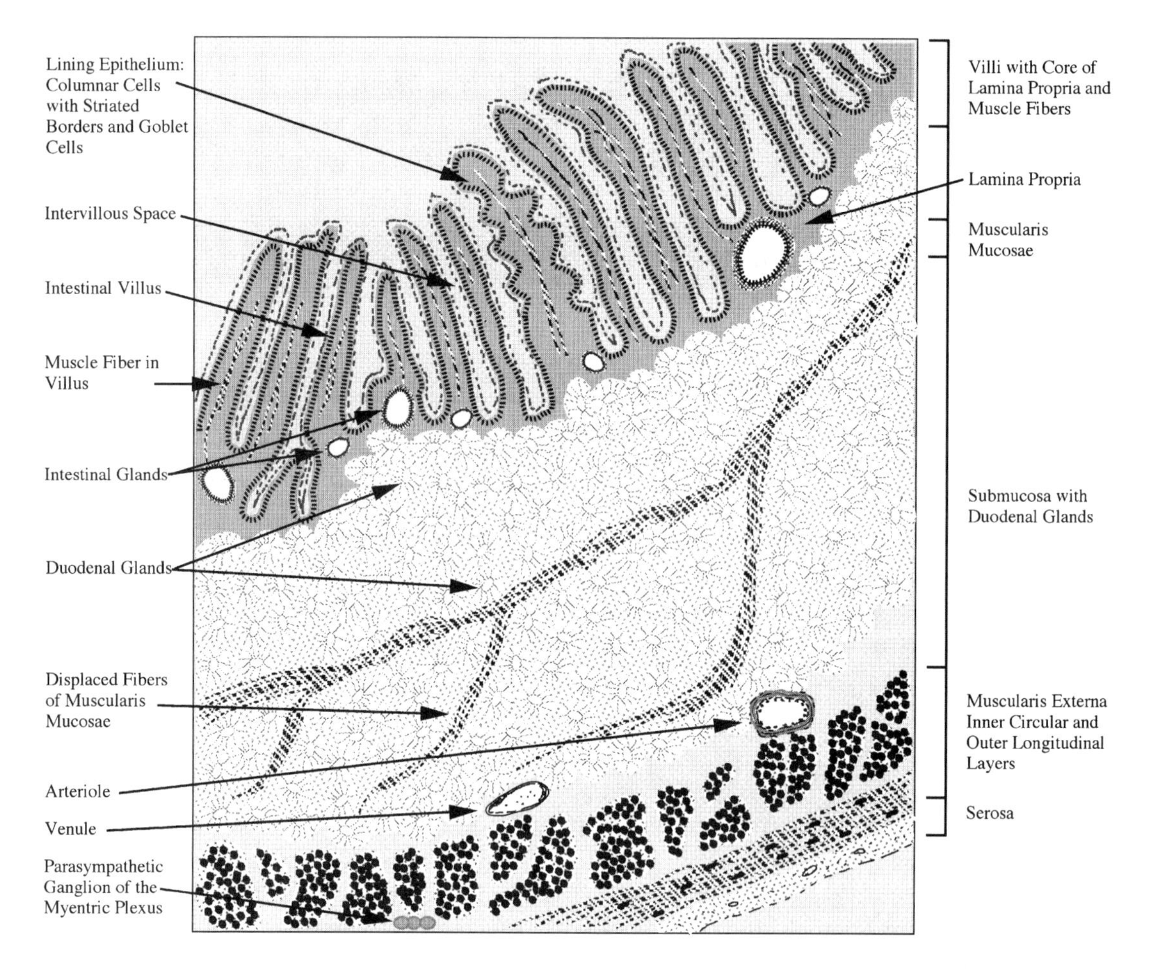

Figure 4.13 Structure of small intestine. In small intestine, the mucosa contains villi or surface projections, lamina propria, and muscle fibers. Submucosa contains duodenal glands, and the muscularis contains muscle fibers.

deeper layer. The muscularis is prominent and contains muscle fibers that are arranged in branching bundles separated by connective tissue. The connective tissue between the muscle fiber bundles merges with the connective tissue of the serosa. The serosa is continuous with the peritoneal lining.

In comparison, the ureter is similar, except the muscularis is less prominent (Figure 4.15). In the ureter, the muscularis consists of an inner longitudinal layer and an outer circular layer of smooth muscle. The outer layer, adventitia, contains fibroelastic and adipose tissue, in which are found blood vessels and nerves.

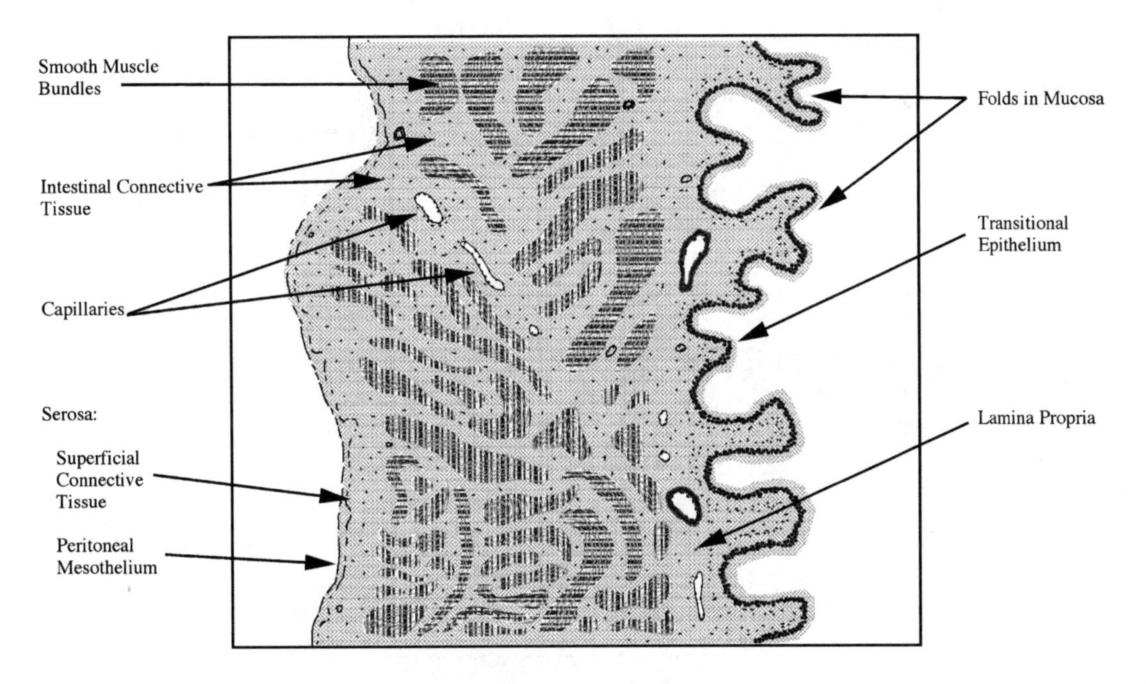

Figure 4.14 Structure of bladder. This structure is composed of mucosa, muscularis, and serosa. In the bladder, the mucosa is infolded when empty and is supported by connective tissue in the lamina propria. Muscularis contains smooth muscle bundles, and the serosa is continuous with the peritoneum.

Parenchymal or Organ-Supporting Structures

Internal organs such as the liver and pancreas have complex structures. They consist of cells that are functional units supported by connective tissue (the parenchyma). The relationship between the cells that make up the functional units vary from organ to organ and so does the geometry of parenchyma. However, in each organ, there is usually an afferent (going into the organ) blood supply and an efferent (going out) blood supply. The location of the blood supplies and the flow of blood through each unit are important to understand. The reader is referred to the excellent histological textbooks in the suggested readings list for further study.

Skeletal Structures

The skeletal tissues of importance include cancellous bone, cortical bone, cartilage, intervertebral disc, skeletal muscle, tendon, and ligament. Histologically,

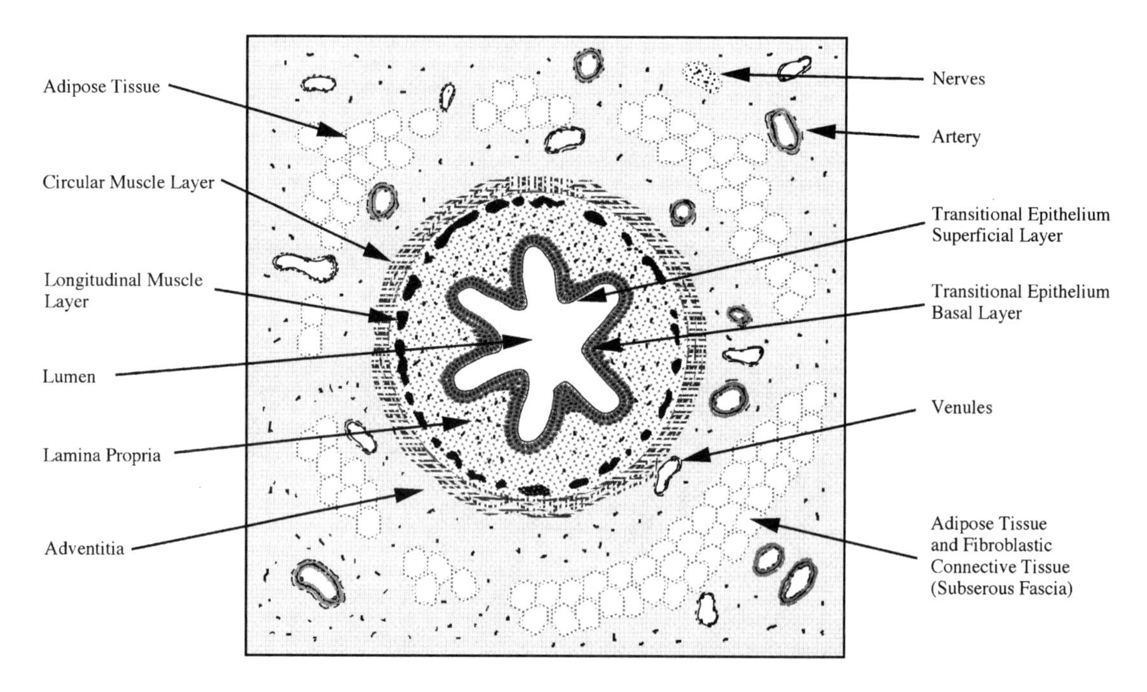

Figure 4.15 Structure of ureter. The structure of the ureter is similar to that of the bladder, except the muscularis is less prominent, and adventitia, the outer layer, contains fibroelastic and adipose tissue in which blood vessels and nerves are embedded.

these are made up of the cellular and noncellular components described in Table 4.3.

Cancellous bone consists of slender trabeculae made up of oriented layers of mineralized collagen fibers that connect and form irregular cavities containing bone marrow (Figure 4.16). The outer layer is composed of periosteum that contains undifferentiated connective-tissue-containing cells that merge with the surrounding tissue. Underneath is peripheral bone, composed of parallel sheets of mineralized collagen fibers in which circumferentially wrapped collagen fibers containing mineral in the form of osteons are inserted. Bone-making cells (osteocytes) are observed in lacunae. The marrow cavity contains adipose tissue and hematopoietic tissue. The endosteum covers the interface between the marrow and the trabeculae.

Compact bone that forms the outer layer of long bones is composed of lamellae, which are thin plates of bony tissue containing osteocytes in lacunae (Figure 4.17). The outer layer of bone beneath the periosteum is formed from

Table 4.3 Cellular and Noncellular components of skeletal tissues

Structure	Cellular layers	Noncellular layers	Function
Cancellous bone	Periosteum		Covers surface
	Osteocytes	Bony trabeculae	Mechanical support
	Osteocytes	Haversian systems	Mechanical support
	Blood cells	Marrow cavities	Source of cells
Compact bone	Periosteum		Covers bone
Circumferential	Osteocytes	Circumferential lamellae of collagen and hydroxyapatite	Mechanical support
Osteon	Osteocytes	Concentric lamellae	Mechanical support
	Endothelium	Connective tissue (haversian canal)	Blood supply
Cartilage	Perichondrium		Covers surface
Hyaline	Chondrocytes	Connective tissue	Mechanical support
	Chondrons	Interterritorial matrix	Connects cells
Intervertebral disc	Annulus fibrosis	Connective tissue	Covers surface
	Nucleus pulposis		
	Chondrocytes	Connective tissue	Shock absorber
Muscle	Fibroblasts	Endomysium (connective tissue)	Holds fibers together
	Muscle cells	Actin and myosin	Contractile elements
Tendon and ligament	Epitendineum		Covers surface
		Fascicle	Separates units
	Fibroblasts	Dense, aligned connective tissue	Supports loads

circumferential layers of mineralized collagen (lamellae) that are parallel to each other and the long axis of the bone. Similarly, the inner layer of bone in the shaft is also circumferential lamellae. Between these two layers are found osteons, which are also known as haversian systems. Each osteon consists of a number of concentric lamellae (plied layers of mineralized collagen containing osteocytes) surrounding a central canal that contains reticular connective tissue and blood vessels. The boundary between each osteon is a thin layer of matrix, the cementing line.

Hyaline cartilage, which is found on the ends of long bones, is surrounded by perichondrium, a connective tissue sheath containing undifferentiated cells (Figure 4.18). Beneath the perichondrium is found a mixture of connective tissue,

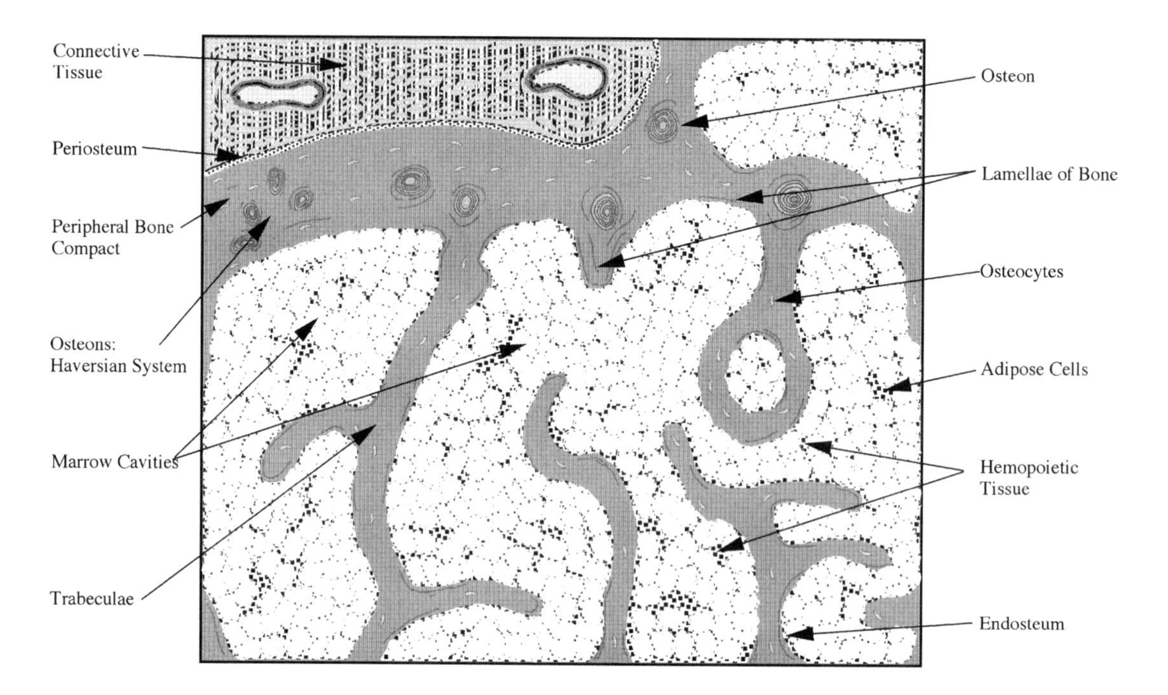

Figure 4.16 Structure of cancellous bone. Cancellous bone is composed of periosteum, an outer layer apposed to the peripheral compact bone. The layer beneath peripheral compact bone contains bone trabeculae, marrow cavities, adipose tissue, hemopoietic cells, and the endosteum.

rich in type II collagen and proteoglycans, and chondrocytes, cells that make matrix. The chondrocytes are formed into chondrons, vertical groups that are surrounded by extracellular territorial matrix (surrounding individual chondrocytes in their lacunae) and interterritorial matrix (separating different chondrocytes and chondrons).

In fibrocartilage such as the intervertebral disc, the outer layer consists of circumferential lamellae of plied collagen and fibroblasts that support an inner layer of chondrocytes that make type II collagen fibers. Small chondrocytes in lacunae usually lie in rows within the collagen matrix. Elastic cartilage is composed of chondrocytes embedded in a matrix containing collagen and elastic fibers (Figure 4.19).

Skeletal muscle is composed of muscle cells, thin and thick filaments, and endomysium, connective tissue containing fibroblasts that holds the fibers together. Interactions between thick (myosin) and thin (actin) filaments result in lines or bands containing one or more of the muscle fiber components. The H band represents overlap of only the thick filaments, whereas the I band represents

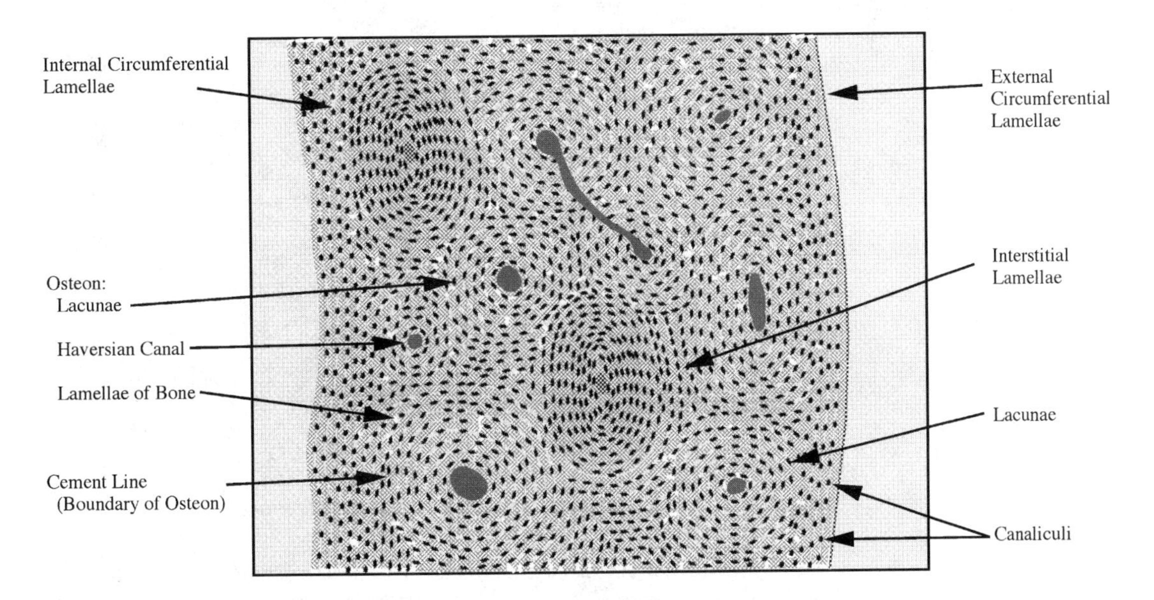

Figure 4.17 Structure of compact bone. Compact bone consists of an outer layer of mineralized lamellae that are wrapped around the shaft of the bone. Beneath the outer layer are concentric rings of mineralized collagenous lamellae. Each concentric unit is termed an osteon and has a vessel running through its center (haversian canal).

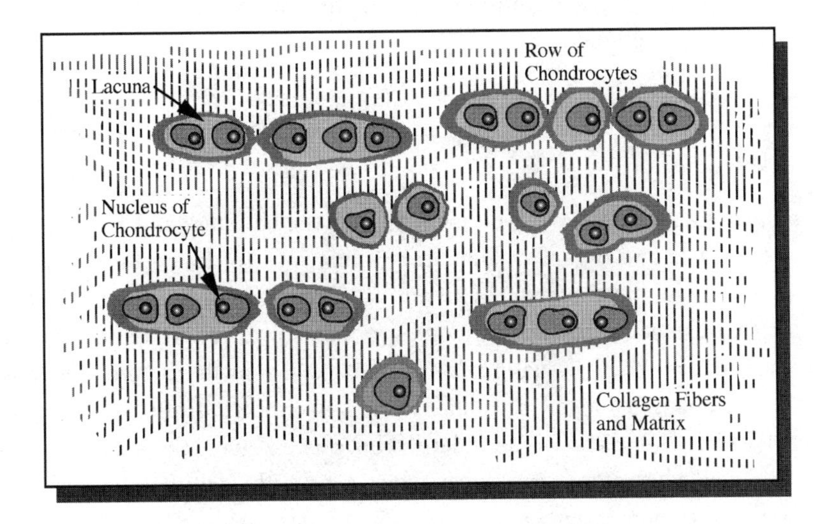

Figure 4.18 Structure of hyaline cartilage. Hyaline cartilage contains chondrocytes within lacunae separated by a matrix containing collagen fibers and proteoglycans.

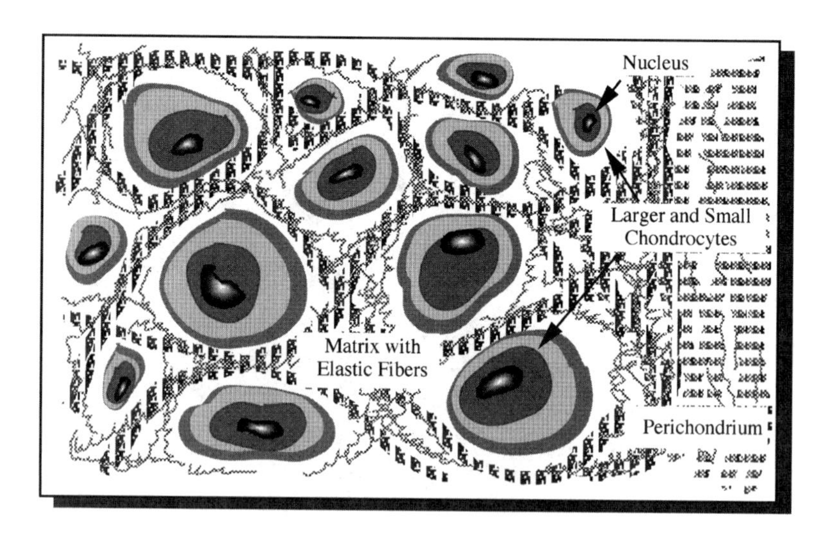

Figure 4.19 Structure of elastic cartilage. Elastic cartilage is composed of chondrocytes in a matrix containing collagen and elastic fibers.

the area of overlap of thin and thick filaments. The Z bands are the points at which the sarcomere repeats itself.

Tendons and ligaments are made up of fascicles, units that are bound into functional units by a sheath called epitendineum (Figure 4.20). Individual fascicles are composed of rows of fibroblasts that alternate with bundles of collagen fibrils parallel to the tendon axis.

Summary

The biomaterials scientist must be familiar with the structure and function of a variety of tissues and organs in animals and humans. To design functional implants, the implant must interface with normal tissue and not cause any structural changes that are pathological. Although gross observations are typically relied on in the industry when evaluating device toxicity in animals, the problem arises when small changes in cell size and shape go unnoticed until pathological changes lead to tissue and organ failure. This process may take years to see at the gross level, whereas microscopic evidence may easily identify such a problem early in the development process. In addition, biomaterials workers must choose animal models to test the efficacy of their designs. To do this, care must be taken to choose an animal that functionally and anatomically is like humans in order

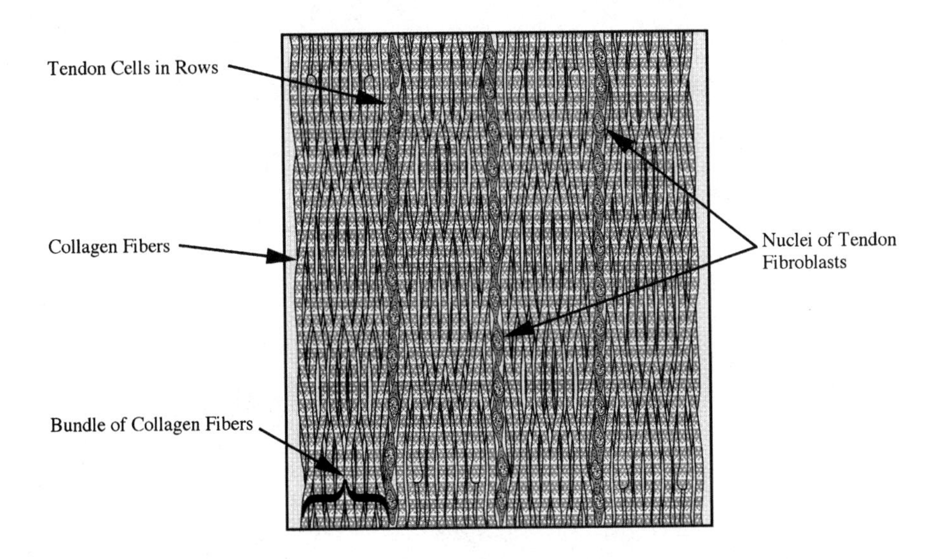

Figure 4.20 Structure of tendon. Tendon is composed of rows of tendon cells that are in columns aligned parallel to the axis and parallel bundles of collagen fibers.

to recognize any changes in tissue and organ function during device implantation. It serves no purpose to use an animal model that under-represents the type of response expected in humans because it takes a lot of time and money to build and test a device. For example, it is well known that device implantation into rodents, even if unsterilized, may not result in injury or death. Therefore, implantation models using rodents may not have much value in predicting the outcome in humans.

Suggested Reading

Bloom W. and Fawcett D.W., A Textbook of Histology, W.B. Saunders Company, Philadelphia, 1975.

di Fiore M.S.H., Atlas of Human Histology, Lea &Febiger, Philadelphia, 1981.

5

Determination of Physical Structure and Modeling

Introduction

Much of our understanding of the way cells and tissues behave reflects to a first approximation the behavior of isolated single macromolecules. For example, the resistance of tendon to deformation reflects the high axial ratio of the collagen molecule and the packing of these molecules into parallel arrays in collagen fibrils and fibers. Although much information is gained from studying molecular structure by X-ray diffraction, from which the average atomic coordinates of molecules in a crystal can be determined, other information is needed to get an exact picture of the molecular structure.

In Chapter 2, the types of structures that are most commonly found in proteins, polysaccharides, and lipids were described. These structures include the α helix, β sheet, collagen triple helix, extended chain structures in polysaccharides, lipids, nucleic acids, and random chain structures. Most macromolecules contain domains that are characterized by one of these structures connected to other types of structural domains. Although the collagen molecule is mostly composed of a triple helix, it contains other regions. Therefore, even though it is largely made up of a rigid triple helix, other domains in collagen make the structure depart from an ideal helical structure. This departure from the ideal makes understanding macromolecular behavior more difficult; however, it also allows us to get a more accurate picture of the dynamics of macromolecular structure.

Many of the methods used to extract information as to the structure of isolated macromolecules come from studying the behavior of macromolecules in solution. These techniques are based primarily on the flow behavior in a velocity gradient, brownian motion, or osmotic effects of individual molecules. The techniques that have been used most extensively include viscometry, light scattering, analytical ultracentrifugation, and electron microscopy.

Viscosity

The property of viscosity is a measure of the size and shape of a particle; long, thin molecules give rise to increased solution viscosity as opposed to small spherical molecules. For example, "maple" syrup typically contains sugar in the form of sucrose, a low molecular weight molecule, and corn syrup, a high molecular weight polysaccharide derived from corn. Because sucrose is a low molecular weight molecule, it can be sweet but it does not effectively make the syrup thick. Therefore, syrup with only sucrose in it is very thin and is not very desirable to put on pancakes. However, corn syrup, which contains higher molecular weight glucose chains, is not only sweet but also has a higher viscosity because of the tendency of poly(glucose) and other polysaccharide molecules with β-(1–4) linkages to adopt an extended conformation very similar to that of hyaluronan. This extended conformation creates a high viscosity, which makes the maple syrup both sweet and thick. Therefore, the viscosity is very sensitive to the conformation and structure of a macromolecular solution. In the next section, we quantitatively analyze the relationship between macromolecule size, shape, and viscosity.

What Is Viscosity?

Viscosity is a measurement of the resistance to flow of a fluid. A simple way by which viscosity (η) is measured is by determining the time required for the fluid to flow from one point to another under the influence of gravity in a capillary tube with a fixed radius and length. A solution with a high viscosity requires a longer time to flow than one with a low viscosity. For example, it is much easier to pour water out of a container than to pour maple syrup; therefore, maple syrup has a higher viscosity than water. During simple fluid flow, layers of molecules slide past each other and experience molecular forces, such as electrostatic interactions, hydrogen bonding, and dispersive forces leading to energy losses that arise from friction. In the simplest case, the shear stress (σ) between adjacent layers of fluid is dependent on the viscosity (η) times the velocity gradient across the tube, dv/dr, as shown by equation 5.1. The fluid layer nearest to the wall moves more slowly than that in the center of the tube because of friction at the wall.

$$\sigma = \eta(dv/dr) \tag{5.1}$$

According to equation 5.1, the viscosity is the proportionality constant between σ and (dv/dt). If η is a constant, then by definition the behavior is termed newtonian. For large macromolecules in solution, the viscosity at a fixed concentration can be a function of the shear rate (dv/dr). This occurs because of alignment of macromolecules with the direction of fluid flow at high shear rates. Therefore, the true viscosity of macromolecular solutions can only be measured at very low shear rates or by extrapolation of measurements at a number of shear rates to zero shear rate.

A further complication of determination of the viscosity of a macromolecular solution is that the viscosity depends on the concentration of macromolecules. Newton developed a formula for predicting the viscosity of a solution of macromolecules and solvent (η') knowing the solvent viscosity (η), shape factor for the macromolecule (ν), and the volume fraction of macromolecules (φ). Newton's law is given in equation 5.2 and shows the relationship between viscosity, shape factor, and volume fraction of a polymer.

$$\eta' = \eta(1 + \nu\varphi) \tag{5.2}$$

Using this formula, it is possible to predict the viscosity of a macromolecular solution knowing the shape factor of a macromolecule and its volume fraction. However, the shape factor must be determined first.

Determination of the Shape Factor

The shape factor of a macromolecule can be determined from theoretical treatments developed by Simha for prolate ellipsoids, cigar-shaped molecules, and oblate spheroids, which are disk-shaped molecules. Einstein determined that the shape factor was 2.5 for spheres. The shape factor for prolate and oblate ellipsoids is given in Figure 5.1 as a function of axial ratio a/b. Values of a and b are the dimensions of the semimajor and semiminor axes for ellipsoids. By knowing the shape factor, the axial ratio can be calculated for the equivalent rod (prolate ellipsoid) or disk. It turns out the shape factor can be estimated by determining the intrinsic viscosity by viscometry.

Determination of Intrinsic Viscosity

The shape factor can be estimated by measuring the time (t) required for a solution of macromolecules to flow through a capillary. If this is measured for a number of solutions with increasing concentrations of macromolecules, then the intrinsic viscosity is obtained by calculating the reduced specific viscosity (η_{sr}),

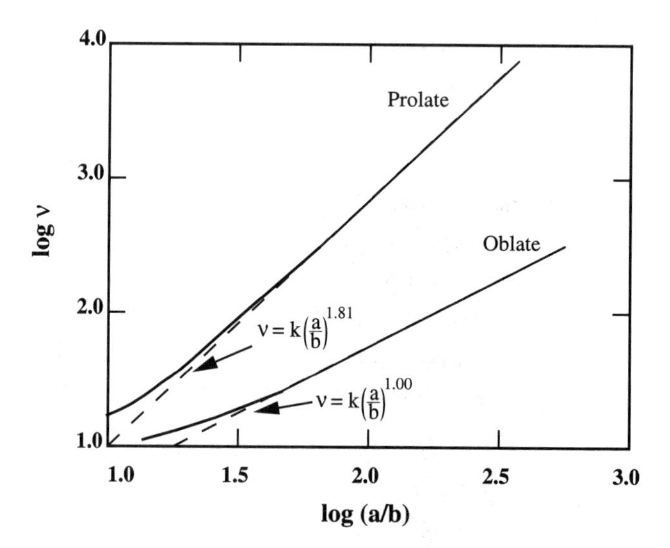

Figure 5.1 Theoretical values of the shape factor for ellipsoids. This is a plot of the shape factor, log ν versus log of the axial ratio (a/b) for prolate and oblate ellipsoids. Note that intrinsic viscosity can be approximated by the shape factor for ellipsoids.

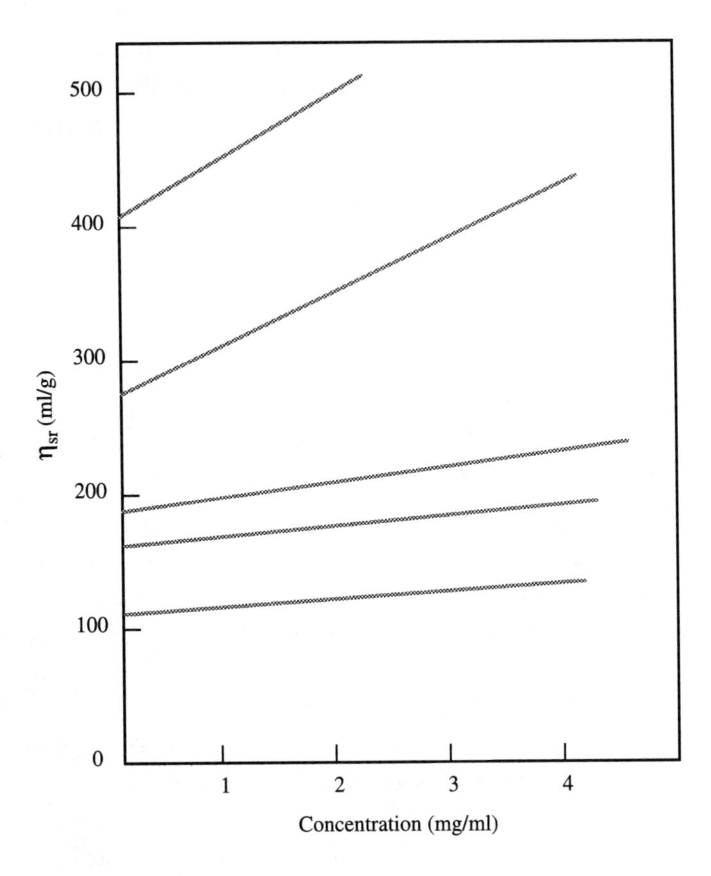

Figure 5.2 Determination of intrinsic viscosity. Intrinsic viscosity is determined by extrapolation of the reduced specific viscosity (η_{sr}) to zero concentration. The reduced specific viscosity is calculated by taking the difference between the flow time in a capillary for polymer in a solvent and that of the pure solvent and dividing the difference by the flow time of the pure solvent times the polymer weight concentration. Data are shown for hyaluronan for various fractions isolated from bovine vitreous.

extrapolated to zero concentration, as shown in Figure 5.2. The intrinsic viscosity $[\eta]$ is defined by equation 5.3 as the limit of the difference between the flow time measured for the macromolecular solution minus the flow time of the solvent divided by the solvent flow time (t_0) times the weight concentration (c).

$$[\eta] = \lim_{c \to 0} \{(t - t_o)/t_o)1/c\} \tag{5.3}$$

Once $[\eta]$ is determined from a plot similar to Figure 5.2, values for *a/b* can be obtained from Figure 5.1 for prolate and oblate ellipsoids. These values must be checked by determining *a* and *b* by other techniques, which are discussed in the next sections.

Light Scattering

Another method by which the size and shape of macromolecules can be determined is based on the ability of large molecules to scatter light. As early as the 1940s, Debye and co-workers recognized that light was scattered by a solution of macromolecules and that the scattered intensity was related to the molecular weight. Because light is an electromagnetic wave characterized by electric and magnetic fields, the interaction of light with the electric and magnetic field around a macromolecule alters these fields. As light moves through space, the magnitude of the electric and magnetic vectors changes as a function of time and position.

The intensity of scattered light (I_s) is proportional to the molecular weight of the macromolecule (M) as well as the scattering angle (ϕ), light wavelength (λ), distance from the scattering source (r), Avogadro's number (No), change in refractive index with weight concentration (dn/dc), the molecular polarizability (α), and the refractive index, (n), as shown in equation 5.4. Note that the intensity of the incident beam is I_o.

$$I_s = \frac{2\pi^2 n^2 (1 + \cos^2\phi)(dn/dc)^2 M}{No\lambda^4 r^2} (c\ I_o) \tag{5.4}$$

A simple way to determine the relationship between molecular weight and intensity of scattered light is to recast equation 5.4 by defining an optical constant K using equation 5.5 and R_θ, the Rayleigh factor (equation 5.6).

$$K = \frac{2\pi^2 (dn/dc)^2 n^2}{No\lambda^4} \tag{5.5}$$

$$R_\theta = \frac{r^2 I_s}{I_o(1 + \cos^2\phi)} = Kc/(1/M) \tag{5.6}$$

In the simplest case, equation 5.6 relates the Rayleigh factor to the molecular weight in the absence of solvent–macromolecule and macromolecule–macromolecule interactions. The molecular weight is determined from the slope of a plot of the Rayleigh factor versus weight concentration of macromolecules.

In the case of macromolecule–macromolecule interactions, the relationship between the Rayleigh factor and weight concentration of macromolecules is complicated by the fact that the coefficient B is needed to account for this effect (equation 5.7).

$$Kc/R_\theta = 1/M + 2Bc \tag{5.7}$$

Operationally, the value of the Rayleigh factor for the solution is subtracted away from that of the solvent plus macromolecule and plotted versus weight concentration to get $1/M$ from the intercept and B from the slope. B is also known as the second virial coefficient, which reflects the state of attraction between macromolecules (negative value of B) or repulsion (positive value of B). Because the weight of each macromolecule contributes to the scattered light intensity, the molecular weight determined by light scattering is a weight average.

For macromolecules that have one or more dimensions larger than 1/20 of the light wavelength, less light actually is measured than is predicted by equation 5.7. The observed scattered light intensity is less than that predicted from theory as a result of interference between light scattered by different portions of a macromolecule. The light waves scattered by different portions of a macromolecule are in some instances one-half wavelength length out of phase from each other and when summed cancel each other. The fraction of the theoretically predicted light that is actually scattered is called the particle-scattering factor, $P(\phi)$. $P(\phi)$ is dependent on the scattering angle and the shape of a macromolecule. Correcting equation 5.7 for the particle-scattering factor results in equation 5.8.

$$Kc/R_\theta = 1/(P(\phi))[1/M + 2Bc] \tag{5.8}$$

The molecular weight of a macromolecular solution is now obtained by plotting $Kc/(R_\theta)$ versus weight concentration at a series of decreasing angles to form a graph referred to as a Zimm plot. The molecular weight is obtained by double extrapolation, that is, to a weight concentration of zero and scattering angle of zero.

This double extrapolation procedure proves very time consuming and requires many hours of measurements in the laboratory. In the case of macromolecules that self-associate, the determination of molecular weight by classical multiangle light scattering is a significant task.

The development of low-angle laser light-scattering devices allowed the measurement of Rayleigh factors at low scattering angles, obviating the need for measurements at multiple angles (Figure 5.3). The particle-scattering factor can be obtained by dividing the Raleigh factor measured at an angle between 90° and 180° by that measured at angles less than 4°.

Theoretical relationships between the particle-scattering factor and molecular shape have been developed for rods, spheres, and random coils. Figure 5.4 illustrates the relationship between the particle-scattering factor and ratio of length to width for rods as a function of scattering angle. The value of rod length L and molecular weight from light scattering can be compared with the value of a or a/b obtained from viscometry to get a better picture of the size and shape of a macromolecule.

Quasi-Elastic Light Scattering

In classical light-scattering theory, the average scattered light intensity is measured, and the molecular weight and particle scattering factor are determined. In addition to the average light intensity, the time dependence of the scattered light intensity gives information concerning the rate of movement (brownian motion) of the macromolecules in solution. Although the average wavelength of the scattered light remains the same, the scattering process broadens the distribution of light wavelengths. In addition, the time dependence of fluctuations of the light intensity are related to the size and shape of the macromolecules in solution. This is because of the diffusion of macromolecules into and out of the control volume illuminated by the light. At any time, the probability that there will be more than or less than the average density of molecules within the volume illuminated by the light is related to the diffusion coefficient and the rate at which molecules diffuse into and out of the light path. The diffusion coefficient can be calculated from the rate of decay of the product of two intensities that are tabulated into a mathematical function called the autocorrelation function.

Mathematically, the autocorrelation function of the scattered light, $G(Ndt)$ in equation 5.9, is related to the product of the two light intensities that are separated by a time interval dt. If dt is chosen correctly, that is, within the time

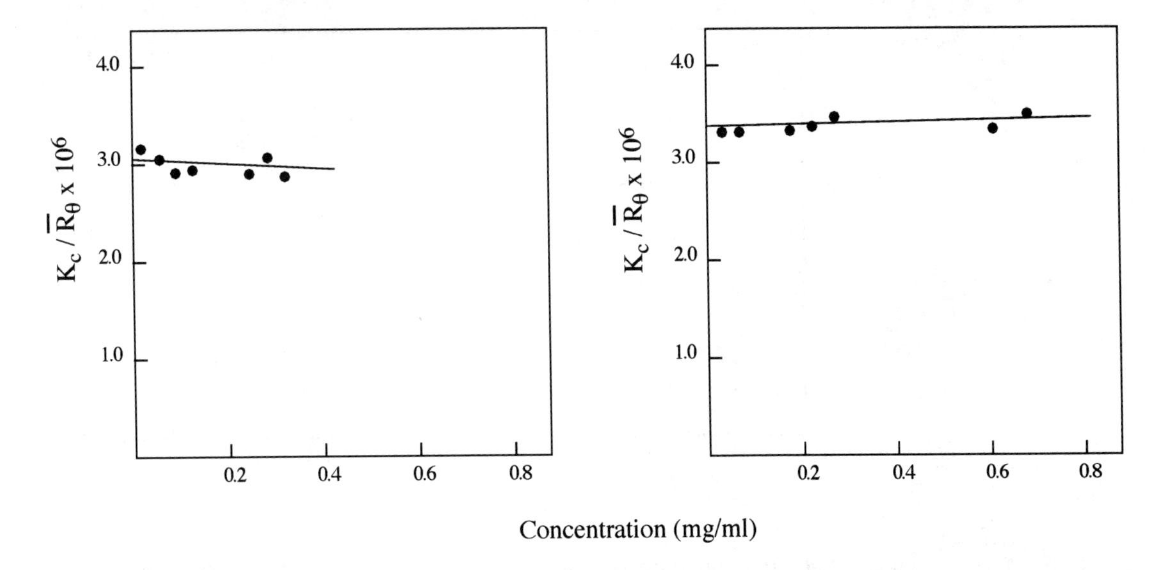

Figure 5.3 Determination of molecular weight from light scattering. The weight average molecular weight is determined from light scattering by determination of the Rayleigh ratio, R_θ, and plotting Kc/R_θ versus concentration. The weight average molecular weight is taken as the reciprocal of the intercept at zero concentration. Plots shown are for type I collagen from rabbit cornea (left) and sclera (right). *Source:* Adapted from Birk and Silver (1983).

period required for diffusion to occur into and out of the control volume, then the diffusion coefficient can be calculated from the decay of the autocorrelation function over an observation interval of *Ndt* (Figure 5.5).

$$G(Ndt) = \sum_{N=1}^{\infty} [I(t)][I(t + dt)]/N \tag{5.9}$$

The autocorrelation function decays to a value of $\langle I^2 \rangle$av, which is the average squared scattered light intensity. Theoretically, $G(Ndt)$ is related to a constant (α), the experimental baseline (B), and the decay constant (Γ), as shown in equation 5.10.

$$G(Ndt) = B(1 + \alpha e^{-2\Gamma t}) \tag{5.10}$$

The decay constant Γ is related to the translational diffusion coefficient D_t and the scattering vector Q, as shown in equation 5.11. A typical normalized autocorrelation function is shown in Figure 5.6.

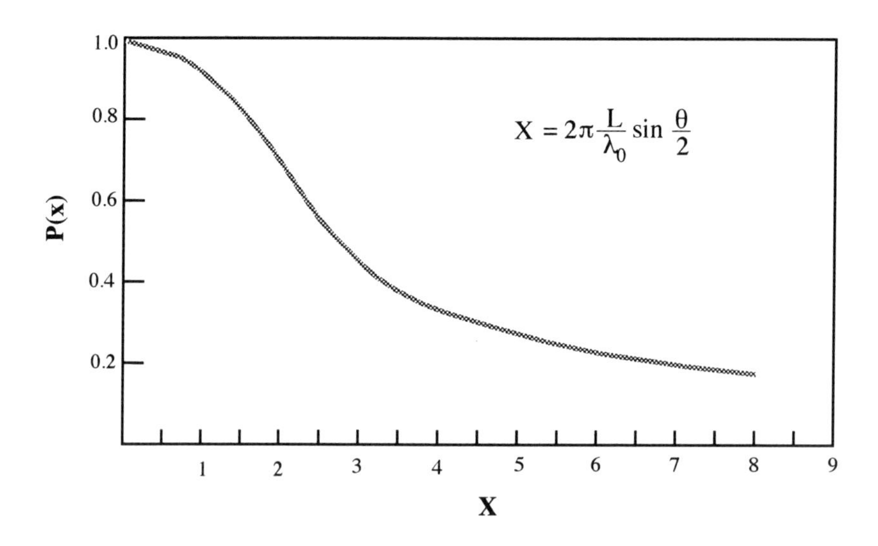

Figure 5.4 Determination of particle-scattering factor. Particle-scattering factor, $P(X)$, as a function of angle (θ) and wavelength of light in solution (λ_o), versus X for rodlike molecules. Note that as the ratio of molecular length to light wavelength increases or the scattering angle increases, the particle-scattering factor decreases.

$$\Gamma = D_t Q^2$$

$$Q = [4\Pi n/\lambda][\sin(\theta/2)] \tag{5.11}$$

Taking the natural logarithm of equation 5.10 and substituting equation 5.11, we find that a plot of $\ln[\,(G(Ndt)/B) - 1]$ versus time has a slope equal to $-2D_t Q^2$, and D_t can be determined knowing the scattering angle, wavelength of light, and solution index of refraction. D_t is determined from such a plot for collagen α chains in Figure 5.7.

The translational diffusion coefficient can be determined this way for macromolecules that are less than 2 μm in their largest dimension at scattering angles less than 4°. Macromolecules larger than this or measurements made at higher scattering angles need to be corrected to get accurate values of D_t.

Once D_t for a macromolecule is determined, it can be compared to theoretical values for prolate ellipsoids and spheres using equations 5.12 and 5.13 for standard conditions of 20 °C and the viscosity of water.

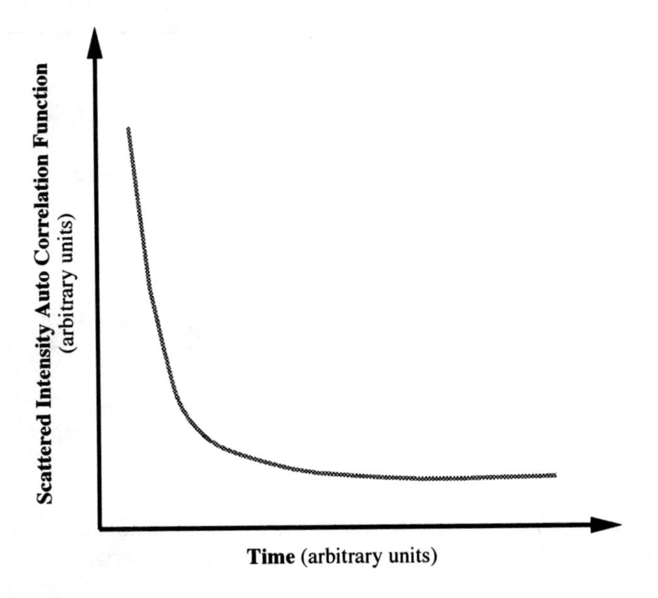

Figure 5.5 Autocorrelation function of scattered light. Schematic diagram showing the decay of the autocorrelation function to a baseline. The autocorrelation function is related to the product of two intensities separated by a time interval. As the time interval increases, the function decays to a baseline. The rate of decay is proportional to the translational diffusion coefficient.

Prolate ellipsoids $Z = a/b > 10$

$$D_{t_{20,w}} = 2.15 \times 10^{-13}\ \text{cm}^3/\text{sec}\ [\ln (2a/b)]/a \tag{5.12}$$

Sphere of radius R (5.13)

$$D_{t_{20,w}} = 2.15 \times 10^{-13}\ \text{cm}^3/R$$

Using equations 5.12 and 5.13 and the experimental measurement of D_t, values of a and b or R can be determined for different macromolecules. These values can be compared to values generated from viscometry and classical light scattering.

Ultracentrifugation

The size and shape of macromolecules in solution can be studied using the techniques of equilibrium and velocity ultracentrifugation. These techniques use

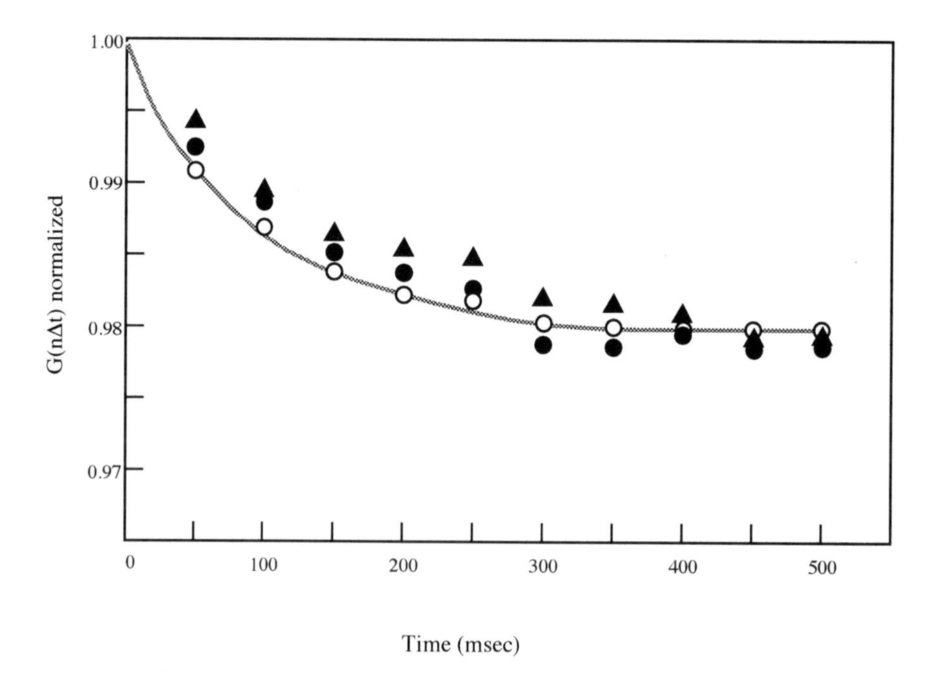

Figure 5.6 Normalized autocorrelation function. Autocorrelation function for collagen single molecules. The autocorrelation function, $G(n\Delta t)$, is normalized by dividing all points by the first experimental point, $G(1)$. The autocorrelation function decays to a value of the average squared intensity of scattered light divided by $G(1)$. The average squared intensity is proportional to the weight average molecular weight, whereas the rate of decay is related to the translational diffusion coefficient. *Source:* Data reproduced from Silver and Trelstad (1980).

an ultracentrifuge to rotate solutions of macromolecules to study their physical properties. The size and shape of the macromolecules can then be determined from the solution's physical properties. The ultracentrifuge is equipped for direct measurement of the solution as it spins at high speed.

At speeds above several thousand revolutions per minute, macromolecules in the cell of the ultracentrifuge sediment toward the periphery of the rotor. At the same time, the molecules tend to diffuse toward the center of the rotor because of the concentration gradient. At equilibrium, the centrifugal force caused by the spinning of the rotor exactly offsets the tendency for the molecules to diffuse toward the center of the rotor.

Measurement of the solution optical density, which is a function of the concentration (c) as a function of the distance (r) from the center of the cell, is

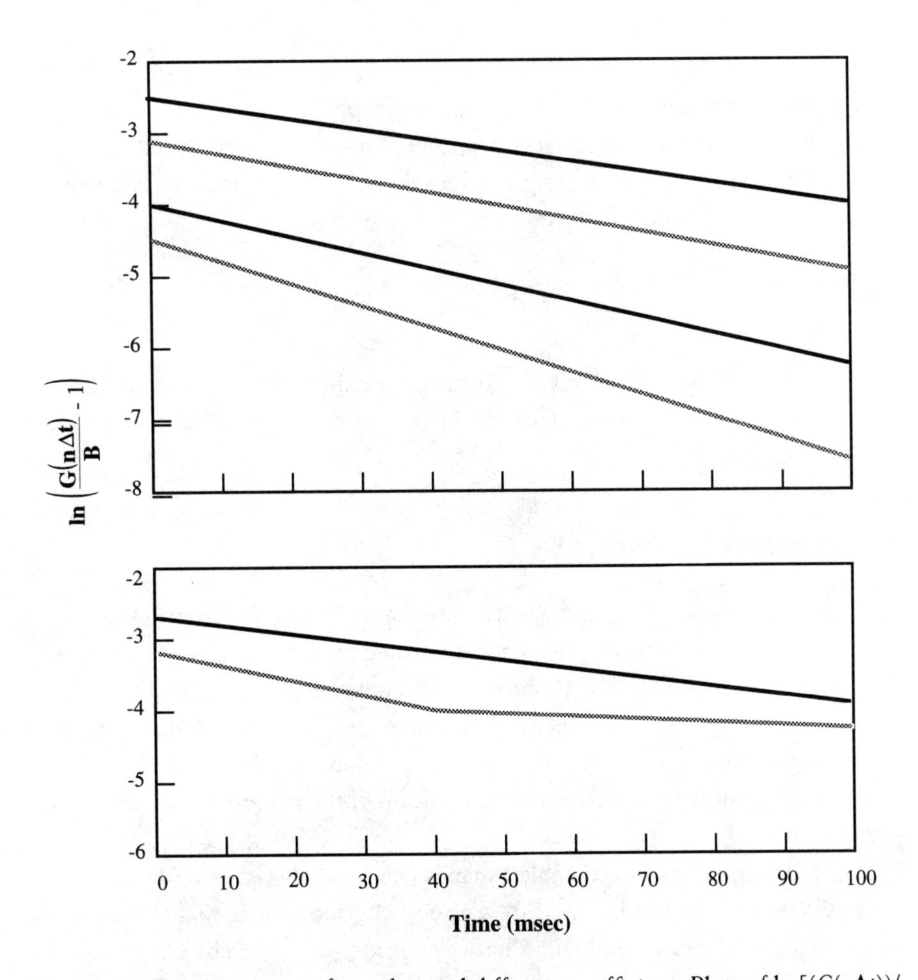

Figure 5.7 Determination of translational diffusion coefficient. Plots of ln $[(G(n\Delta t))/B) - 1]$ versus time for collagen α chains (top) and mixtures of α chains and β components (bottom). The translational diffusion coefficient is obtained by dividing the slope of each line by $-2Q^2$, where Q is the scattering vector. B is the baseline of the autocorrelation function. Note for single molecular species, the slope is constant, and for mixtures with different molecular weights, the slope varies (molecular weight for α chains is 95,000 and for γ components is 285,000). *Source:* Data reproduced from Silver and Trelstad (1980).

then expressed as a function of the apparent molecular weight (M_{app}) by means of equation 5.14. V_2 is the volume fraction of macromolecule, R is the gas constant, ω is the angular velocity, ρ is the solution density, and T is temperature.

$$\ln c(r) = (1 - V_2\rho)/2RT\ [\omega^2 M_{app} r^2] + \text{constant} \qquad (5.14)$$

At high speeds greater than 40,000 rpm in the ultracentrifuge, macromolecules are sedimented toward the rotor periphery. Under these conditions, the sedimentation coefficient (s) is determined from the speed of sedimentation divided by the angular acceleration. The sedimentation coefficient is related to molecular weight using equation 5.15.

$$M = RTs/[D_t(1 - V_2\rho)] \qquad (5.15)$$

Using M and s from ultracentrifugation experiments, the D_t can be calculated and compared to the D_t determined from quasi-elastic light scattering.

Electron Microscopy

The identification of size and shape can most easily be done by direct observation in the electron microscope. This is accomplished by depositing single molecules (if they can be obtained) directly on polymer-coated copper grids and shadowing them with heavy metals or making a replica of the molecular surface on mica. The sample can then be viewed in the transmission electron microscope, and photographs can then be taken after calibration of the magnification factor (Figure 5.8).

From these images, complex shapes can be analyzed using bead models (Figures 5.9) to calculate Dt and $P(\theta)$. The equations for calculating Dt for a series of N beads of diameter d and radius between beads r_{ij} are given by equations 5.16 and 5.17. In equation 5.16, the summations are made for N from 1 to j and i from 1 to N as long as i does not equal j. In equation 5.17, k is the Boltzmann constant.

$$f = 6\pi\eta\ d/2N/[1 + d/N\ \Sigma\Sigma < r_{ij} - 1 >] \qquad (5.16)$$

$$D_t = kT/f \qquad (5.17)$$

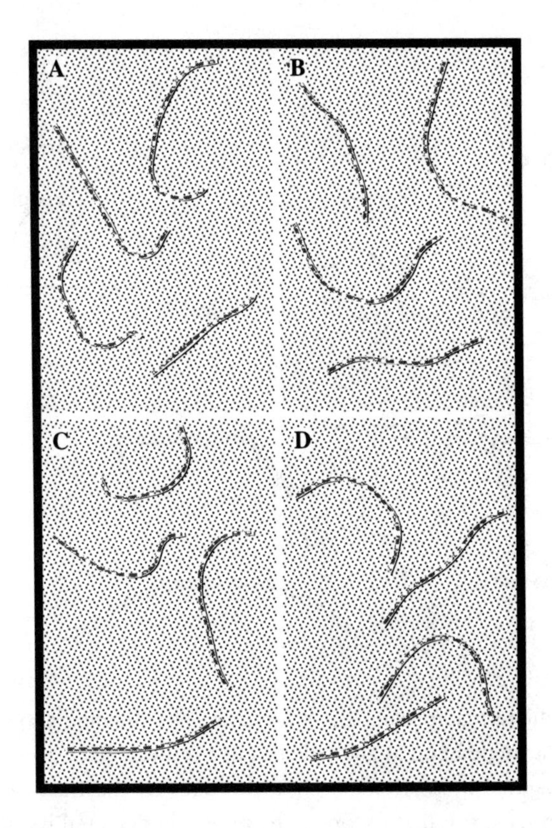

Figure 5.8 Electron microscope images of single molecules. Schematic images of collagen molecules obtained by shadowing the molecules with heavy metals and viewing either the molecules or replicas of the molecules under the electron microscope. Molecules shown include collagen types I (A), II (B), III (C), and V (D), and each molecule is roughly 300 nm in length.

Determination of Physical Parameters for Biological Macromolecules

The values of intrinsic viscosity, molecular weight, translation diffusion coefficient, and sedimentation coefficient have been determined for a variety of different macromolecules and are listed in Table 5.1. As we can see from these data, molecules with high values of the intrinsic viscosity, that is, greater than 1,000 mL/g, and relatively low translational diffusion coefficients (1×10^{-7}) and sedimentation coefficients (3×10^{-13}) are rigid or semiflexible rods. Molecules with low translational diffusion coefficients and intrinsic viscosity and high sedimen-

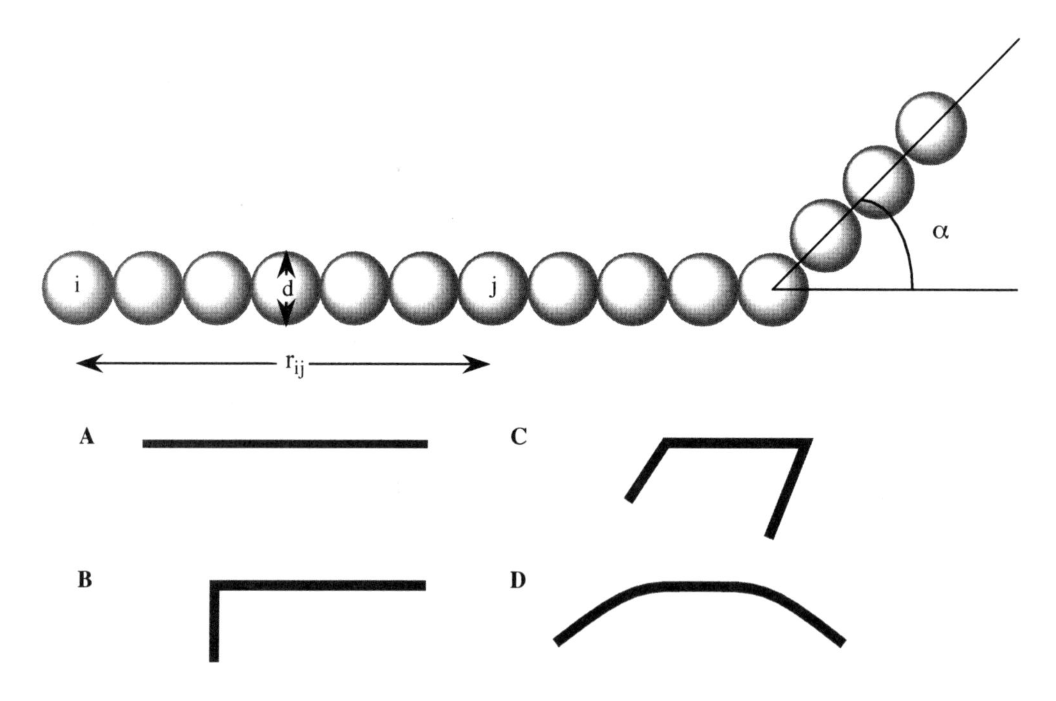

Figure 5.9 Bead models used to simulate macromolecular structures. The drawing shows a model used to simulate a collagen molecule with a single bend at an angle α. Each molecule is represented by a series of N beads of diameter d that are separated by a distance r_{ij}. (A) through (D) illustrate the typical shapes observed for collagen type I by rotary electron microscopy. The molecule depicted has a total length of 300 nm.

tation coefficient are flexible rods. Molecules with high translational diffusion coefficients and low intrinsic viscosity are roughly spherical.

Measurement of physical parameters in solution for isolated macromolecules provides a manner by which the shape of a macromolecule can be determined. The approximate dimensions and axial ratio or radius can be calculated by applying equations 5.3 through 5.17.

Summary

The size and shape of a macromolecule can be determined by measuring the physical properties of isolated macromolecules in solution. Large, rigid macromolecules that are derived from extended structures, including the collagen triple helix, result in rodlike rigid or semirigid structures. This is important because

Table 5.1 Physical constants for biological macromolecules

Macromolecule	M	$D_{20,w}$	$S_{20,w}$	[η]	Shape
Collagen I	285,000	0.85	2.96	1,100	Semiflexible rod
II	285,000	0.85		1,530	Semiflexible rod
IV	535,000	0.66		75	Flexible rod
Fibrinogen	330,000	2.02	7.9	25	Flexible rod
Hyaluronan (vitreous)	178,000	0.86	3.10	718	Wormlike coil
Myosin	570,000	1.0	6.4	217	Flexible rod
Serum albumin	66,000	5.94	4.31	3.7	Roughly spherical
Keratin	110,000	4.3	4.3	154.1	Flexible rod
Tubulin	110,000	4.7	6.0		Roughly spherical
G-Actin	42,000	7.9–8.13		3.0	Roughly spherical

Note: $D_{20,w}$, $S_{20,w}$ = translational diffusion constant and sedimentation coefficient back calculated to a temperature of 20 °C and the viscosity of water.

these molecules tend to have high moduli and are packed into strong stiff networks in tissues.

Flexible molecules tend to form from structures that have helixes connected by flexible segments. These molecules are not only mobile in solution but form flexible low modulus structures in tissues.

Suggested Reading

Birk D.E. and Silver F.H., Corneal and Scleral Type I Collagens: Analyses of Physical Properties and Molecular Flexibility, Int. J. Biol. Macromol. 5, 209, 1983.

Silver F.H. and Trelstad R.L., Type I Collagen Structure in Solution and Properties of Fibril Fragments, J. Biol. Chem. 255, 9427, 1980.

Silver F.H., Biological Materials: Structure, Mechanical Properties, and Modeling of Soft Tissues, NYU Press, New York, chapter 4, 1987.

6 Assembly of Biological Macromolecules

Introduction

The study of animal tissues is complex because water, ions, cells, macromolecules, tissues, and organs exist in equilibrium. From a structural point of view, biological tissues contain highly ordered arrays of macromolecules. One might wonder why biological structures need to be made up of highly ordered arrays of proteins, polysaccharides, and lipids. The reason is that individual polymer molecules cannot sustain the weight of gravity without rearranging. For example, if your skin were made of just collagen molecules without being cross-linked into crystalline fibers, it would sag. This is because individual molecules, in a similar manner to water molecules, can move around or diffuse. In the case of water molecules, a container is needed to shape them. In the case of tissues, the molecules need to assemble into ordered structures and be cross-linked for the shape of the tissue to be maintained. In some cases, assemblies of macromolecules are purposely not cross-linked so that shape can be changed quickly. For example, cytoskeletal actin filaments are rapidly assembled and disassembled to allow for changes in cell shape. In this example, cross-links prevent rapid shape changes; however, actin filaments by themselves are rigid enough to maintain cell shape at any one instant. In contrast, collagen fibers in the skin must be cross-linked to form force-bearing units to prevent tearing when skin is stretched. This comparison is used to underscore the complexity of biological tissue structure and its relationship to physical properties. In some cases, rigidity is sacrificed for structural flexibility. In other cases, structural flexibility is sacrificed for permanence.

Biological macromolecules are largely found in animal tissues in the form of helixes (α and collagen triple helix), β structures, amorphous or flexible chains, and combinations of these structures. It is the assembly of these units that make up the crystalline arrays that are the noncellular components of tissues. It is amazing how the assembly of a few types of structures gives rise to many different tissue structures.

The assembly of biological macromolecules into fibrous and other supramolecular structures has been studied extensively since the early 1900s. Oosawa and Kasai (1962) pointed out that various biological macromolecules form intramolecular helixes, that is, helixes that are formed from hydrogen bonds within a single polymer chain, but no one considered the possibility of intermolecular helical structures that had hydrogen bonds between chains. The purpose of their paper was to present a simple theory of the helical aggregation of macromolecules and to compare theoretical predictions with experimental results. The theory developed suggested that an equilibrium distribution existed between monomers (single polymer chains) and linear and helical aggregates containing more than one polymer chain. When the concentration of macromolecules is increased, the helical aggregates begin to appear at the critical concentration determined by solvent conditions. Above the critical concentration, very long helical aggregates coexist in equilibrium with a constant concentration of dispersed monomers. They used this theory to explain the equilibrium and kinetic features of the transformation of globular monomeric actin (G-actin) to fibrous aggregated actin (F-actin).

Another approach for thinking about the mechanism of assembly of biological macromolecules involved considering it as a phase transition involving a solid phase (aggregate) in equilibrium with a solution phase containing single isolated polymer chains. Flory developed free-energy calculations to understand the energy required to transfer molecules between the liquid and solid phases. The free-energy change required to mix solute molecules and solvent molecules is related to the change in the number of bonds formed or broken as well as the associated entropy change.

Specifically, Flory developed relationships for the free energy of mixing of solvent and solute (polymer) molecules with different solute axial ratios (length/width) and solute interaction parameters. In the absence of interactions between solute molecules, the free energy of the system lowers when rodlike molecules precipitate out of solution and form a separate solid phase. This is because small water molecules must order themselves around large rodlike macromolecules in solution, so the system is most stable when the water molecules and rodlike molecules are separated in space.

Flory's mathematical relationships for the free-energy change that occurs when a solute is mixed with a solvent for the solution phase and the solid phase are discussed in the work by Silver (1987). The free energy of mixing of solute and solvent per mole of molecules is equivalent to the change in chemical potential of a desired state (μ) from that of the reference state ($\mu_1{}^{\circ}$). These relationships are given by equations 6.1 through 6.4 where the subscripts 1 and 2 represent solvent and solute, respectively; R is the gas constant; T is the absolute temperature; V is the volume fraction of solute; Z is the axial ratio of the solute (half length/half width); y is the disorientation index (the amount of disorientation of the solute); and χ_1 is the interaction parameter.

Isotropic or liquid phase $$\left(\frac{\mu_1 - \mu_1^0}{RT}\right)_l = \ln(1 - V_2) + \left(1 - \frac{1}{Z}\right) V_2 + x_1 V_2^2 \qquad (6.1)$$

$$\left(\frac{\mu_2 - \mu_2^0}{RT}\right)_l = \ln(V_2/Z) + (Z - 1)V_2 - \ln Z^2 + x_1 Z(1 - V_2)^2 \qquad (6.2)$$

Anisotropic or solid phase $$\left(\frac{\mu_1 - \mu_1^0}{RT}\right)_s = \ln(1 - V_2) + \frac{(y - 1)}{Z} V_2 + \frac{2}{y} + x_1 V_2^2 \qquad (6.3)$$

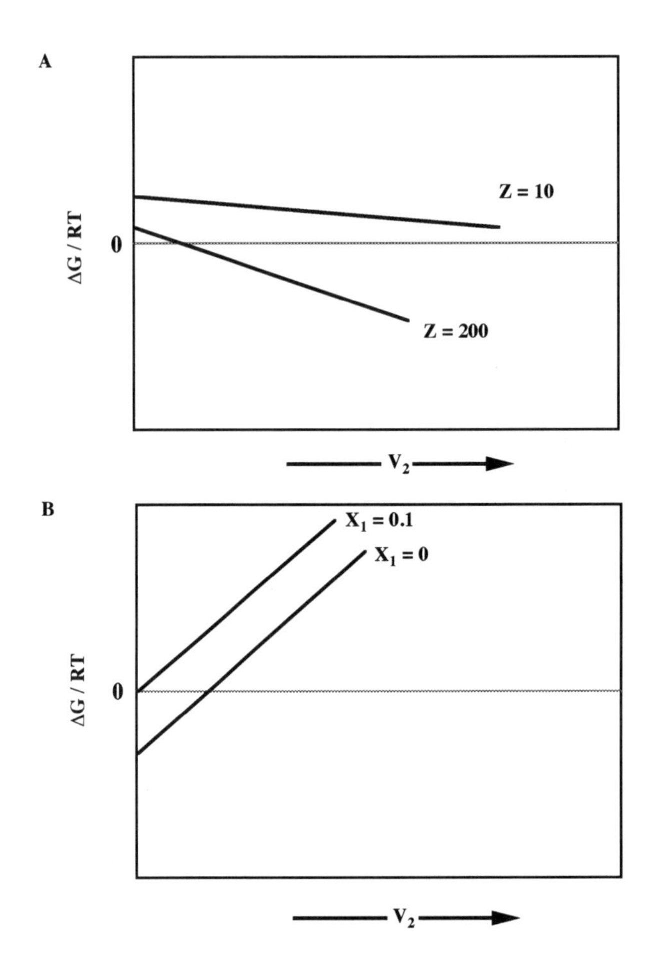

Figure 6.1 Free-energy change of mixing for rods and solvent molecules. Free-energy change ($\Delta G/RT$) of (A) solid phase associated with transfer of a solute molecule (macromolecule) from the liquid to the solid state as a function of solute volume fraction (V_2) for low ($Z = 10$) and high ($Z = 200$) axial ratios; and (B) liquid phase as a function of solute volume fraction in the presence ($X_1 = 0.1$) and absence ($X_1 = 0$) of interactions between solute molecules. The diagrams show that separation of solute and solvent molecules occurs spontaneously for high axial ratios above a critical volume fraction and that the free energy of the solvent is raised by intermolecular interactions.

$$\left(\frac{\mu - \mu_2^0}{RT}\right) = \ln(V_2/Z) + (y - 1)V_2 + 2 - \ln y^2$$
$$+ x_1 Z (1 - V_2)^2 \, \Delta G = \# \text{ moles}(\Delta\mu) \quad (6.4)$$

What these equations tell us is that for molecules with large axial ratios (i.e., 100 or larger), the free energy is lowered (the lower the free energy, the more stable the system) for the solute by undergoing a phase transition from the liquid to the solid states, and the solute precipitates out. The phase separation is further supported by increasing the solute concentration (Figure 6.1). This observation suggests that polymer molecules that form α or collagen triple helixes or β structures in solution tend to spontaneously phase separate because of their long, thin profiles and form a new solid phase. If the molecules exhibit attractive forces, they are likely to self-assemble into higher-order structures. In the next section, Flory's equations are used to understand collagen self-assembly.

Several self-assembling macromolecules are of interest in this text, including (1) collagen, the primary structural material found in the extracellular matrix; (2) actin, a component of the cell cytoskeleton that is involved in cell locomotion and forms the thin filaments of muscle; (3) microtubules, which are involved in cell mitosis, movement, and organelle movement; and (4) fibrinogen that forms fibrin networks that minimize bleeding from cut vessels. Self-assembly is important in these systems because the function of these macromolecules can be modified via processes that increase the molecular axial ratio and hence the solubility.

Methods for Studying Self-Assembly Processes

Several methods have been developed for studying self-assembly processes, including light scattering, ultracentrifugation, and electron microscopy. Although these methods have some associated problems when applied to studying self-assembly, a great deal of information has been obtained by applying several of these methods and then comparing the results.

Light Scattering

Phenomenologically, it was observed many years ago that when a solution of macromolecules self-assembled, it became turbid, and the resulting turbidity-time curve could be used to characterize this process. Typically, the starting solution was transparent to light at wavelengths between 313 and 500 nm, and once assembly proceeded, the process was characterized by a sigmoidal curve containing lag, growth, and plateau phases, as diagrammed in Figure 6.2. This type of curve was seen with collagen, and it was hypothesized that during the lag phase, nuclei assembled that later grew rapidly during the growth phase.

The mathematics of analysis of the events that occur during the turbidimetric lag and growth phase are discussed elsewhere (Silver and Birk, 1983). The turbidity per unit path length (τ) is equivalent to 2.303 times the absorbance (measured using a spectrophotometer) and is proportional to an optical constant (H) times several factors, including the molecular weight (M), weight concentration of macromolecule (c), and particle dissipation factor (Q) (equation 6.5). The particle dissipation factor is related to the largest dimension of the macromolecule; it is 1 if the macromolecule is small with respect to the light wavelength in solution and 0 if the largest dimension is infinite. The optical constant is given by equation 6.6, where n is the index of refraction of the solution, dn/dc is the refractive index increment of the macromolecular component, No is Avogadro's number, and λ is the wavelength of light in solution.

$$\tau = HMcQ \tag{6.5}$$

$$H = \frac{32\pi^3 n^2 (dn/dc)^2}{3No\lambda^4} \tag{6.6}$$

Equation 6.5 is analogous to the equation for the scattered light intensity at an angle 90° to the transmitted beam, and Q is related to the particle-scattering factor at a scattering angle of 90°. Using this analogy, the increased turbidity reflects the increase in molecular weight. At the same time, if the particle formed is getting longer, then Q may cause the turbidity to decrease because of destructive interference. Therefore, it is evident from equation 6.5 that increases in turbidity are a result of changes in size (M) and shape (Q).

Q and H can be calculated for rodlike assembly, and the results depend on wavelength. At wavelengths of 300 and 650 nm, the turbidity per unit concentration (τ/c) is related to the mass per unit length (M/L), as indicated in equation (6.7).

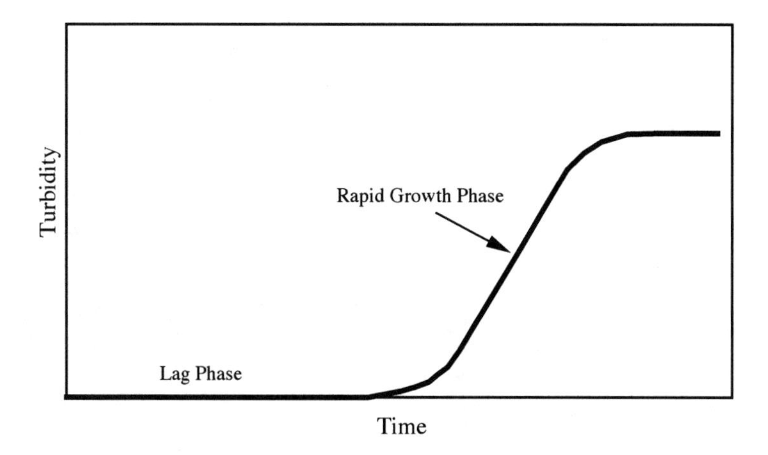

Figure 6.2 Turbidity-time curve illustrating collagen self-assembly. Turbidity-time curve illustrating lag phase, during which small linear and lateral aggregates form, and growth phase, during which unit fibers form that rapidly grow into fibers. The plateau is characteristic of termination of fibril growth.

$$\tau/c = 6.57 \times 10^{-3}\ (M/L)\ \lambda = 300 \text{ nm} \tag{6.7}$$

$$\tau/c = 4.49 \times 10^{-4}\ (M/L)\ \lambda = 650 \text{ nm}$$

As one can see from reviewing the mathematics, the turbidity per unit concentration is proportional to the molecular weight per unit length. This implies that the turbidity lag phase can be characterized either by a constant molecular weight or a molecular weight per unit length that does not increase. The latter case occurs when self-assembly proceeds in a linear fashion, and the length doubles every time the molecular weight doubles. This mathematical analysis of turbidity-time curves suggests that turbidity measurement is not the recommended way to follow linear growth of macromolecular assemblies. However, rapid turbidity increases during the growth phase of turbidity-time curves are characteristic of systems that grow linearly and laterally at the same time.

Another manner to characterize self-assembly using light-scattering techniques is to measure the light intensity scattered at a fixed angle. Although the problems associated with measuring turbidity changes during self-assembly also affect measurement of the intensity of light scattered at an angle of 90°, these problems can be averted by measurement of the scattered light intensity at very small scattering angles. Figure 6.3 illustrates that measurement of the scattered

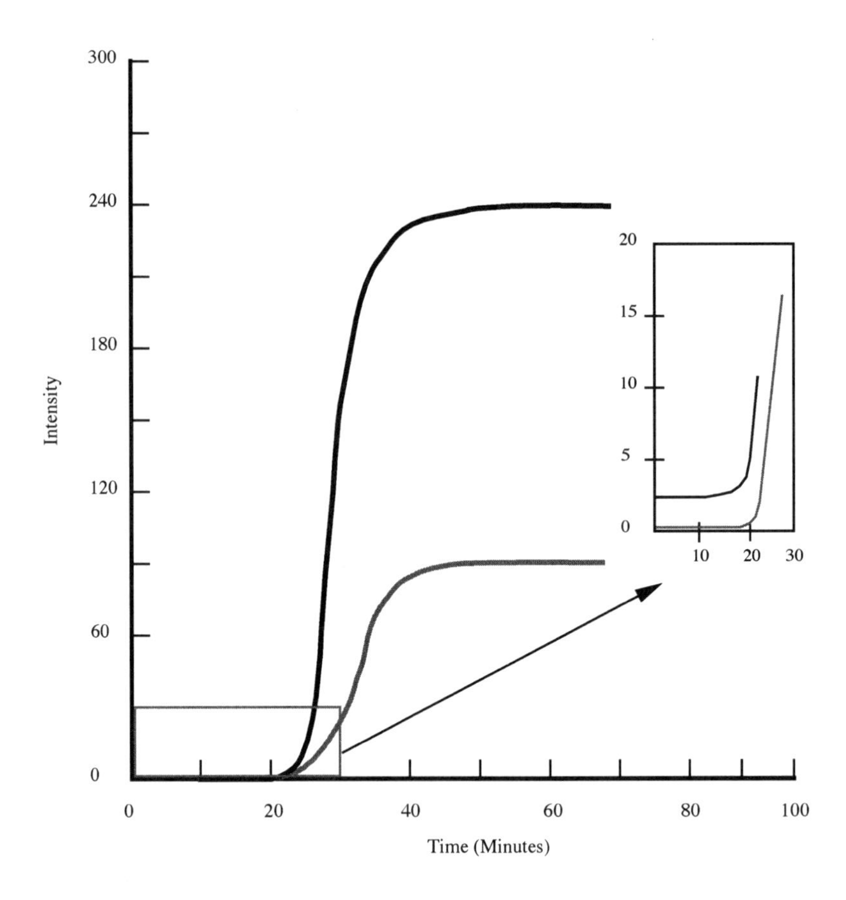

Figure 6.3 Scattering intensity during initiation of collagen self-assembly. Relative scattering intensity versus time at a scattering angle of 90° (upper curve) and extrapolated to 0° (lower curve) during the initiation of collagen self-assembly; changes are observed at a scattering angle of 0° before they are measured at 90°.

light intensity extrapolated to 0° during self-assembly is more sensitive than that at a scattering angle of 90°. Therefore, measurement of the light intensity at low scattering angles can be used to follow the early events associated with self-assembly.

In addition, by measuring the intensity fluctuations at low scattering angles, changes in the translational diffusion coefficient can be measured during the early phases of self-assembly, as illustrated by Figure 6.4. The translational diffusion coefficient decreases as the ratio of the length to width increases as linear assembly occurs.

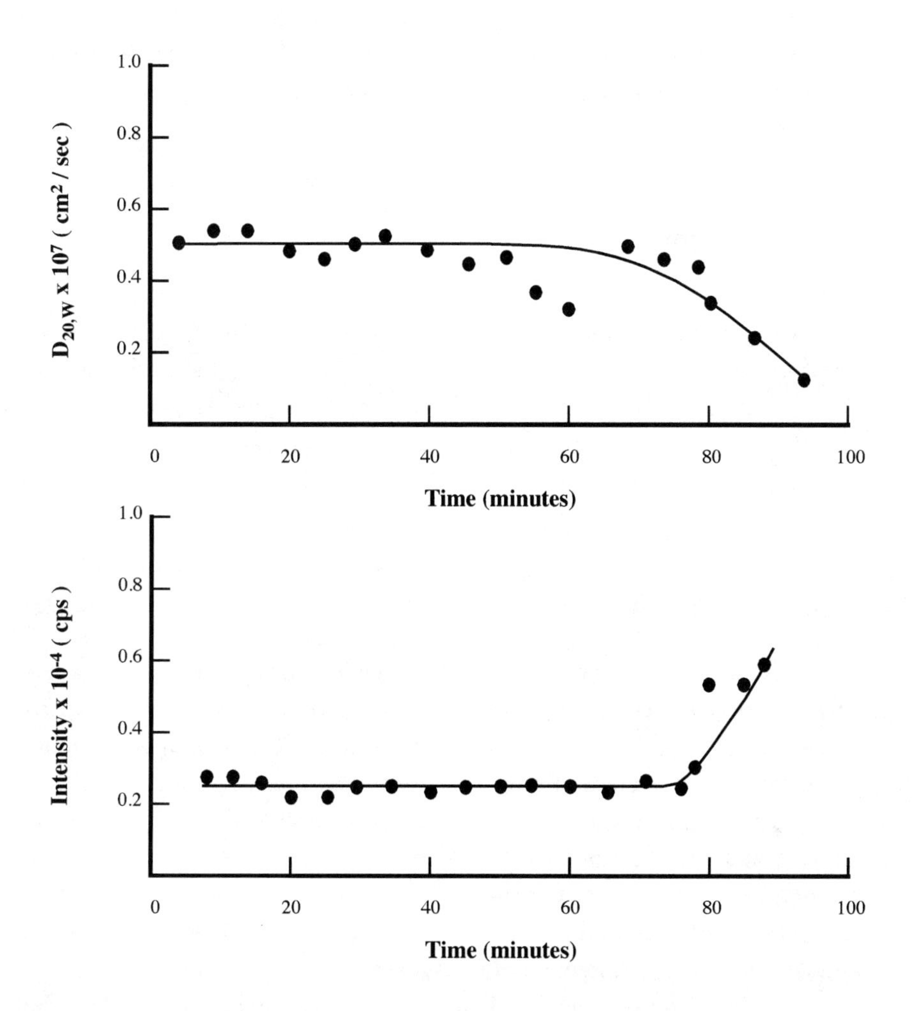

Figure 6.4 Measurement of physical properties during initiation of collagen self-assembly. Translation diffusion coefficient ($D_{20,w}$) (top) and intensity of scattered light at 90° (bottom) versus time for type I collagen. Note translational diffusion constant decreases, whereas intensity of scattered light remains initially unchanged.

Equilibrium Ultracentrifugation

Another possible method for studying self-assembly has been to measure the concentration gradient as a function of position and as a function of time during equilibrium centrifugation. The major criticism of this approach is that if the products that form during self-assembly are very large, they can be sedimented out and removed from the sampling window. However, even with these criticisms, it is worth using this approach to observe small aggregates that form during early self-assembly steps.

Electron Microscopy

Although electron microscopy is always interpreted with care and caution, because of the possibility of artifacts, a picture is always of value in developing models of self-assembly products. In self-assembly experiments, it is important to stop the process of self-assembly so that direct visualization can be achieved without modifying the process.

Collagen Self-Assembly

The self-assembly of collagen to form rigid gels was observed about 30 years ago by Jackson and Fessler and also by Gross and co-workers. In their experiments, collagen was solubilized and then heated to 37 °C in a buffer containing a neutral salt solution. Under these conditions, it was recognized that collagen molecules and aggregates of molecules would spontaneously assemble, forming fibrils that had the characteristic 67-nm repeat distance when viewed in the electron microscope. Subsequent studies showed that ions, alcohols, and other substances that affected both electrostatic interactions and hydrophobic bonds were able to modify the assembly of collagen molecules.

The mechanism by which collagen molecules assemble into fibrils is still a subject of intense interest. Cassel and associates (1962) studied the kinetics of the transition of collagen from the solution to the solid phase by raising the temperature over a wide range of pHs and ionic strengths. Their results suggested that collagen self-assembly involved a phase transition with no change in molecular conformation and that it was controlled by the rate of addition of molecules at the fibril surface at temperatures above 16 °C and limited by the rate of diffusion of collagen molecules at temperatures less than 16 °C. Thermodynamic studies by Cooper (1970) in which the solubility of collagen was measured in the temperature range of 20 to 37 °C indicated that native collagen fibril formation was an endothermic process (required energy) made thermodynamically favorable by the large increase in mobility (entropy) of the water molecules obtained by separating them from the collagen molecules. Other electron microscopic studies suggested that linear growth of fibrils appeared to occur by tandem addition of aggregates to each other and subsequently to the ends of a subfibril; lateral growth of these subfibrils occurs by ropelike entwining (Trelstad et al., 1976). These papers as well as other studies were recently reviewed (Veis and George, 1994; Kadler et al., 1996) to understand how fibrils are formed from precursor molecules.

Several observations help us understand the factors that are important in the assembly process. Under typical solution conditions, collagen single molecules with low pH and low salt content are in equilibrium with larger aggregates (Yuan and Veis, 1975; Silver and Birk, 1984). The aggregate in equilibrium with single molecules has been estimated to be between 1.5 million (Yuan and Veis, 1975) or 5 molecules and 5 million (Silver and Birk, 1984)or about 17 molecules. Later studies with procollagen type I indicated that the propeptides appear to limit association, and the aggregate in equilibrium with single molecules contained about 6 molecules (Berg et al., 1986).

The aggregates that form in solution may range in length from 4-D staggered dimers about 570 nm long (Silver et al., 1979; Kobayashi et al., 1985; Ward et al., 1986) to about 700 nm long (Bernengo et al., 1978). This corresponds to between 2 and 3 collagen molecules long. Taken together, the length and width suggest that aggregates that form during the initial phases of self-assembly contain 5 to 17 molecules and are 2 to 3 collagen molecules long. Estimates of the diameter of the first aggregates formed are initially 1 to 2 nm and then 2 to 6 nm (Gale et al., 1995). From these observations, it appears that initiation of assembly involves both linear and lateral growth and occurs by formation of a unit containing about 5 molecules packed laterally that is 2 or 3 collagen molecules long, as diagrammed in Figure 6.5. These units grow lengthwise and therefore do not result in turbidity increases during the lag phase. The initial phases have been shown to involve the nonhelical end regions (Bensusan

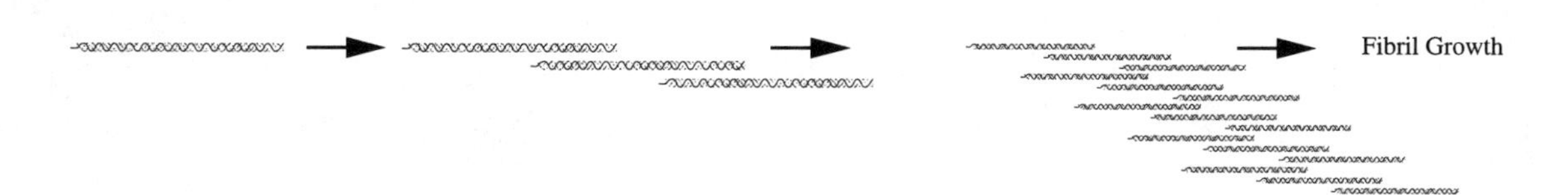

Figure 6.5 Collagen self-assembly. The diagram models the initiation of collagen self-assembly via formation of linear aggregates containing about three molecules that then laterally associate. The lateral assembly step may require a supramolecular twist, explaining why linear aggregation precedes lateral aggregation.

and Scanu, 1960), and the lag time is prolonged by removal of the nonhelical end regions with enzymes (Comper and Veis, 1977). Prevention of Schiff-base-mediated cross-links makes the aggregation step reversible (Suarez et al., 1985). Cross-linking during aggregation in the telopeptide region explains the memory effect that has been observed during the initial aggregation phase. The role of the nonhelical end regions that contain cross-link sites has been well established.

The initial phases of collagen self-assembly appear to involve the formation of linear aggregates of molecules that laterally aggregate to form thin filaments; however, the later phases are quite different. During the turbidimetric growth phase, thin collagen fibrils appear to fuse laterally (Trelstad et al., 1976), which causes rapid increases in turbidity that is correlated with fibril diameters (Birk and Silver, 1984), as diagrammed in Figure 6.5.

The role of the telopeptides (nonhelical ends) in self-assembly of collagen has been studied extensively. These studies demonstrate that enzymes (pronase) that cleave both the amino and carboxylic nonhelical ends arrest self-assembly. Pepsin, which removes portions of the amino and carboxylic nonhelical ends, slows aggregation down and results in wide fibrils. Leucine aminopeptidase removes the amino-terminal nonhelical end and leaves the carboxylic end intact, resulting in a lengthened lag phase. Carboxypeptidases A and B remove the carboxyl nonhelical end, leaving the amino end intact and decreasing lateral growth without affecting lag phase. From these studies, it was concluded that the amino-terminal nonhelical end is involved in formation of the initial thin fibrillar unit, and the carboxy-terminal nonhelical end is involved in fusion of thin fibrillar units. Lowering the pH from 7.5 to about 5.5 increases the lateral diameter of fibrils, and ionic strength changes fibril diameter, which suggests that fibrillar fusion involves electrostatic interactions. In tissues, collagen fibril diameter may in part be controlled by removal of extrahelical propeptides by specific enzymes (Miyahara et al., 1989) or by association with other macromolecules, such as decorin, a proteoglycan (Scott, 1996).

Assembly of Cytoskeletal Components

Actin and tubulin are two important cellular components that are involved in cell shape and movement. Actin is present in all mammalian cells and is involved in cellular transport and phagocytosis (eating of extracellular materials). It provides rigidity to cell membranes, and when bonded to tropomyosin and troponin, forms the thin filaments of muscle. Tubulin is the subunit from which microtubules are self-assembled. Microtubules are most commonly known for their role in cell division. The mechanisms of self-assembly of these macromolecules have been well studied and are important models of biological assembly processes.

Actin exists in two states within the cytoplasm of the cell: in the monomeric form (G-actin) and in the self-assembled or fibrous form (F-actin). G-actin has been shown to exist as an oblate ellipsoid by electron microscopy, and F-actin exists as a double helical structure. G-Actin and F-actin exist in cells in equilibrium with actin-binding proteins. These proteins are involved in the polymerization and depolymerization of F-actin.

Actin Self-Assembly

Actin is one of the most abundant cellular proteins and is present in all mammalian cells. F-Actin is physically cross-linked to form the cellular cytoskeleton, and its contraction allows cell deformation and movement during processes such as phagocytosis. The G-actin to F-actin assembly and disassembly sequence in the cell cytoplasm allows shape changes during cell movement. Gelsolin is an actin-binding protein that inhibits assembly of F-actin by binding to G-actin. G-actin is globular in shape (see Figure 2.39) and consists of α-helical regions connected by aligned regions containing β structures. The 3-D structure shows that G-actin has large and small domains. In the cleft between the domains lies the ATP-binding site, and there is also a binding site for calcium.

In the presence of filamin, an actin-binding protein, G-actin is polymerized into helical F-actin. Step one of polymerization involves formation of a G-actin–calcium complex. Step two involves replacement of the calcium with a magnesium cation, forming an altered G-actin; the altered form of G-actin binds an additional magnesium cation and then a dimer of two altered G-actin monomer forms. Binding of an additional altered G-actin monomer makes a trimer, the nucleus, with one pointed end and one barbed end. The addition of altered F-actin molecules to the nucleus causes elongation, and the barbed end grows faster than the pointed end (Figure 6.6).

G-Actin is soluble in water and is transformed into F-actin in neutral salt solutions. Conversion of G-actin to F-actin is associated with hydrolysis of ATP, leading to the formation of filaments that are up to several micrometers long and 7 to 10 nm wide. The process has been characterized as containing two steps, nucleation and growth (Korn et al., 1987), and, like polymerization of other

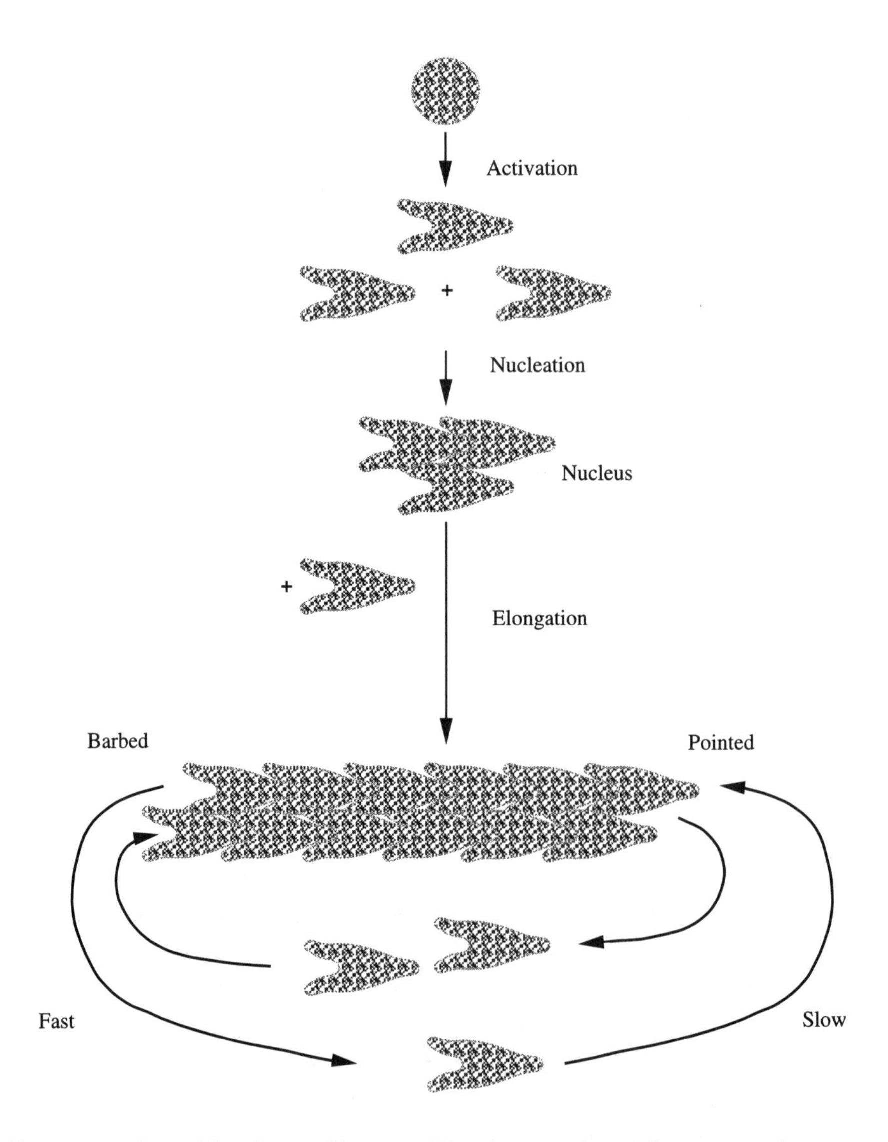

Figure 6.6 Assembly of actin filaments. The diagram shows the steps in the transition of actin monomer, G-actin, to actin filaments, F-actin. Monomers are activated by binding calcium and then exchanging calcium for magnesium, leading to nucleus formation. The nucleus consists of a trimer of G-actin with one pointed end and one barbed end. The addition of activated G-actin monomer to the nucleus causes elongation of the barbed end faster than the pointed end.

macromolecules, is entropy driven. In muscle, actin-filament tropomyosin binds to F-actin and mechanically stabilizes the muscles.

Tubulin

Tubulin is the building block from which microtubules are formed. Microtubules are essential in many cellular processes, including mitosis, cell-shape changes, and internal organelle movement of cilia and flagella. Tubulin is made up of α and β subunits that contain two components (thus they are dimers); each subunit has a MW of 55,000. Assembly of these subunits occurs longitudinally, forming a protofilament, and then 13 protofilaments close into a hollow cylinder, forming the basic structure of the microtubule. The microtubule has an outer diameter of 30 nm and an inner diameter of 14 nm, as shown in Figure 6.7 (Snyder and McIntosh, 1976).

Tubulin polymerization has been followed turbidimetrically. The shape of the turbidity-time curve is similar to that observed for collagen; it is S-shaped and contains lag, growth, and plateau phases (Gaskin et al., 1974). This has led to the interpretation that the lag phase involves linear growth of subunits into protofilaments, similar to that of collagen, which associate laterally into microtubules during the growth phase, as diagrammed in Figure 6.8. During assembly, the β subunit binds guanosine 5′-triphosphate (GTP) and hydrolyzes GTP to guanosine 5′-diphosphate (GDP) when the dimer adds to a microtubule. In addition, microtubule-associated proteins (MAPs) help to stabilize the structure and also help link the microtubules to other structures in the cell.

Actin–Myosin Interaction

The interaction of actin and myosin is key to the generation of contractile forces in skeletal muscles. Skeletal muscles are composed of thick and thin filaments; the thick filaments are composed primarily of myosin, and the thin filaments contain actin. The interaction between the two is a reversible self-assembly process that is followed by disassembly that separates actin from myosin. This separation initiates movement of myosin with respect to actin and results in shortening of the muscle, causing force generation. It is the assembly and disassembly processes that interest us in this section.

The process involves binding of ATP to myosin. The myosin molecule has a length of more than 16.5 nm and is approximately 6.5 nm wide and

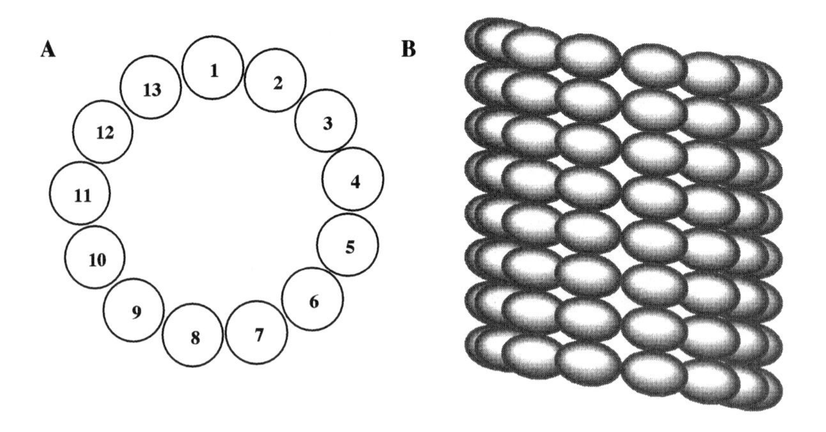

Figure 6.7 Schematic diagram of microtubular structure. A cross-section (A) and longitudinal section (B) of a microtubule are shown.

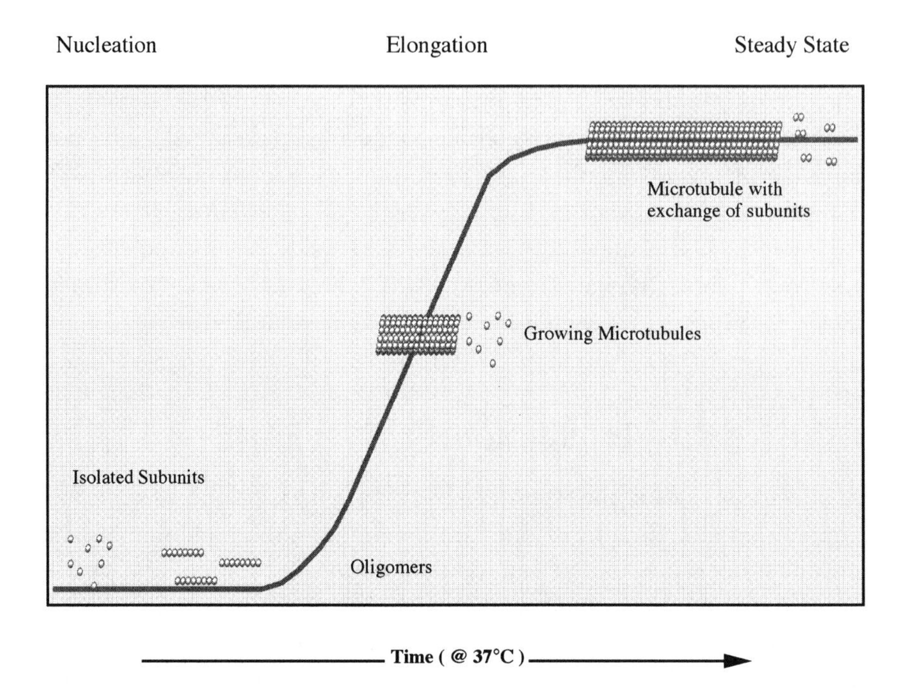

Figure 6.8 Turbidity-time curve for microtubule assembly. The diagram illustrates the turbidity-time changes that occur during microtubule assembly. Initially during the turbidity lag phase, tubulin monomers form rings of tubulin subunits that cause microtubule elongation during the growth phase.

4.0 nm deep at its thickest end (Rayment and Holden, 1994) (see Figure 2.40). The secondary structure contains several long α helixes. The actomyosin complex is a very large macromolecular complex. The starting point in the contractile cycle is taken as the point when myosin is tightly bound to actin in the absence of ATP. In this state, it is believed that the space between the two parts of the head region is closed. The first step in the contractile cycle involves opening of the space between the components of the head when ATP binds, which disrupts

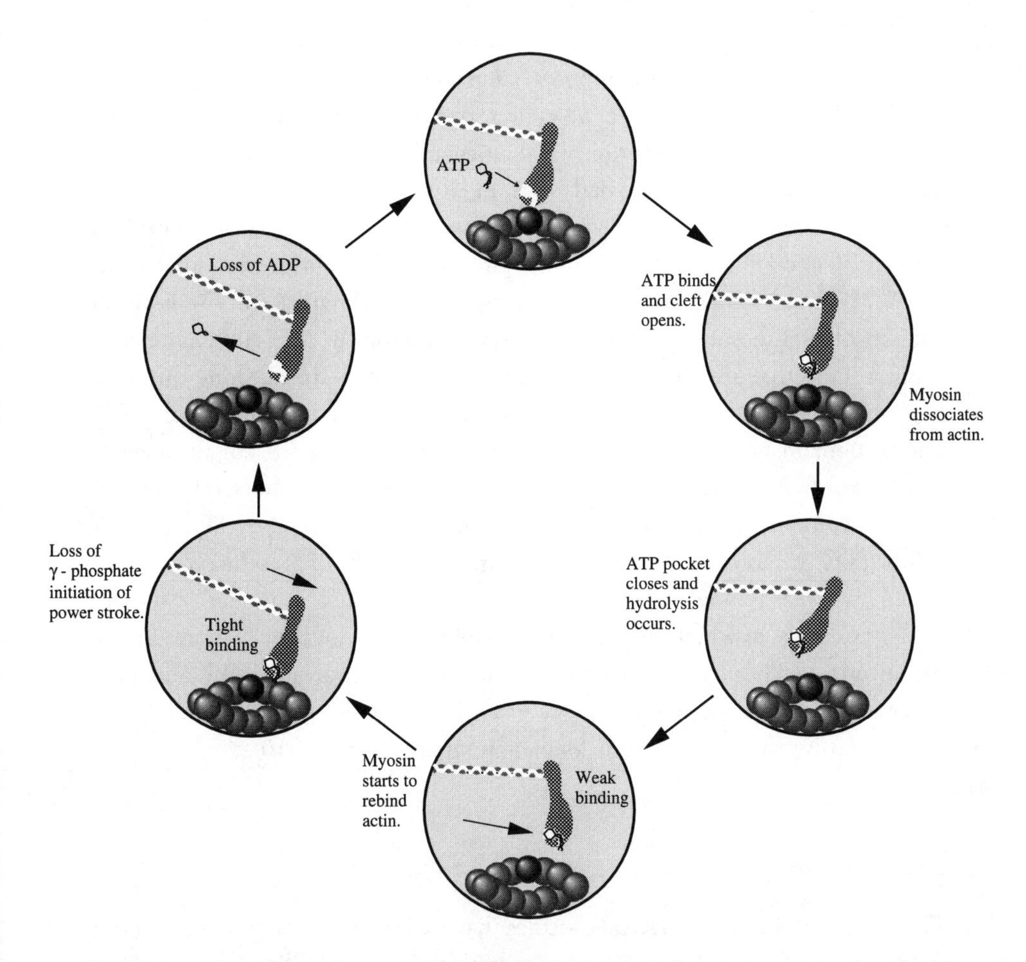

Figure 6.9 Summary of contraction cycle. Interaction between myosin head and F-actin involves the binding of ATP to a cleft in the myosin head (top) that disassociates myosin head from actin, causing hydrolysis of ATP into ADP. This is followed by binding of myosin head to actin, loss of γ-phosphate, and then loss of ADP and repeat of the cycle. *Source:* Diagram modified based on Rayment and Holden (1994).

the actin-binding site on myosin (Figure 6.9). After ATP binding occurs and the space between the two head components narrows, hydrolysis occurs. This is followed by myosin starting to rebind to actin, which is followed by loss of γ-phosphate and initiation of the power stroke. The power stroke is completed as ADP is lost from the site between the two head components.

Fibrinogen

Fibrinogen is a protein that composes 3% of the weight of blood. It is found in blood cells and platelets. It is involved in formation of fibrin networks that make up the noncellular component of blood clots and is converted into fibrin via the intrinsic and extrinsic clotting pathways. In either of these pathways, prothrombin is converted to thrombin, which proteolytically removes fibrinopeptides A and B (FPA and FPB) from the fibrinogen molecule. The fibrinogen molecule has a MW of about 330,000; its structure is diagrammed in Figure 2.37. The molecule consists of two terminal domains with MWs of about 67,200 and diameters of about 6.0 nm that are connected to a central domain by 16-nm-long threadlike domains. These domains have a coiled-coil structure and contain α-helixes. The central domain has a MW of about 32,600 and is about 5.0 nm in diameter. Removal of FPA and FPB result in only a small decrease in molecular weight and allow assembly to proceed by the interaction of sites blocked by FPA and FPB. Historically, it has been proposed that removal of FPB leads to linear polymerization, whereas removal of FPA leads to lateral growth. This has led to a model of self-assembly in which initiation of assembly by removal of FPA and dimerization of two fibrinogen molecules is followed by removal of FPB and linear polymerization. Further assembly occurs by lateral packing of elements that are formed during the previous step, as diagrammed in Figure 6.10.

Summary

The ability of biological macromolecules that are composed of helixes, including the α helix and the collagen triple helix, to assemble into higher-order structures has some common elements. First, many of these processes are characterized by turbidimetric lag and growth phases, suggesting that linear growth of a stable element or nucleus must occur before lateral growth. In many instances, linear growth involves formation of dimers and trimers in an end-to-end or overlapped fashion, possibly because aggregation in this manner favors formation of a new

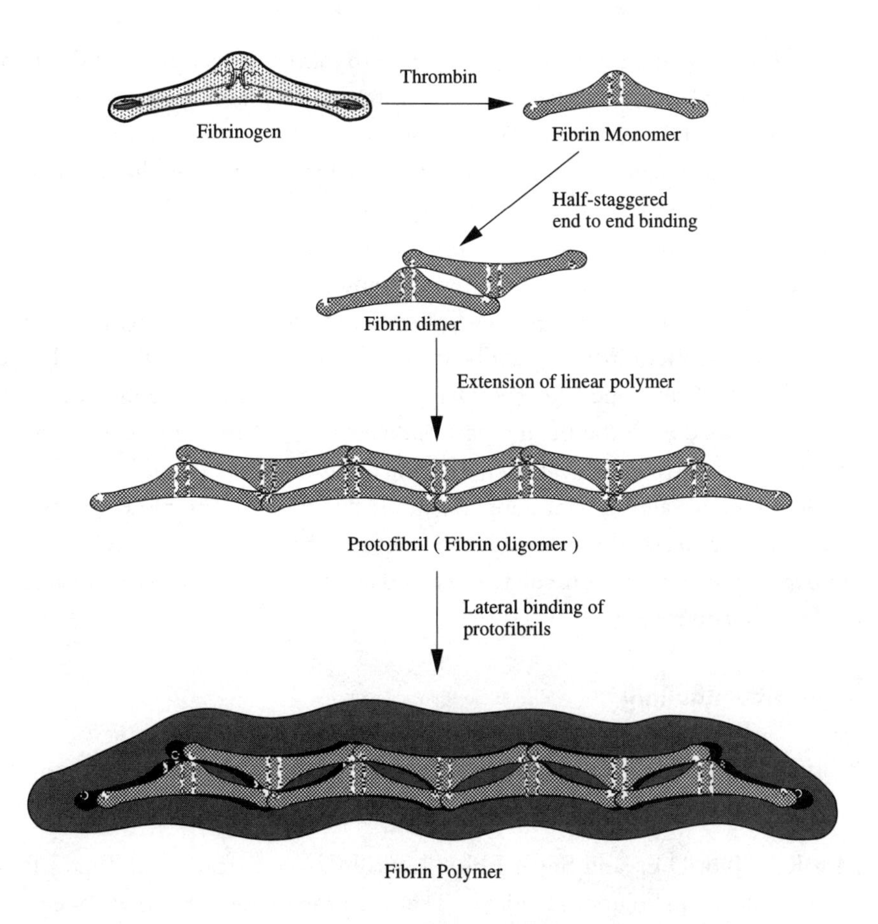

Figure 6.10 Conversion of fibrinogen to fibrin. The diagram illustrates the assembly of fibrin monomer after cleavage of fibrinopeptides A and B by thrombin, causing end-to-end binding of monomers followed by lateral growth.

phase (solid) in equilibrium with the liquid phase. From another perspective, linear aggregation decreases the probability for the molecules to tumble, thereby stabilizing later lateral growth steps. Whether linear growth stabilizes a nucleus as in classical nucleation and growth kinetics is unclear; however, linear growth provides an element that has more surface area, whereas lateral growth minimizes the surface area exposed for new growth. Maximization of the surface area would allow rapid growth of large numbers of macromolecules, especially if each successive step in the assembly process used larger and larger building blocks. In addition, it would also provide the beginning of a twisted unit if growth involves formation of a superhelix.

Turbidity increases during the growth phase reflect the rapid lateral growth of large units. This favors interpretation of biological assembly as multistep processes, because formation of large structures can only occur rapidly if high molecular weight units grow in diameter by fusion. The ultimate consequence of self-assembly at either the cellular or extracellular levels is the formation of structures that can transmit or resist large forces.

A practical example of the need for biological structures to grow rapidly over a small time frame is observed when one characterizes the strength of tissues just before and after birth. Prenatally, the extracellular matrix is rather weak, and the collagen fibrils appear to be discontinuous, at least in tendons. Just after birth, associated with the beginning of movement, collagen fibril diameters increase as does the mechanical strength. The only way that the fibrils could grow rapidly in length and in width appears to involve fibril fusion. However, as discussed in Chapter 7, the mechanical properties of collagenous tissues, at least, are dependent not only on collagen fibril diameters but on how the molecules are cross-linked together.

Suggested Reading

Bensusan H.B and Scanu A.J., Fiber Formation From Solutions of Collagen. II. The Role of Tyrosyl Residues, J. Am. Chem. Soc. 82, 4990, 1960.

Berg R.A., Birk D.E., and Silver F.H., Physical Characterization of Type I Procollagen in Solution: Evidence That the Propeptides Limit Self-Assembly, Int. J. Biol. Macromol. 8, 177, 1986.

Bernengo J.C., Herbage D., Marion C., and Roux B., Intermolecular Interactions: Studies on Native and Enzyme-Treated Acid-Soluble Collagen, Biochem. Biophys. Acta 532, 305, 1978.

Birk D.E. and Silver F.H., Collagen Fibrillogenesis In Vitro: Comparison of Types I, II and III, Arch. Biochem. Biophys. 235, 178, 1984.

Cassel J.M., Mandelkern L., and Roberts D.E., The Kinetics of the Heat Precipitation of Collagen, J. Am. Leather Chem. Assoc. 51, 556, 1962.

Comper W.D. and Veis A., Characterization of Nuclei in In Vitro Collagen Fibril Formation, Biopolymers 16, 2133, 1977.

Cooper A., Thermodynamic Studies of the Assembly In Vitro of Native Collagen Fibrils, Biochem. J. 118, 355, 1970.

Gale M., Pollanen M.S., Markiewicz P., and Goh M.C., Sequential Assembly of Collagen Revealed by Atomic Force Microscopy, Biophys. J. 68, 21248, 1995.

Gaskin F., Cantor C.R., and Shelanski M.I., Turbidimetric Studies of the In Vitro Assembly and Disassembly of Porcine Neurotubules, J. Mol. Biol. 89, 737, 1974.

Gross J., Highberger J.H., and Schmitt F.O., Some Factors Involved in the Fibrogenesis of Collagen In Vitro. Proc. Soc. Exp. Biol. Med. 80, 462–465, 1952.

Jackson D.S. and Fessler J.H., Isolation and Properties of a Collagen Soluble in Salt Solution at Neutral pH. Nature 176, 69–70, 1955.

Kadler K.E., Holmes D.F., Trotter J.A., and Chapman J.A., Collagen Fibril Formation, Biochem. J. 316, 1, 1996.

Kobayashi K., Ito T., and Hoshino T., Electron Microscopic Demonstration of Acid-Labile, 4D-Staggered Intermolecular Association of Collagen Formed In Vitro, Collagen Rel. Res. 5, 253, 1985.

Korn E.D., Carlier M.-F., and Pantaloni D., Actin Polymerization and ATP Hydrolysis, Science 238, 638, 1987.

Miyahara M., Hayashi K., Berger J., Tanzawa K., Njieha F.K., and Trelstad R.L., Formation of Collagen Fibrils by Enzymatic Cleavage of Precursors of Type I Collagen In Vitro, J. Biol. Chem. 259, 9891, 1989.

Oosawa F. and Kasai M., A Theory of Linear and Helical Aggregations of Macromolecules, J. Mol. Biol. 4, 10, 1962.

Rayment I. and Holden H.M., The Three Dimensional Structure of a Molecular Motor, Trends Biochem. Sci. 19, 129, 1994.

Scott J.E., Proteodermatan and Proteokeratan Sulfate (Decorin, Lumincan/Fibromodulin) Proteins Are Horseshoe Shaped. Implications for Their Interactions with Collagen, Biochemistry 35, 8795, 1996.

Silver F.H., Self-Assembly of Connective Tissue Macromolecules, in Biological Materials: Structure, Mechanical Properties, and Modeling of Soft Tissues, NYU Press, New York, chapter 5, pp. 150–153, 1987.

Silver F.H., Biomaterials, Medical Devices and Tissue Engineering: An Integrated Approach, Chapman & Hall, London, chapter 1, 1994.

Silver F.H. and Birk D.E., Kinetic Analysis of Collagen Fibrillogenesis: I. Use of Turbidity-Time Data, Collagen Rel. Res. 3, 393, 1983.

Silver F.H. and Birk D.E., Molecular Structure of Collagen in Solution: Comparison of Types I, II, III and V, Int. J. Biol. Macromol. 6, 125, 1984.

Silver F.H. and Trelstad R.L., Linear Aggregation and the Turbidimetric Lag Phase: Type I Collagen Fibrillogenesis In Vitro. J. Theor. Biol. 81, 515–526, 1979.

Snyder J.A. and McIntosh J.R., Biochemistry and Physiology of Microtubules, Annu. Rev. Biochem. 45, 699–720, 1976.

Suarez G., Oronsky A.L., Bordas J., and Koch M.H.J., Synchrotron Radiation X-Ray Scattering in the Early Phases of In Vitro Collagen Fibril Formation, Proc. Natl. Acad. Sci. U.S.A. 82, 4693, 1985.

Trelstad R.L., Hayashi K., and Gross J., Collagen Fibrillogenesis: Intermediate Aggregates and Suprafibrillar Order, Proc. Natl. Acad. Sci. U.S.A. 73, 4027, 1976.

Veis A. and George A., Fundamentals of Interstitial Collagen Self-Assembly, Extracellular Matrix Assembly, Academic Press, New York, pp. 15–45, 1994.

Ward N.P., Hulmes D.J.S., and Chapman J.A., Collagen Self-Assembly In Vitro: Electron Microscopy of Initial Aggregates Formed During the Lag Phase, J. Mol. Biol. 190, 107, 1986.

Yuan L. and Veis A., The Self-Assembly of Collagen Molecules, Biopolymers, 12, 1437, 1975.

7
Mechanical Properties of Tissues

Introduction to Analysis of Tissue Mechanical Properties

Tissues are composed of macromolecules, water, ions, and minerals, and therefore their mechanical properties fall somewhere between that of random chain polymers and that of ceramics. Table 7.1 gives the physical properties of cells, soft and hard tissues, metals, polymers, ceramics, and composites. The properties of biological tissues are wide ranging, from cell membranes with a modulus of about 10^{-4} MPa to bone with ultimate tensile strength (UTS) and modulus of about 200 MPa and 15 to 20 GPa, respectively. This range of properties is achieved largely by use of protein building blocks in association with mineral. As discussed in Chapter 2, the mechanical properties of macromolecules are intimately related to the 3-D structures of the polypeptide chains. Because

Table 7.1 Mechanical properties of tissue and synthetic materials

Material	Ultimate tensile strength (MPa)	Modulus (MPa)
Soft tissue		
Arterial wall	0.5-1.72	1.0
Hyaline cartilage	1.3-18	0.4-19
Skin	2.5-16	6-40
Tendon/ligament	30-300	65-2,500
Hard tissue (bone)		
Cortical	30-211	16-20 GPa
Cancellous	51-93	4.6-15 GPa
Polymers		
Synthetic rubber	10-12	4
Glassy	25-100	1.6-2.6 GPa
Crystalline	22-40	0.015-1 GPa
Metal alloys		
Steel	480-655	193 GPa
Cobalt	655-1,400	195 GPa
Platinum	152-485	147 GPa
Titanium	550-860	100-105 GPa
Ceramics		
Oxides	90–380 GPa	160-4,000 GPa
Hydroxyapatite	600	19 GPa
Composites		
Fibers	0.09–4.5 GPa	62-577 GPa
Matrixes	41–106	0.3-3.1
Aortic cell	—	90×10^{-6}[a]

Source: Adapted from Silver (1994).

[a]Data from Sato et al. (1990).

most proteins are composed of α helices, β structures, and collagen triple helixes, the mechanical properties of these structures are of interest. In this chapter, we introduce the terminology used to describe and measure mechanical properties, followed by a discussion of the mechanical properties of α helices, β structures, and collagen triple helixes.

Elastic and Viscous Behavior

When a force is placed on a body, there is an instantaneous elastic response and a time-dependent response. Consider as an example placing a weight on the end of a rubber band (Figure 7.1). The instantaneous deformation is the elastic response, and the time-dependent deformation is the viscous response. If a material does not have a time-dependent response, it is considered to exhibit purely elastic behavior; if it has a time-dependent behavior, it is considered to exhibit viscoelasticity. Polymers behave as viscoelastic materials and exhibit both time-independent and time-dependent behavior; in addition, their mechanical behavior is also temperature dependent. Liquids are viscous materials because they exhibit only time-dependent behavior.

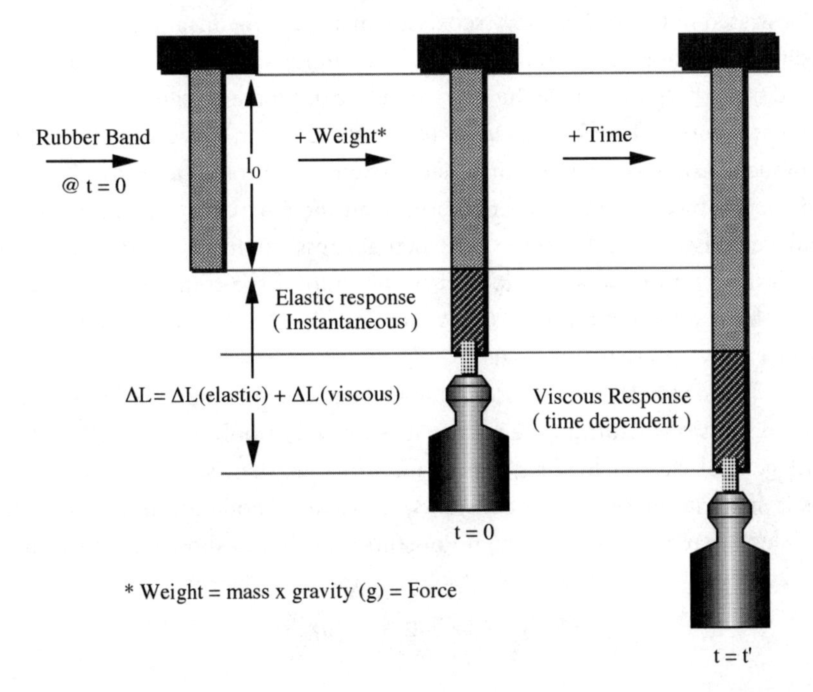

Figure 7.1 Definitions of elasticity and viscoelasticity. If a weight is placed on a rubber band at time zero and the initial length is l_o, then instantaneously the weight causes a length increase that is followed by a continual slow length increase until the rubber band reaches some new equilibrium length $l_o + \Delta L$. The elastic response is the length increase that occurs instantaneously, and the viscous response is the time-dependent length component.

When a force or stress (stress is defined as force per unit cross-sectional area) is applied to an elastic body and a strain (strain is defined in this book as change in length divided by original length) results, we say the material is *hookean* if the loading curve is identical to the unloading curve with no permanent deformation of the material (Figure 7.2A). If the unloading curve is not identical to the loading curve and there is permanent deformation in the material, the behavior is *plastic* (Figure 7.2B). When a viscoelastic material is placed under a constant load, the strain is time dependent compared to an elastic material, which exhibits a time-independent strain (Figure 7.2C).

Material properties are characterized by measuring the stress required to stretch a material until it fails. In a constant rate-of-strain experiment, the specimen is stretched at a constant strain rate, and the force per unit area is measured at different time intervals. By knowing the original specimen length and the rate at which the specimen is stretched, a stress-versus-strain curve can be plotted, as illustrated in Figure 7.3. For viscoelastic materials, the magnitude of the stress at different strains is strain-rate dependent. Important parameters that are obtained from this plot include the stress at failure (ultimate tensile strength [UTS]), strain at failure (ε_f), and the tangent to the stress–strain curve (tangent modulus or modulus). If a viscoelastic material is stretched through a series of intervals and then allowed to relax to equilibrium, an incremental stress–strain curve is produced (Figure 7.4). From the incremental stress–strain curve, we can calculate the elastic fraction, which is defined as the ratio of the total force (F_o) divided by the force at equilibrium (F_e). This ratio is 1 for elastic materials and between 0 and 1 for viscoelastic materials.

This characterization assumes that the material being tested is uniform in three dimensions (isotropic) and that it is loaded in only one direction. In the most general case of loading, the relationship between stress and strain in an elastic material involves two constants, E (Young's modulus) and G (the shear modulus), that are related through Poisson's ratio (v), as shown in equation 7.1.

$$G = E/2(1 + v) \tag{7.1}$$

In the case of bone and materials that are not isotropic (anisotropic), the relationship between stress and strain is more complex, and there is more than one value of Young's modulus and the shear modulus, which is dependent on the direction of loading. The values of these parameters and the values of Poisson's ratio are found by determination of the constants in the elastic compliance matrix, as described in the work by Cowin (1989).

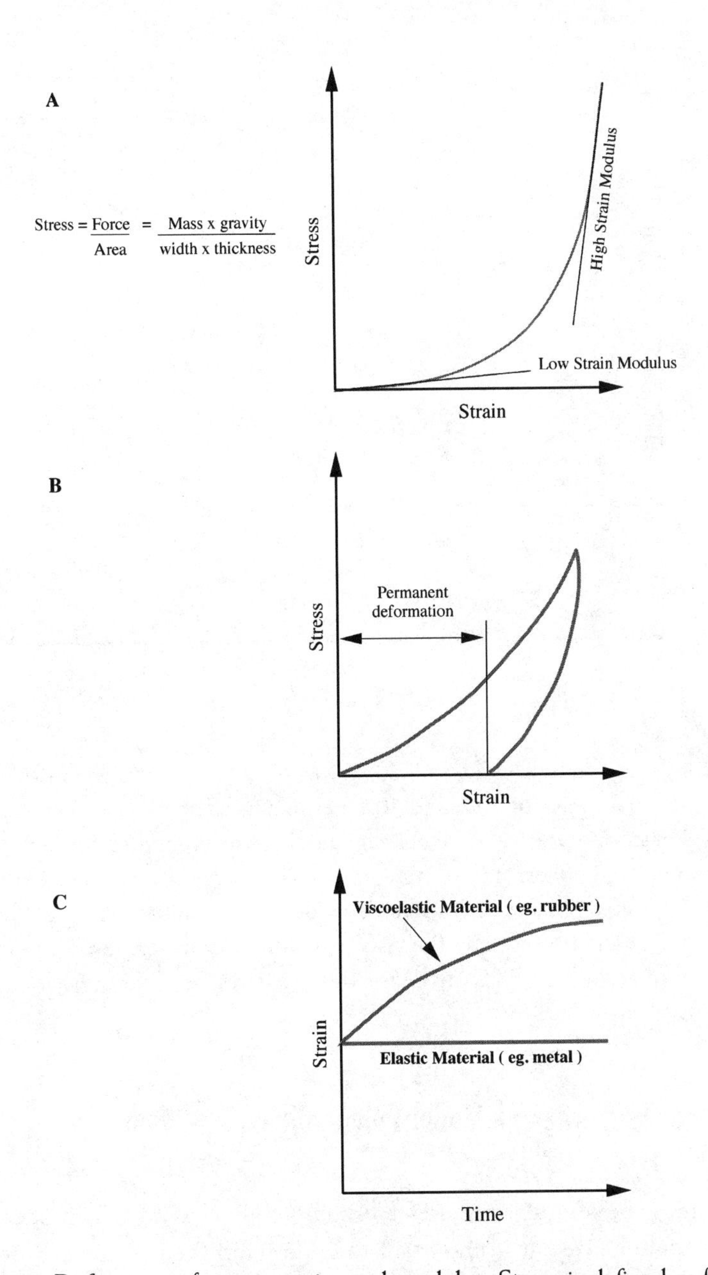

Figure 7.2 Definitions of stress, strain, and modulus. Stress is defined as force per unit area, and strain is the change in length divided by the original length. When stress is plotted versus strain, then the slope is the modulus (**A**). When the load is removed, any strain remaining is called permanent or plastic deformation (**B**). When elastic materials are loaded, they are characterized by a constant strain as a function of time, whereas viscoelastic materials have strains that increase with time (**C**).

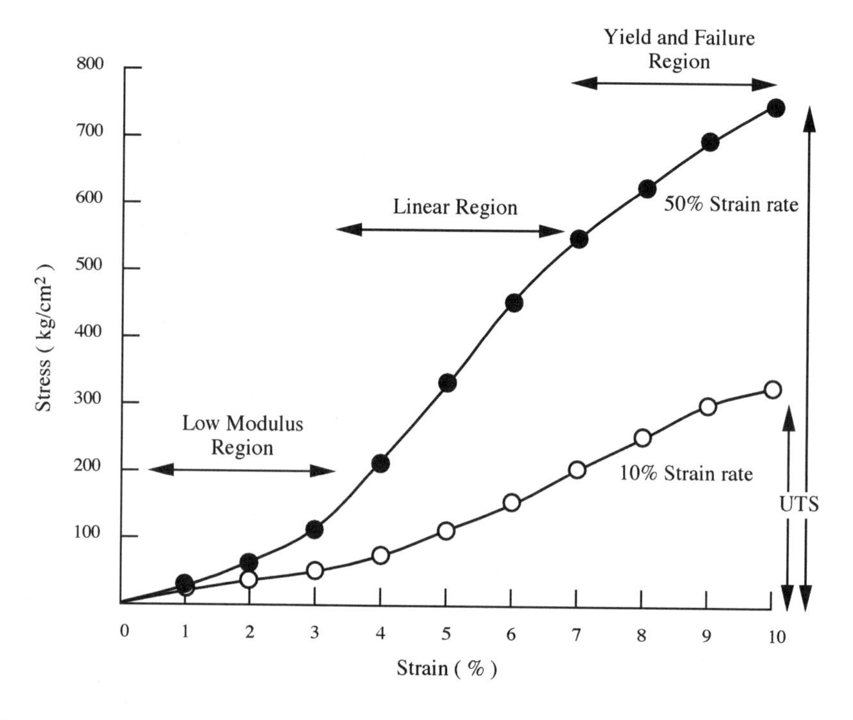

Figure 7.3 Stress–strain curve for tendon. Stress–strain curves for wet rat-tail tendon in tension at strain rates of 10% and 50% per minute. The curve is made up of low modulus, linear, and yield and failure regions that correspond to the straightening of crimped collagen fibers, the stretching of collagen fibrils, and the disintegration of fibrils, respectively. The strain-rate dependence is a characteristic of the viscoelastic behavior and is caused by the viscous sliding of collagen fibrils by each other during deformation. The UTS is defined as the stress at failure. *Source:* Diagram adapted from Silver (1987).

Mechanical Properties of Model Polypeptides

Because most biological structures are composed of α helices, β sheets, and collagen triple helixes, it is important to understand the mechanical properties of these basic structures. The mechanical properties of collagen fibers from tendon are thought to reflect the mechanical properties of the collagen triple helix. The stress–strain curve of tendon is S-shaped, and tendon fails at stresses of above 50 MPa and strains of 10 to 20% (Figure 7.5). The slope of the linear portion of the curve at high strains is 2 GPa, suggesting that the modulus of the collagen molecule has that value. Note that these properties are for wet collagen

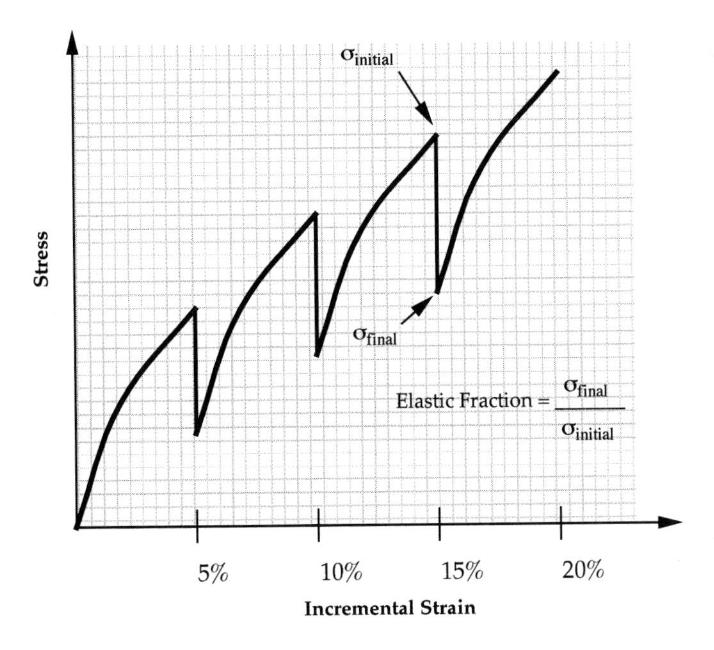

Figure 7.4 Determination of elastic and viscous components. Incremental stress–strain curve constructed by stretching a specimen in strain increments of 2 to 5% and allowing the specimen to relax to an equilibrium stress before an additional strain increment is added. The elastic fraction is defined as the equilibrium stress divided by the initial stress. *Source:* Adapted from Silver (1987).

fibers that have about 50% water by weight. Therefore, the properties of the fibers would need to be corrected for the water content. As we discussed in Chapter 2, the conformation of the collagen molecule is such that it is highly extended, suggesting that highly extended conformations lead to polypeptide structures with high stiffness (about 4 MPa after correction for water content) and low extensibility (about 10%). Fibrous collagen is believed to prevent premature mechanical failure and shape changes of biological tissues, and therefore, its extended structure and high stiffness are properties that have evolved to perform these functions.

The second example of mechanical properties of a standard polypeptide structure is that of the β structure of silk. The stress–strain curve of the dragline silk of a spider (Figure 7.6) shows a UTS of about 800 MPa, a modulus of about 7.0 GPa, and a strain at failure of 30%. The UTS and modulus values are greater than those for collagen, which reflects the greater axial rise per residue in the β structure (0.36 nm) compared to collagen (0.29 nm).

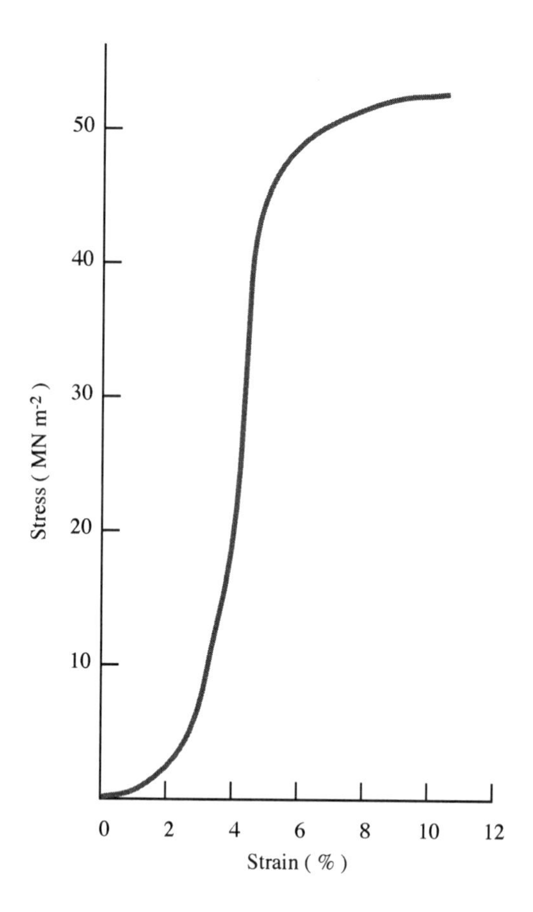

Figure 7.5 Typical stress–strain curve for tendon. The diagram illustrates the stress–strain curve for an isolated collagen fiber from tendon. Note that collagen fibers from tendon fail at UTS values above 50 MPa and at strains between 10 and 20%. The slope of the linear portion of the curves at high strains is 2 GPa.

The stress–strain curve of stratum corneum, which is made of several different keratins with helical structures, is shown in Figure 7.7. Although the structure of keratin contains more than a simple arrangement of α helixes, at low relative humidity the stress–strain curve is almost linear and has a UTS of about 1.8 MPa and a modulus of about 120 MPa. This is consistent with the observations that the UTS and modulus are lower than for the collagen triple helical and β structures because the rise per residue along the α helix is only 0.15 nm.

The mechanical properties of standard polypeptide chains suggest that the UTS and modulus are related to the axial rise per residue along the helix. The more extended the structure is in space, the more rigid it is and the higher

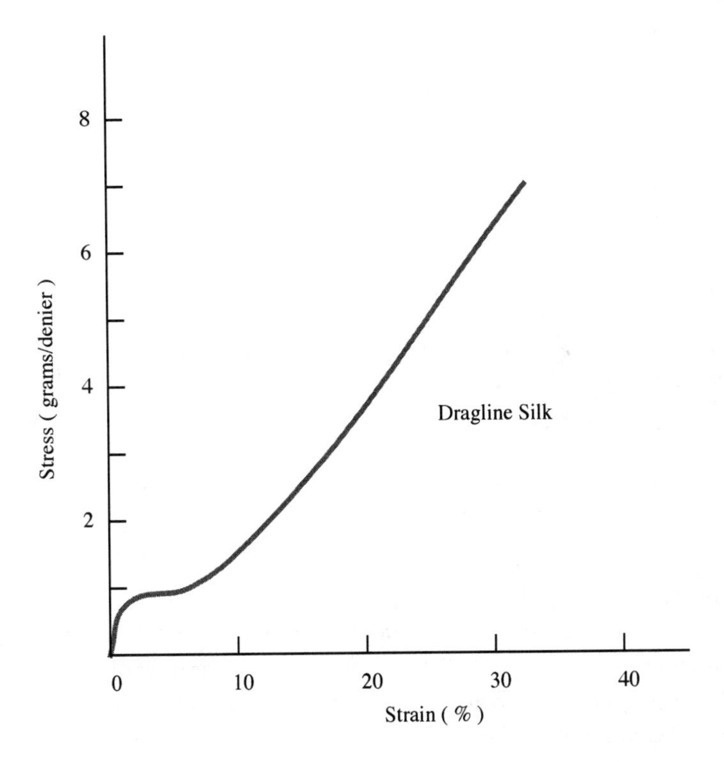

Figure 7.6 Stress–strain curve for silk. The ultimate strength of dragline silk is about 800 MPa, with a modulus of about 7 MPa and strain at failure of 30%. Note that these values are higher than those for collagen, reflecting the higher axial rise per residue along the molecule.

its tensile strength. This suggests the high strength of collagen and silk is a consequence of their extended structures in space. A corollary to this rule is that any process that causes chain extension in space strengthens and stiffens a protein. In comparison to the helical macromolecules, the mechanical behavior of random polypeptide chains can be determined from the behavior of elastin. The UTS for elastin is about 0.25 MPa, the modulus is about 0.15 MPa, and the strain at failure is about 80% (Bush et al., 1982). This demonstrates that a random chain structure that has an axial rise per residue less than that of helical chain structures is more extensible but also has lower values of strength and modulus. For this reason, elastin is believed to play a role in energy storage in the arterial wall at low strains, but it does not significantly contribute to the strength or high strain stiffness.

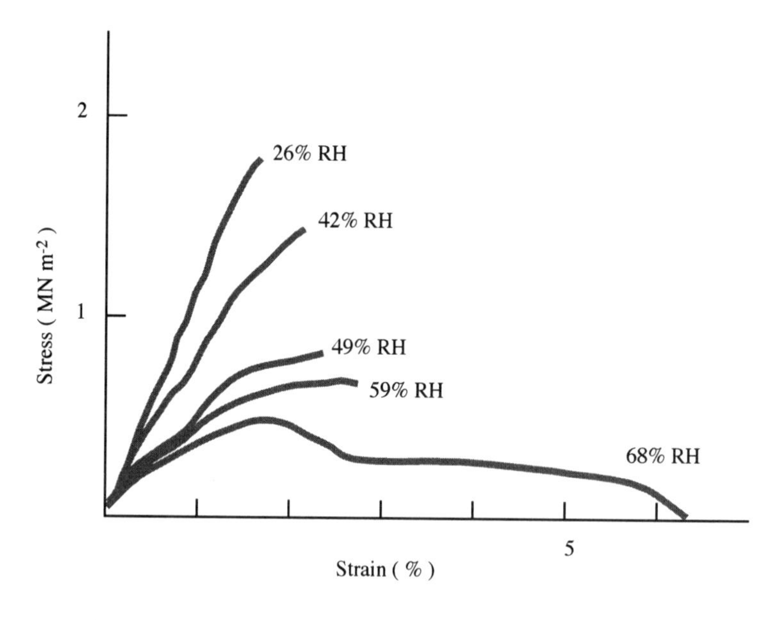

Figure 7.7 Stress–strain curve for stratum corneum. At low moisture contents, the stress–strain properties of stratum corneum are almost linear, with a UTS of about 1.8 MPa and a modulus of about 120 MPa. Note that the properties are lower than those of collagen and silk.

Mechanical Properties of Collagenous Tissues

Connective tissue functions to maintain shape, transmit and absorb loads, prevent premature mechanical behavior, partition cells and tissues into functional units, and act as a scaffold that supports tissue architecture (Silver, 1987). The primary structural element in mammalian connective tissue is fibrillar type I collagen. Type I collagen molecules form fibrillar elements, 20 to several hundred nanometers in diameter, which in turn pack into fibril bundles or fibers, fascicles, and higher-level tissue architectures, as discussed in Chapter 4. The structural hierarchy and mechanical properties of fibrillar type I collagen vary from tissue to tissue; however, the UTS of tissues containing type I collagen has been correlated with fibril (Parry, 1988) and fiber (Doillon et al., 1985) diameters.

Although the mechanical properties of connective tissue are dependent on the properties of the collagen fibrils, their diameters, and the content of collagen in the tissue, they are also dependent on several other factors, including the content of other components, such as elastic tissues and proteoglycans; ori-

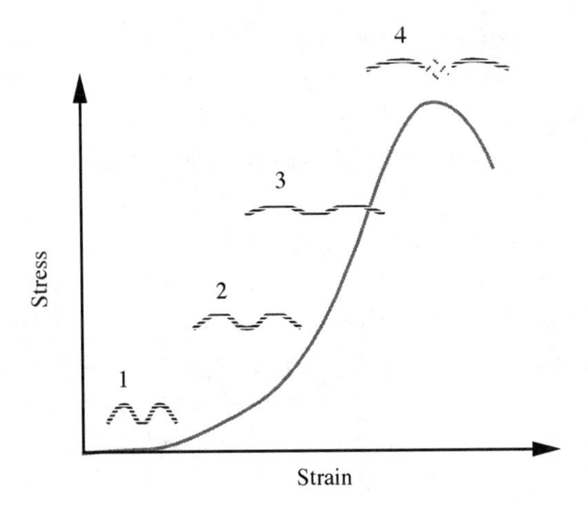

Figure 7.8 Generalized uniaxial stress–strain curve for connective tissue. The stress–strain curve for unmineralized connective tissue consists of two nonlinear regions and a linear region. The initial nonlinear region (region 1) involves uncrimping or geometrical alignment, or both, of the collagen fibers or expression of fluid. The linear region (regions 2 and 3) is a result of stretching and sliding of collagen fibers that fail by defibrillation in the final nonlinear region (region 4).

entation of the collagen fibers; the direction of loading; the degree of cross-linking; the extent of tissue hydration; the age and sex of the host; the anatomical location; the extent of mineralization; and the cellular contribution. The generalized uniaxial stress–strain curve for collagenous tissue is shown in Figure 7.8. It consists of a nonlinear region (region 1); a linear region (regions 2 and 3), and a final nonlinear region (region 4). The initial nonlinear region involves both uncrimping and geometrical alignment of collagen fibers and the expression of fluid. The linear region is a result of the direct loading of collagen fibers that fail in region 4 by defibrillation.

The mechanical properties of isolated collagen fibers 50 to 100 mm in diameter have been studied in uniaxial tension. Values of UTS as high as 40 MPa, strains up to 10%, and high strain moduli in excess of 500 MPa have been reported (Kato and Silver, 1990). These properties reflect the stiffness of the collagen triple helix, which has been estimated to have a stiffness as high as 4 GPa, based on solution properties (Nestler et al., 1983). In contrast, the mechanical behavior of elastic fibers have been modeled by a rubberlike state; however, some nonideality of the stress–strain behavior has been observed and may be the result of the interaction of water with the hydrophobic regions of this protein. The UTS of elastic tissue is 0.1 MPa or only a small fraction of that of collagen fibers, as discussed earlier, whereas the modulus and the strain at failure are about 0.1 MPa and 300%, respectively (Mukherjee et al., 1976). It is believed that elastic fibers contribute to the low strain mechanical behavior of connective tissue, whereas collagen limits high strain deformation and prevents failure. Rheological studies on proteoglycans indicate that the complex shear modulus is about 1×10^{-5} MPa, suggesting that proteoglycans do not mediate stress transfer directly between collagen fibers, but may facilitate sliding or dissipate energy by blocking charged sites on the collagen molecule (Pins et al., 1997).

Because collagen dominates the high strain behavior of connective tissue, for simplification, the mechanical behavior of connective tissue is classified by the type of collagen network. The types of collagen networks found in tissues include parallel aligned collagen networks, alignable networks, or composites containing aligned and alignable networks. In the next section, each of these classifications is discussed individually.

Mechanical Properties of Oriented Collagen Networks

Much of our understanding of the mechanical behavior of collagen has developed by studying the mechanical properties of oriented collagen networks, including tendons and ligaments. Although the microscopic structure of these tissues varies from location to location throughout the body, the basic structural units, including aligned collagen fibrils and fibers, are found throughout these tissues, as discussed in Chapter 4. Tendons and ligaments are multicomponent cablelike elements that cyclically transmit force while limiting their deformation. Tendons and ligaments are loaded under physiological conditions along or close to the tissue axis; therefore, uniaxial mechanical testing in vitro gives a reasonable approximation of the mechanical properties in vivo. The stress–strain curve for tendon and ligament follows the generalized curve for collagenous tissues (see Figure 7.3). It is, however, very difficult to predict the level of prestress existing in a tissue (all tissues are under some finite load even when they are not transmitting forces to cause locomotion), so the exact range of positions on the stress–strain curve over which the tissue operates is debatable.

The stress–strain curve of tendon (see Figure 7.3) is typically divided into three regions: (1) a low modulus toe region; (2) a linear region; and (3) a yield and failure region. The toe region is characterized by a very low modulus and has been attributed to the straightening of the crimp, a planar zigzag within collagen fibers. However, the toe region in tendon may also be from the geometric alignment of the fibers with the tensile axis, because other tissues contain a low modulus region without exhibiting crimped collagen fibers. As discussed in the next paragraph, molecular stretching and sliding as well as subsequent fibrillar slippage occur during the linear region. Once fibrillar slippage leads to breakage of cross-links between fibrils, fibril failure occurs.

On a molecular basis, during loading collagen molecules stretch and slide past each other as loads are transferred between fibrils. Because fibrils branch and are discontinuous, the transfer of loads between fibrils occurs by molecular extension. The exact mechanism by which mechanical energy is translated into molecular and fibrillar deformation is still unclear; however, up to a macroscopic deformation of about 2%, molecular stretching predominates (Sasaki and Odajima, 1996). Beyond 2%, increases in the D-period are a result of molecular slippage (Folkhard et al., 1987; Sasaki and Odajima, 1996). The exact magnitude of the strain at which molecular deformation of the triple helix becomes small compared to fibrillar slippage depends on the strain rate and tissue studied; however, it is becoming clear that lateral interactions between fibrils are an important aspect of collagen mechanical behavior.

In addition to collagen, another highly prevalent macromolecule in connective tissue is decorin, a small dermatan sulfate proteoglycan found attached to collagen fibrils. Decorin is regularly and specifically associated with the surface of fibrillar type I collagen in extracellular matrixes (Scott, 1984). It has been hypothesized that decorin limits collagen fibril diameters by inhibiting lateral fusion of fibrils and inhibits mineralization of fibrillar collagen (Scott and Orford, 1981). Decorin filaments appear to connect adjacent collagen fibrils, and it has been proposed that it plays a role in maintaining the mechanical integrity of aligned, fibrillar connective tissue (Vogel, 1993; Cribb and Scott, 1995). A recent report indicates that decorin incorporation increases the UTS of collagen fibers, and it is hypothesized that decorin facilitates fibrillar slippage during deformation and thereby improves the tensile properties of collagen (Pins et al., 1997).

Because connective tissue is viscoelastic, the stress at a fixed strain is time dependent. For this reason, the tabulation of a value for Young's modulus is not meaningful because the modulus and other material properties are dependent on load history and the method of testing. Therefore, the tangent to the stress–strain curve is normally substituted for Young's modulus in describing connective tissue, and it usually refers to a particular level of strain and strain rate.

Dependence of the modulus and UTS on the deformation rate of whole tissues is usually significant for connective tissue, particularly if the strain rate is changed by more than an order of magnitude. Poisson's ratio for tendons that

Table 7.2 Mechanical properties of soft tissues

Tissue	Maximum strength (MPa)	Maximum strain (%)	Elastic modulus (MPa)
Arterial wall	0.24–1.72	40–53	1.0
Tendons/ligaments	50–150	5–50	100–2,500
Skin	2.5–15	50–200	10–40
Hyaline cartilage	1–18	10–120	0.4–19

Source: Adapted from Silver et al., 1992.

do not extrude water during testing is equal to about 0.5. Values of UTS as high as 120 MPa and strains of about 10% have been reported for tendons; for ligaments, these values are 147 MPa and 70%, respectively (Table 7.2)

Although the static mechanical properties and the properties of collagenous tissues at different strain rates tell us that these tissues are viscoelastic, this information does not help us understand the full implications of tendon and ligament behavior. One of the ways that we study the viscoelastic behavior of connective tissue is by measuring the incremental stress–strain curve (see Figure 7.4). By determination of the ratio of the equilibrium stress divided by the total stress, we can determine at each strain what fraction of the force and stress is stored in the tendon (elastic component) and what part is lost as heat (viscous component). Tendons are characterized by an elastic fraction of about 0.75 that is independent of strain. If tendons are cycled many times, the collagen fibers that are not perfectly aligned will subsequently improve their alignment with the tendon axis after about 10 cycles, and the elastic fraction will increase from 0.75 to above 0.80. Therefore, the energy lost in cycling tendons through a series of tensile loads followed by relaxation steps decreases as the number of cycles increases.

Dura mater and pericardium, the collagenous layers that protect the brain and heart, are made up of multiple layers, each with aligned collagen networks. In the case of dura mater, the networks of collagen fibers are aligned parallel and perpendicular to the midline of the brain, whereas in pericardium there are three networks that can be superimposed by rotation through a 120° angle. When loaded along the collagen fiber direction, dura and pericardium have a similar response to loading when compared to tendon. Because fewer fibers are aligned with the load direction under uniaxial loading, the stiffness at any strain and UTS are lower for these tissues as compared to tendon. When loaded along the fiber direction, dura and pericardium have elastic fractions equal to 0.75 that are independent of strain (Figure 7.9). The similarity between the elastic fractions of tendon, pericardium, and dura indicate that aligned collagen networks efficiently transmit stress along the fiber axis.

Mechanical Properties of Orientable Collagen Networks

Some tissues, such as tendon, function to transmit loads along a single direction and have collagen fibers oriented in only one direction. Other tissues, such as skin, bear loads within a plane and need to have collagen fibers oriented in the plane. Because skin normally bears loads within the plane of the skin, the collagen fibers become oriented with the direction of loading and reorient during mechanical deformation. In this manner, skin is able to bear loads in any direction within its plane. For this reason, the collagen network in skin must be loose enough to be able to rearrange.

The stress–strain curve for skin is shown in Figure 7.10. During the first region (strains up to 0.3), the collagen network offers little resistance to deformation, which is similar to the toe region of tendon. Collagen fibers are initially wavy and become straight on initiation of loading. Once the collagen fibers are aligned (strains between 0.3 and 0.6), they begin to offer resistance to increased extension. Once the strain reaches 0.6, the fibers begin to yield and fail. At high strains, the tangent to the stress–strain curve gives a stiffness of about 22.4 MPa, and the UTS is about 14.00 MPa, based on the true cross-sectional area. The elastic fraction of skin increases from about 0.50 to 0.73 with increasing strain, as shown in Figure 7.9. A summary of the mechanical properties of skin are tabulated in Table 7.3.

The concept of true engineering stress becomes important when evaluating the mechanical properties of skin. Engineering stress is defined as the force divided by the initial cross-sectional area. In tissues that deform more than about 10%, the cross-sectional area decreases as the sample is stretched, so the actual cross-sectional area is smaller than the original one. If the material is incompressible, that is, the volume does not change during deformation, then the true stress can be obtained from the engineering stress by multiplying by the extension ratio, L/L_o, where L is the extended length and L_o is the original length.

The mechanical behavior of skin is time and strain-rate dependent, and for this reason, the elastic fraction is less than 1. The time dependence of the mechanical properties and the strain dependence of the elastic fraction imply an interesting structural transition that occurs in skin with strain. Although skin

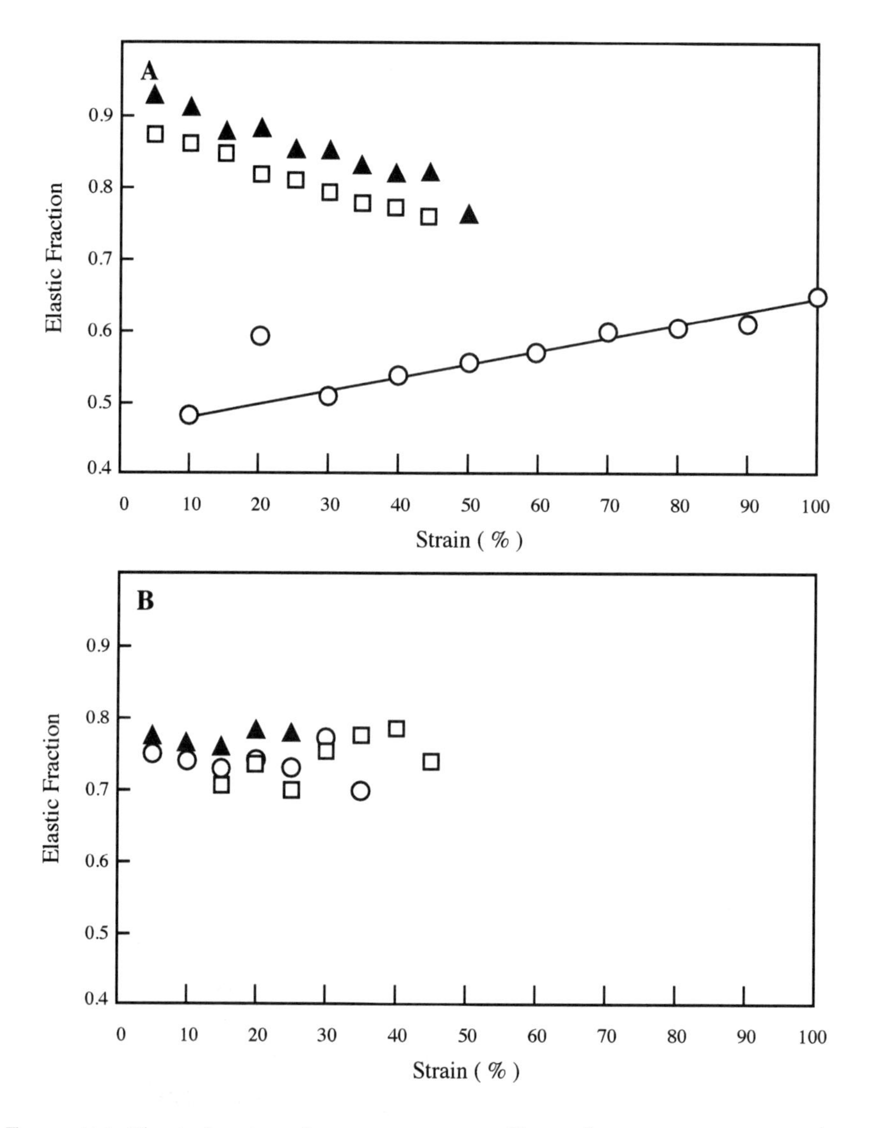

Figure 7.9 Elastic fraction of connective tissue. Elastic fractions versus strain determined from incremental stress–strain curves for (A) aorta in the circumferential (solid triangles) and axial (open squares) directions and skin (open circles) and (B) pericardium (open squares), psoas major tendon (solid triangles) and dura mater (open circles). *Source:* Adapted from Silver (1987).

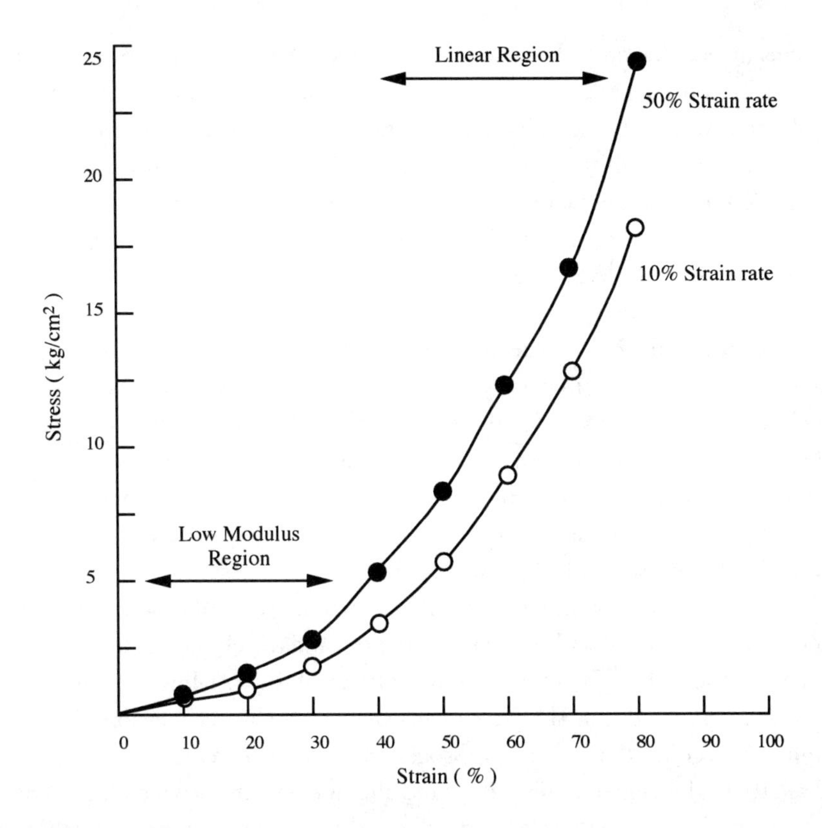

Figure 7.10 Stress–strain curve for skin. Stress–strain curves for wet back skin from rat at strain rates of 10 and 50% per minute. The low modulus region involves the alignment of collagen fibers along the stress direction that are directly stretched in the linear region. Disintegration of fibrils and failure occurs at the end of the linear region. *Source:* Adapted from Silver (1987).

Table 7.3 Properties of skin

Tissue	Failure strain	Failure stress (MPa)	Tangent modulus (MPa)
Abdominal skin (human)	0.8–1.00	4–14	1–24
Back skin (rat)	0.9–1.0	5–12.5	6–44

Source: Adapted from Silver (1987).

contains both collagen and elastic fiber networks, the mechanical behavior at all strains is a consequence of the collagen network, which is discussed in more detail later. At low strains, the elastic fraction is low and appears to involve the viscous realignment of the collagen fibers with the stress axis. At higher strains, the elastic fraction approaches 0.75, the value for tendon. Several types of connective tissue studied exhibit an elastic fraction approaching 0.75 at high strains. We interpret this observation as suggesting that the collagen network limits the high strain deformation in all types of all connective tissue.

Although the elastic fibers compose about 5% of the protein in skin, they appear to not affect the mechanical properties of this tissue. It is believed that the elastic fibers contribute to the recoil of the skin, which gives it the ability to be wrinkle-free when external loads are removed. As humans age, the elastic fiber network found in the skin is lost, and the skin begins to wrinkle. The mechanical role of the elastic fiber network in skin is very different than its role in vascular tissues, which is discussed later.

Successive mechanical cycling of skin in uniaxial tension shifts the stress–strain curve to the right within the linear region, and the elastic fraction increases and approaches 0.85. Therefore, one must be careful to establish a well-defined protocol to precondition skin specimens before mechanical testing. The preconditioning effect that occurs is associated with an increased elastic fraction and relates to the degree of unfolding and alignment of the wavy collagen fibers.

Complicating matters even further is the fact that the mechanical properties of skin depend on the location from which the specimen is obtained and the direction of loading. The contribution of the epithelial layer to the mechanical properties depends on its thickness. For most regions, the stratum corneum is thin and does not contribute significantly to the mechanical properties of skin; however, in places where it is thick, such as the balls of the feet and the palms of the hand, it needs to be considered.

The directional dependence of the mechanical properties of the skin was first observed by Langer in 1861. Langer pierced small circular holes in the skin of cadavers and observed that these holes became elliptical with time. By joining the long axes of the ellipses, he produced a series of lines that subsequently were named after him. It is well known to surgeons that incisions made along the direction of Langer's lines close and heal more rapidly. Histological sections made parallel and perpendicular to Langer's lines show that collagen fibers in skin are oriented along these lines.

Another example of an alignable collagen network is found in cardiovascular tissue, such as the aorta. This tissue contains three layers: the intima, media, and adventitia. The mechanical properties are dictated primarily by the media (Silver et al., 1992). The media in humans is about 2.5 μm thick and contains 50 to 65 concentric layers of elastic lamellar units, consisting of smooth muscle cells, elastic fibers, and collagen fibrils. Smooth muscle contains a layer of cells surrounded by a common basement membrane and a closely associated group of collagen fibrils. Wavy collagen fibers are woven between the layers of elastic fibers and smooth muscle cells, and a higher collagen content is found in the abdominal aorta. The elastic fiber content is higher in the thoracic aorta.

Selective enzymatic digestion of the aorta suggests that the collagen and elastic networks in this tissue are independent. Measurement of elastic fractions for the aorta along the vessel axis (longitudinal direction) and in the hoop direction (circumferential direction) indicates that it ranges from about 0.9 at low strains to about 0.75 at high strains. At high strains, this number is similar to that found for other types of connective tissue and represents the stretching of the collagen fiber network. At low strains, the elastic fraction is higher and represents the contribution of the elastic fiber network to the mechanical behavior of the tissue. This behavior is consistent with the function of the aorta as an auxiliary pump that expands and contracts to maintain blood pressure as the heart is filling. At high strains, the collagen network prevents overdilation and aneurysm formation. For this to occur, the collagen fiber network must be folded or wavy at low strains and only become stretched when the elastic network is fully expanded.

Stress–strain behavior of aorta is qualitatively similar to that of other types of connective tissue, such as skin, as shown in Figure 7.11. At low strains, the slope of the curve is small and reflects the unfolding of the elastic and collagen fiber networks. At high strains, the resistance to deformation increases because of loading of the collagen network. The exact stress–strain behavior will depend on the location from which tissue is extracted. The aortic stiffness in vivo varies from about 0.100 to about 1.66 MPa at 100 mm Hg pressure, depending on the testing direction and method, as shown in Table 7.4. Aortic wall mechanical properties include an ultimate tensile strength of less than 42 kPa, a modulus of about 1 MPa, and strain at failure of about 50% (Silver et al., 1992). The UTS of this tissue is lower than that observed for tendon and skin, which reflects the lower collagen content and degree of organization of cardiovascular tissue.

Mechanical Properties of Composite Collagen Networks

The mechanical properties of some types of connective tissue are complicated by the types of loading patterns and by the geometries of the collagen networks.

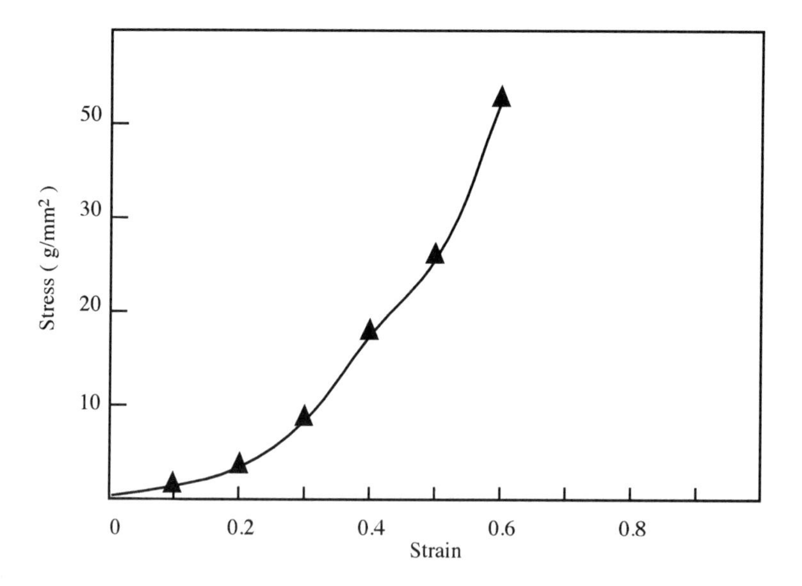

Figure 7.11 Stress–strain curve for aorta. Tensile stress–strain curve for human thoracic aorta in the circumferential direction obtained at a strain rate of 50% per minute. At strains less than 0.2, the elastic fibers dominate the behavior, whereas above 0.2, alignment of collagen fibers occurs. *Source:* Adapted from Silver (1987).

Table 7.4 Mechanical properties of cardiovascular tissue

Site	Modulus (MPa)
Aorta	
Cow	0.15–0.19
Dog	0.7–1.05
Carotid artery (dog)	0.41
Thoracic aorta	
Dog	0.1–1.6
Human	0.1–1.66

Source: Adapted from Silver (1987).

Cartilage is a tissue containing aligned collagen fibers on the surface and fibers perpendicular to the bone interface. The layers between contain collagen fibers that are in other directions. When a tensile or compressive load is applied to cartilage, the stress–strain curve is qualitatively similar to that for connective tissue (see Figure 7.3). Because the superficial zone contains aligned collagen fibers similar to those of tendon, it behaves mechanically in tension similar to tendon. When tested in tension along the direction of the collagen fibers, the superficial zone has UTS values from 15 to 25 MPa and strains at failure from 24 to 80%. These values fit into the approximate range of values expected for an aligned collagen network (Silver et al., 1992).

The mechanical properties of intact cartilage range from 1.0 to 18 MPa (UTS), 8 to 120% (maximum strain), and 0.4 to 16 MPa (modulus). In contrast to the superficial and deepest layers, the intermediate zone of cartilage has a system of bracing fibers that has properties more closely approaching those of skin. When a force is applied to cartilage, fluid is forced out of it into the surrounding joint space. As the negatively charged proteoglycans are compressed, they repel each other, preventing further deformation and causing the forces required for further compression to increase exponentially.

Mechanical Properties of Hard Tissue

Mechanical properties of unmineralized connective tissue depend to a first approximation on the orientation of the collagen fibers. Other factors that play a secondary role in determining mechanical properties include collagen fibril diameters, collagen content, strain rate, loading direction, degree of cross-linking, and the presence of other components (i.e., cells and macromolecules). In contrast, mechanical properties of mineralized tissues are somewhat more complicated, because factors such as time dependence, preservation, cutting and machining, density, mineral content, fat content, water content, and specimen orientation are important. Because bone is a composite structure containing lamellar and osteonic components, the properties depend to a high degree on the specimen orientation during testing.

In an attempt to simplify the discussion, we will ignore the fact that the modulus of bone is dependent on the testing direction and mineral content. A typical stress–strain behavior for cortical bone is illustrated in Figure 7.12. Mineralized connective tissue shows a much higher UTS and modulus, and the strain

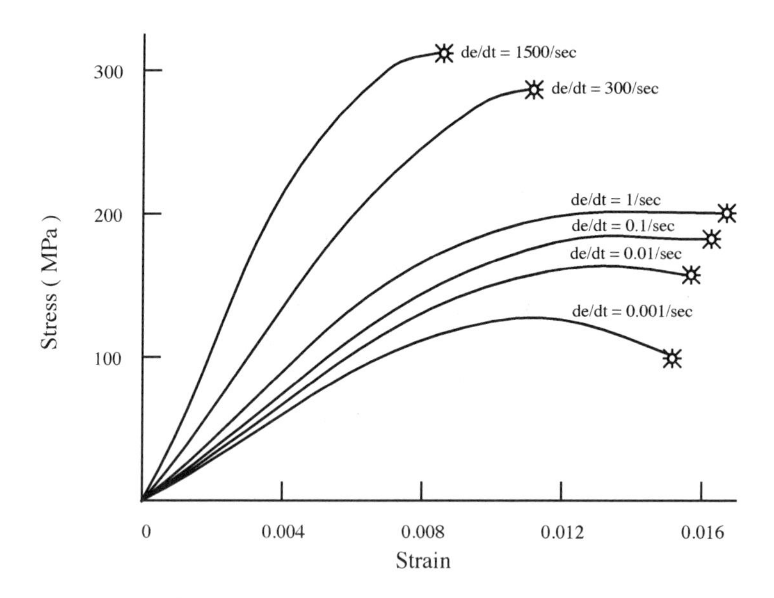

Figure 7.12 Strain-rate dependence of stress–strain curve for bone. The diagram illustrates dependence of the stress–strain curve on the strain rate. Note that at higher strain rates the curve moves to the left and becomes more elastic. *Source:* Adapted from Cowin (1989).

at failure is markedly decreased. In the same manner that increased cross-linking increases the UTS of unmineralized tissue, mineral deposition acts as cross-links and improves the tensile strength of connective tissue. However, the mineral drastically limits the deformation that bone can undergo before it fails in tension. Mineral deposition does not limit the viscoelasticity of bone because the strain-rate dependence is still evident, as seen in Figure 7.12. The UTS for cortical bone varies from 100 to 300 MPa, the modulus varies from several to more than 20 GPa, and the strain at failure is only 1 to 2% (Cowin, 1989).

Mechanical properties of cancellous bone are dependent on the bone density and porosity, and therefore the strength and stiffness of spongy bone is much lower than that of cortical bone. The axial and compressive strength is related to the square of the bone density, and moduli are reported to be in the range of 1 to several GPa.

Cellular Biomechanics

Although it well known that cells change their morphology when stimulated by mechanical stress, the mechanism of mechanochemical transduction is unclear.

Endothelial cells are normally hexagonal, but under shear stress, they change morphology to torpedo shaped, and the long axis is aligned with the flow direction (Dewey et al., 1981). Accompanying this change is a change in intracellular ion flux, gene regulation, transcription and translation, and cytoskeletal structure (Satcher and Dewey, 1996).

The endothelial cell cytoskeleton contains F-actin, intermediate filaments, and microtubules. During shear-induced structural changes to the endothelial cell, prominent changes occur in microfilaments containing cross-linked F-actin. Microfilaments are grouped together with myosin and other actin-binding proteins to form *stress fibers*, 20- to 50-nm diameter bundles. The elastic modulus of F-actin has been estimated to be between 10^4 and 10^5 dynes/cm^2 (Janmey et al., 1991). Stress fibers can theoretically increase the modulus by a factor of 2 to 10, depending on whether they are in series or parallel to the network in transmitting force, as discussed by Satcher and Dewey (1996). This modulus is many times smaller than the modulus of mineralized or unmineralized collagen fibers, which leads us to question how cells are arranged within tissues under tension and compression so that they are not destroyed during mechanical loading.

The cell cytoskeleton is linked via integrins to extracellular matrix collagen. Therefore, tensile and compressive forces in the extracellular matrix are transduced into cytoskeletal changes via shear of the cell membrane and stress fibers associated with the membrane. Because the modulus of the cell cytoskeleton is small in comparison to the stiffness of the extracellular matrix, small changes in deformation of the extracellular matrix lead to large changes in shear stress applied to the cell. In this manner, changes in tension and compression of the extracellular matrix can lead to changes in cell shape and biosynthetic activity.

Summary

Nature has provided strong ductile materials for forming the musculoskeletal system and other systems of vertebrates. This is accomplished with a fixed number of molecules arranged in a variety of ways. The variation in the properties of materials, from liquids or gel-like materials within the cell membrane to hard mineralized bone, is impressive.

The mechanical properties of tissues are intimately related to the molecular structure and packing of biological macromolecules. The stiffness, strain at failure, and UTS are all related to the average axial rise per residue along the largest dimension of the molecule. The rise per residue is smallest for the random chain structure, and as a result, the strain to failure is highest and the modulus and UTS are lowest. In the α helix, the rise per residue is 0.15 nm, which is smaller than that for the collagen triple helix (0.29 nm) and for the β-extended conformation (0.36 nm). Surprisingly, the modulus and UTS go up as the value of the axial rise per residue increases, whereas the strain at failure decreases. This is an oversimplification because the extent of macromolecular alignment, the number and type of macromolecular–macromolecular interactions, and the location and number of cross-links are all important.

Suggested Reading

Bush K., McGarvey K.A., Gosline J.M., and Aaron B.B., Solute Effects on the Mechanical Properties of Arterial Elastin, Connect. Tissue Res. 9, 157, 1982.

Cowin S.C., Mechanics of Materials, in Bone Mechanics, edited by S.C. Cowins, CRC Press, Boca Raton, FL, chapter 2, pp. 15–42, 1989.

Cribb A.M. and Scott J.E., Tendon Response to Tensile Stress. An Ultrastructural Investigation of Collagen: Proteoglycan Interactions in Stressed Tendon, J. Anat. 187, 423, 1995.

Dewey C.F., Busslorai M.M., Gimbrone M., and Davies P.F., The Dynamic Response of Vascular Endothelial Cells to Fluid Shear Stress, J. Biomech. Eng. 103, 177, 1981.

Doillon C.J., Dunn M.G., Bender E., and Silver F.H., Collagen Fiber Formation In Vivo: Development of Wound Strength and Toughness, Collagen Rel. Res. 5, 481, 1985.

Folkhard W., Mosler E., Geercken W., Knorzer E., Nemetschek-Gansler H., Nemetschek Th., and Koch M.H.J., Quantitative Analysis of the Molecular Sliding Mechanism in Native Tendon Collagen: Time-Resolved Dynamic Studies Using Synchrotron Radiation, Int. J. Biol. Macromol. 9, 169, 1987.

Janmey P.A., Hvidts S., Kas J., Levche D., Megg A., Sackman E., Schliwa M., and Stossel T.P., The Mechanical Properties of Actin Gels, J. Biol. Chem. 269, 32503, 1991.

Kato Y.P. and Silver F.H., Formation of Continuous Collagen Fibers: Evaluation of Biocompatibility and Mechanical Properties, Biomaterials 11, 169, 1990.

Mukherjee D.P., Kagan H.M., Jordan R.E., and Franzblau C., Effect of Hydrophobic Elastin Ligands on the Stress–Strain Properties of Elastin Fibers, Connect. Tissue Res. 4, 177, 1976.

Nestler F.H., Hvidt S., Ferry J.D., and Veis A., Flexibility of Collagen Determined from Dilute Solution Viscoelastic Measurements, Biopolymers 22, 1747, 1983.

Parry D.A.D., The Molecular and Fibrillar Structure of Collagen and Its Relationship to Mechanical Properties of Connective Tissue, Biophys. Chem. 29, 195, 1988.

Pins G.D., Christiansen D.L., Patel R., and Silver F.H., Self-Assembly of Collagen Fibers: Influence of Fibrillar Alignment and Decorin on Mechanical Properties, Biophys. J. 73, 2164, 1997.

Sasaki N. and Odajima S., Elongation Mechanism of Collagen Fibrils and Force-Strain Relationships of Tendon at Each Level of Structural Hierarchy, J. Biomech. 9, 1131, 1996.

Satcher R.L. Jr. and Dewey C.F. Jr., Theoretical Estimates of Mechanical Properties of the Endothelial Cell Cytoskeleton, Biophys. J. 71, 109, 1996.

Sato M., Theret D.P., Wheeler L.T., Ohshima N., and Nerem R.M., Application of the Micropipette Technique to the Measurement of Cultured Porcine Aortic Endothelial Cell Viscoelastic Properties, J. Biomech. Eng. 112, 263, 1990.

Scott J.E. and Orford C.R., Dermatan Sulfate-Rich Proteoglycan Associates with Rat Tail-Tendon Collagen at the D Band in the Gap Region, Biochem. J. 197, 213, 1981.

Scott J.E., The Periphery of the Developing Collagen Fibril, Biochem. J. 218, 229, 1984.

Silver F.H., Mechanical Properties of Connective Tissues, in Biological Materials: Structure, Mechanical Properties, and Modeling of Soft Tissues, NYU Press, New York, chapter 6, 1987.

Silver F.H., Kato Y.P., Ohno M., and Wasserman A.J., Analysis of Mammalian Connective Tissue: Relationship Between Hierarchical Structures and Mechanical Properties, J. Long-Term Effects Med. Implants 2, 165, 1992.

Torp S., Baer E., and Friedman B., Effects of Age and of Mechanical Deformation on the Ultrastructure of Tendon, in Structure of Fibrous Biopolymers, Colston Papers No. 26, edited by E.D.T. Atkins and A. Keller, Butterworths, London, pp. 223–250, 1975.

Vogel K.G., Glycosaminoglycans and Proteoglycans, in Extracellular Matrix Assembly and Structure, edited by P.D. Yurchenco, D.E. Birk, and R.P. Mecham, Academic Press, San Diego, pp. 243–280, 1993.

8

Pathobiology and Response to Tissue Injury

Introduction

The biomaterials scientist must understand the normal cell functions in tissues that are injured or diseased. To do this, one must understand gross and microscopic anatomical structures, which were introduced in Chapter 4, and cellular and tissue changes that are observed in injured or diseased tissue. Once the anatomical and microscopic structure and biochemistry of normal tissues is well understood, it is then possible to understand how injury or disease affects cells and tissues.

Before we look into the changes that are associated with injury and disease, it is necessary to briefly review the normal components of the cell as observed under light and electron microscopy. This is necessary to be able to identify the changes in cell shape and number that are associated with cellular adaptation and injury. These changes include induction, secretion, adaptation, and cell injury and death. Changes occur when foreign stimuli come in contact with normal cells and tissues and dictate whether the outcome of contact is stimulation of changes in cell size and shape, attraction of inflammatory cells, or cellular injury and death.

Analysis of the pathobiological responses to implants has been hindered by a lack of understanding of how implants affect cellular metabolism and systemic responses. Although most implant materials, including poly(dimethylsiloxane) (Silicone), poly(ethylene terephthalate)(Dacron), and poly(tetrafluoroethylene) (Teflon), have been used for decades as implant materials with no widespread adverse effects, specific individuals may have very negative responses to any of these materials. In fact, one of every several thousand individuals who have surgery will have a hypersensitive response to any polymeric material used as an implant. For example, latex (natural rubber) surgical gloves were used almost exclusively until latex allergies began to appear in the general public. Now there are many patients who are allergic to latex products, and their skin and mucous membranes swell uncontrollably when exposed to this material. The exact mechanism of latex-induced hypersensitivity is unknown, but all foreign materials induce cellular responses in some hosts. The nature and degree of the response must be considered when assessing the relative safety of the material.

Cellular Components

Cells are composed of proteins, nucleic acids, lipids, ions, and water (see Chapter 1). Cells are organized into functional compartments that include the cell and nuclear membranes, cytoplasm, nucleus, and organelles (Figure 8.1). The organelles that are commonly studied include endoplasmic reticulum (ER), mitochondria, ribosomes, Golgi apparatus, and lysosomes (Table 8.1). Cellular interactions with most implants are studied by evaluation of the cell structure before and after contact with a surface. Changes in cellular components are the first indication that a material will induce cell injury and perhaps death.

The normal cell has a lipid bilayer in the form of a cell membrane that separates the fluid and tissue outside a cell from the organelles inside of the cell. The cell or plasma membrane is a lipid bilayer that contains proteins on the inner and outer surfaces (Figure 8.2). The function of the cell membrane is to (1) maintain ionic and chemical concentration gradients; (2) carry specific surface

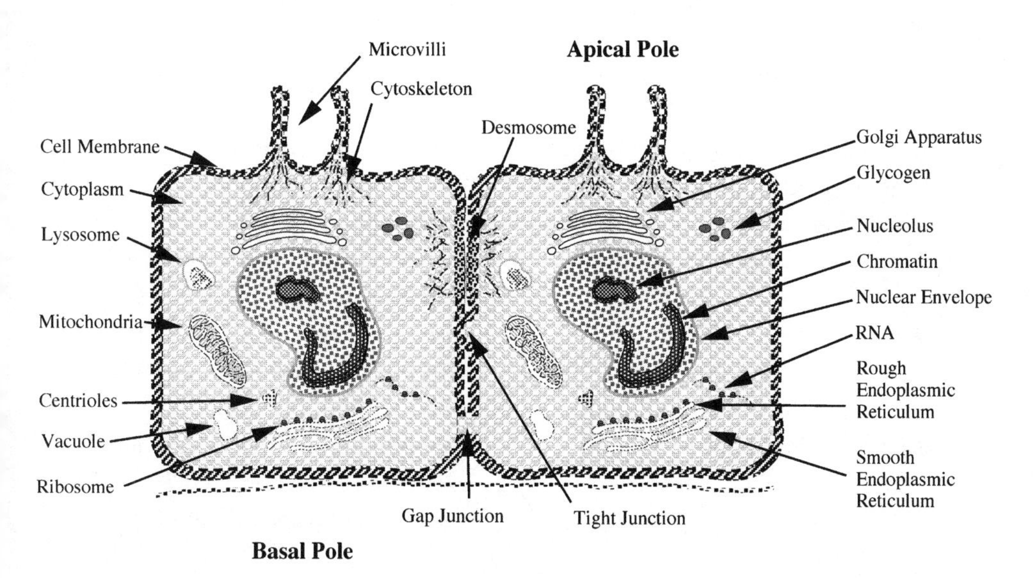

Figure 8.1 Diagram of cell components. Generalized diagram showing components of cell including ER, mitochondria, ribosomes, Golgi apparatus, lysosomes, and other components.

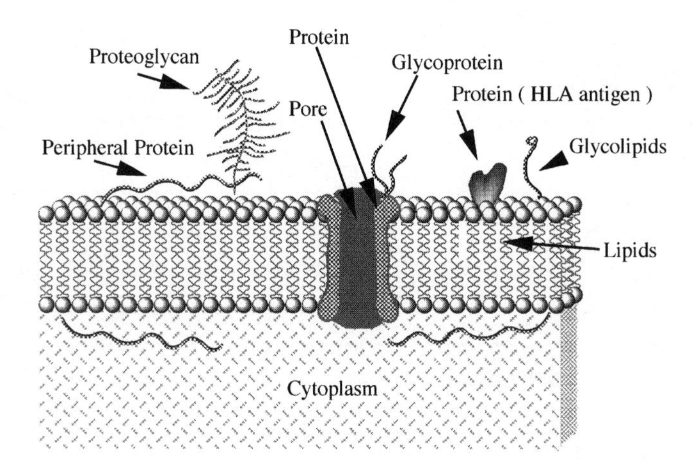

Figure 8.2 Diagram of cell membrane. The plasma membrane is a lipid bilayer containing peripheral proteins, including HLA antigens, proteoglycans, and integral proteins, such as pore-forming proteins that transverse the cell membrane.

Table 8.1 Components of the normal cell and their functions

Cell component	Composition	Function
Cell membrane	Lipid bilayer, containing surface proteins (peripheral proteins), proteins totally embedded in the membrane (intregal proteins), and glycoproteins partially embedded in the membrane	Maintains ionic and chemical concentration gradients, cell-specific markers, intercellular communication, regulates cell growth and proliferation
Cytosol or cytoplasm	Water, ions, soluble proteins	Contains enzymes and structures for generation of energy (ATP) in the absence of oxygen (TCA cycle), activation of amino acids, carrying out specialized cell functions
Endoplasmic reticulum	Membrane-enclosed channels	Involved in transport of proteins for extracellular secretion and modification or detoxification of chemicals
Mitochondria	Contains membrane-lined channels to which enzymes are attached that generate ATP from glucose	Involved in TCA cycle, respiratory chain, and oxidative phosphorylation
Ribosomes	Small and large subunits composed of ribosomal RNA; strands of messenger RNA form a complex with large subunits	Involved in synthesis of enzymes and structural components and proteins for extracellular release
Golgi apparatus	Membrane-lined tubular system that forms stacks	Packaging of proteins in vesicles for extracellular release
Lysosomes	Membrane-lined vescicles containing hydolytic enzymes	Involved in breakdown of intracellular and extracellular material
Nucleus	Membrane-limited area of cell containing nucleolus and chromatin	Site of synthesis of RNA and chromatin, involved in cell division

Note: ATP = adenosine triphosphate; TCA = tricarboxylic acid cycle.

markers and receptors, such as human leukocyte antigens, growth factors, and hormone receptors; (3) participate in intracellular communication; and (4) regulate cell growth and differentiation.

Figure 8.2 is a diagram of the fluid mosaic model of the cell membrane, which incorporates a discontinuous lipid bilayer composed of phospholipids and cholesterol. Proteins and glycoproteins are embedded or attached to the membrane in this model. Peripheral proteins, such as the glycoprotein fibronectin, are found attached to the cell membrane, and integral proteins, including the proteoglycan syndecan, are found at least partly embedded in the cell membrane. Peripheral proteins are involved in maintaining cellular shape and transducing mechanical forces across the cell membrane, and integral proteins are involved in cell recognition and transport of molecules across the cell membrane. Transplantation of tissues from one unrelated host to another involves matching the cell surface antigens to avoid stimulating immune responses that lead to rejection.

The cell membrane is involved in interaction among cells, allowing for the flow of ions and electrical impulses between neighboring cells. Cell-to-cell attachment via specific types of specialized junctions allows for the exchange of proteins and ions. One example is that of epithelial cells that exhibit gap junctions, tight junctions, and desmosomes. Typically, the integrity of the cell membrane is evaluated by identifying what percentage of cells take up a vital dye called trypan blue. Cells that take up trypan blue have defective cell membranes; therefore, materials that cause cells to have faulty membranes are likely to cause cell injury and death. Cell cytotoxicity is evaluated by contacting a surface of a material or the extract of a material or implant with cells in culture. If more than 5% of the cell population stains with dye after contact with the material or an extract, then it is considered cytotoxic.

Cells normally exhibit polarity; they have top, bottom, left, and right sides. Most cells, such as the epithelia, present an apical pole that is characterized by many microvilli, which are ruffles in the cell membrane, and a basal pole that is in contact with a basement membrane. This polarity is important because normal cell function can only be expressed if the cell has the correct orientation.

The gel-like material within the cell membrane is the cytoplasm or cytosol. It is composed of ions, water, soluble proteins, and enzymes that are involved in generation of energy in the form of ATP by the tricarboxylic acid (TCA) cycle (Krebs cycle) in the absence of oxygen (Figure 8.3). It is also involved in activation of amino acids for protein synthesis (see Table 8.1 and Figure 8.3).

In addition to organelles, the cell cytoplasm contains actin filaments that make up the cellular cytoskeleton that control shape. Myosin and α-actinin are

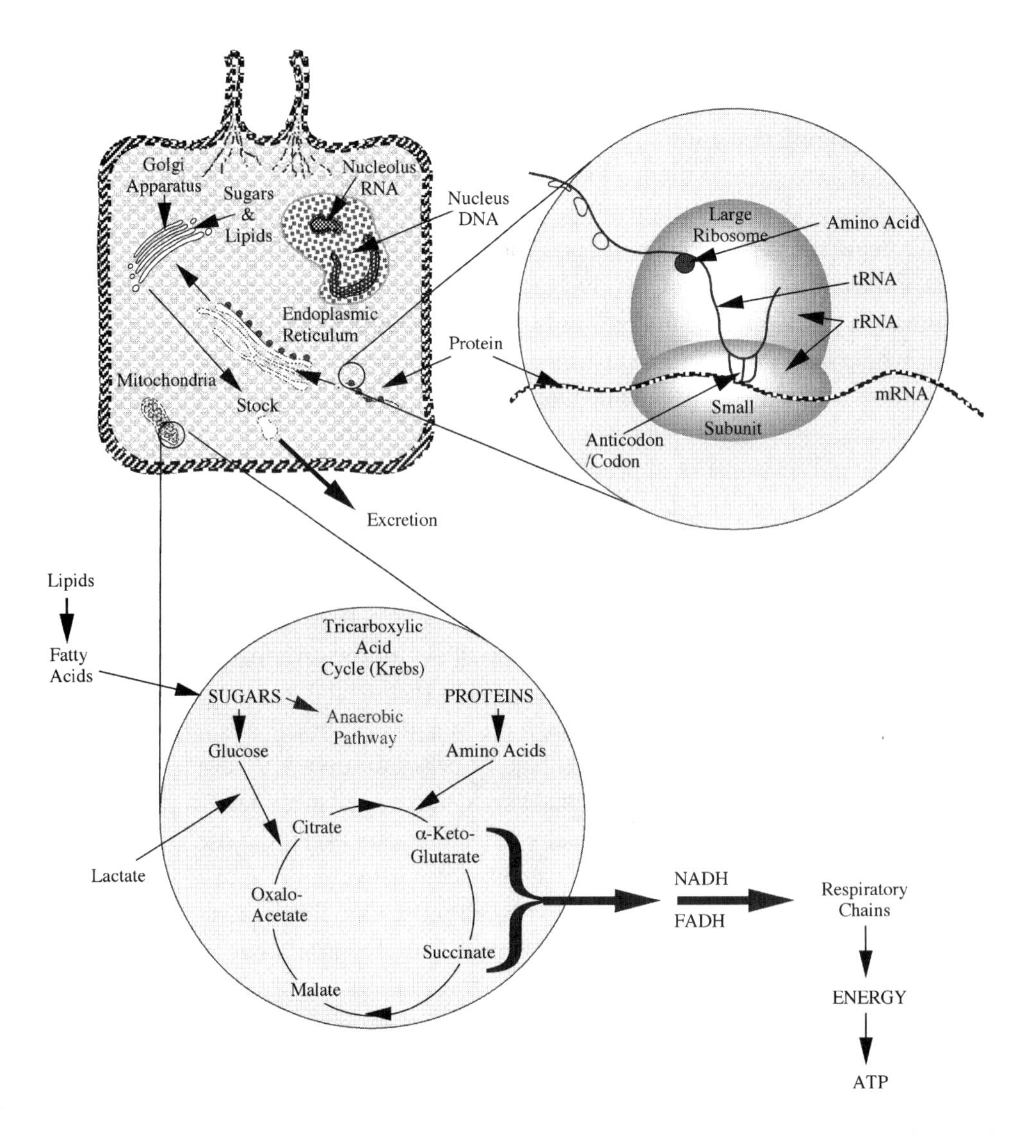

Figure 8.3 Krebs cycle and protein synthesis. This diagram illustrates the generation of ATP through the Krebs cycle and the protein synthesis that occurs within the cell cytoplasm.

also found in the cytoplasm and are believed to be involved in cell contraction. Other filaments, including intermediate filaments, tubulin, calmodulin, and spectrin, form networks within the cytoplasm that modify cell and organelle mobility and shape.

The ER is a branching system of membrane-limited channels that are found within the cytoplasm (see Table 8.1 and Figure 8.3). These channels are

40 to 70 nm wide and are enclosed by a membrane similar to the plasma membrane. In rough ER, the channels are covered with spheres 20 nm in diameter. The spheres are ribosomes and are the site of the synthesis of proteins. The ER serves as a conduit for the synthesized proteins for export extracellularly. Endoplasmic reticulum that has no ribosomes is called smooth ER (SER). Proteins, such as collagen, are synthesized on the ribosomes, pass into the center of these channels, and are transported into the Golgi apparatus, where they are packaged into vesicles for release from the cell. Membranes that line the SER are involved in modification and detoxification of low molecular weight materials that are released by implants. Implants or biomaterials that stimulate proliferation of the SER lead to an increase in cell size. Increased cell size in the presence of an implant is an indication that the implant may cause irreversible cell changes. Active proliferation of the ER alone is not a sufficient indicator of whether implant-induced cell changes occur.

Mitochondria are cigar-shaped organelles (see Figure 8.1) that are separated from material found in the cell cytoplasm by a double membrane. Both inner and outer mitochondrial membranes are similar to plasma membranes. The inner membrane is connected to a series of folded channels (cristae) on which the enzymatic reactions of the tricarboxylic acid cycle (TCA), respiratory chain, and oxidative phosphorylation occur (Figure 8.3). These reactions are required to generate ATP from glucose in the presence of oxygen. Mitochondria are prevalent in cells that are actively secreting proteins, such as collagen, to form a fibrous capsule and in cells that are synthesizing enzymes required for modification or detoxification of chemicals.

Proteins are synthesized on ribosomes that are made of small and large subunits containing nucleic acids. The small subunit (see Figure 8.3) is shaped like a donut that has been cut in half. The large subunit is spherical and contains a notched groove on the top surface. A strand of messenger ribonucleic acid (mRNA) is found between the notched groove of the large subunit and the hole in the center of the small subunit; mRNA in association with the large and small ribosomal subunits acts as a template for protein synthesis. Synthesis of enzymes and structural proteins used within the cell occurs on free ribosomes found within the cytoplasm. Proteins that are synthesized for release from the cell are synthesized on ribosomes attached to the ER. Release of these proteins extracellularly involves the Golgi apparatus.

The Golgi apparatus (Figure 8.1) is a membrane-bound system of tubes that is connected to the ER. Individual tubules make up a winding system of stacked cisternae that form a dictyosome. Proteins synthesized on ribosomes

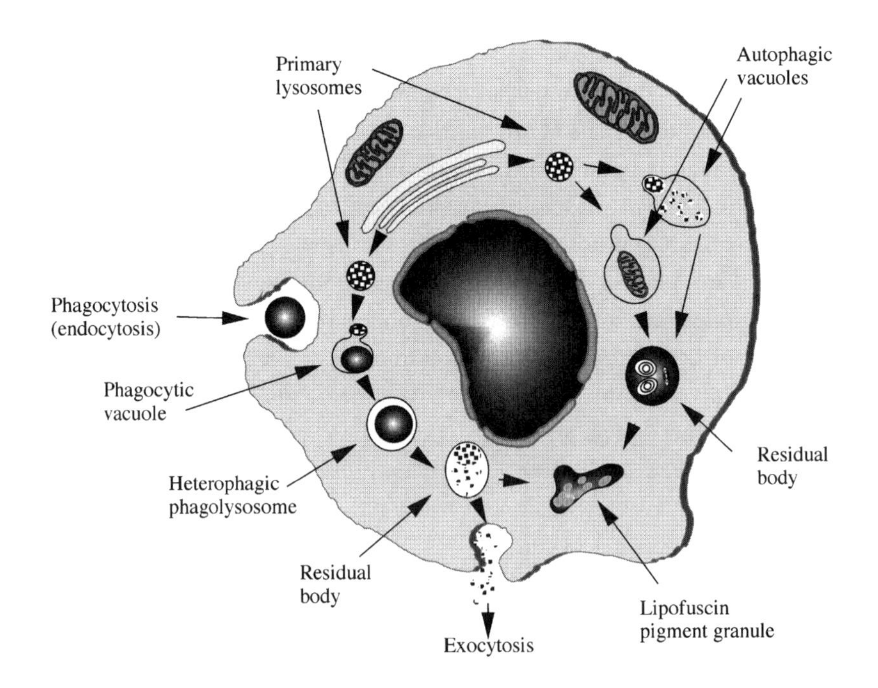

Figure 8.4 Lysosomes and phagocytosis. Dead and foreign material are removed via phagocytosis and degradation of macromolecular material using lysosomal enzymes.

attached to the inner membrane of the ER are transported into the Golgi apparatus where they are packaged into vesicles. Vesicles 40 to 80 nm in diameter at the outside of the cisternae are used to release these proteins extracellularly during wound healing. Extracellular matrix cells synthesize large quantities of collagen and other proteins, and so the Golgi apparatus is prominent. Coincidental with synthesis of new proteins is the removal of old proteins via phagocytosis.

During phagocytosis, foreign or old autologous proteins are ingested by the cell and then removed by fusion with lysosomes, vesicles containing hydrolytic enzymes. Lysosomes (Figure 8.4) are membrane-bound vesicles that are used to break down proteins, nucleic acids, sugar polymers, and other extracellular or intracellular materials. Lysosomes are 0.1 to 0.8 μm in diameter and contain a variety of hydrolytic enzymes. Cells rich in lysosomes include polymorphonuclear leukocytes (neutrophils), monocytes, and macrophages. These white cells are involved in the inflammatory process to remove dead cellular debris and damaged extracellular matrix. Implantation of medical devices leads to trauma and cell death that stimulates migration of inflammatory cells and phagocytosis.

Inflammatory cells clean up the debris by attempting to eat the material, including the implant. The implant can break down into particles and be taken into lysosomes within the cell; this is observed with polyethylene wear particles from hip joints. The implant can also remain intact and be dormant or continue to stimulate inflammation. If the implant causes death to inflammatory cells, extracellular release of lysosomal enzymes causes tissue destruction and further trauma.

The nucleus of the cell (see Figure 8.1) is composed of a porous nuclear membrane, a nucleolus, and soluble materials. The nucleolus contains RNA and chromatin, genetic material that codes for the proteins synthesized on the ribosomes in the cell cytoplasm. The nuclear membrane is continuous with the outer membrane of the ER. Messenger RNA synthesized in the nucleus is transported across the nuclear membrane and is involved in protein synthesis. It fits into the groove between the large and small rRNA subunits (see Figure 8.3) where it acts as a template for transfer RNA (tRNA) to add the appropriate amino acids to the growing protein chain. Messenger RNA is synthesized off of a DNA template that is in the form of a double helix (Figure 8.5). In turn, DNA is associated with proteins called histones, and the complex makes up the chromosomes or genetic material. There are 23 sets of chromosomes containing all the genetic material (DNA) required to synthesize all the proteins found within the human body. Chromosomes within the cell are characterized by the size and shape of their long and short legs. They all look like the letter X with varying lengths and geometries.

Ribosomal RNA is synthesized and packaged within the nucleolus of the cell. The nucleic acids that are synthesized within the nucleus pass through pores in the nuclear membrane into the cytoplasm. The nucleolus appears as a dense spot in the nucleus under the light microscope after staining with special dyes. The staining characteristics of the chromatin in the nucleus are somewhat different. Chromatin is acidic and stains darkly with basic dyes. Loosely coiled chromatin (euchromatin) stains lightly with basic dyes, and tightly coiled DNA (heterochromatin) stains darkly with basic dyes. Stimulation of cell replication results in increased amounts of heterochromatin, whereas resting cells exhibit increased amounts of euchromatin. The nuclear staining characteristics can be useful in determining the activity of cells in the capsule around an implant. Division of cells within the capsule surrounding an implant indicates that the capsule thickness is likely to increase and that the implant is causing a reaction.

During cell division (M phase, see Figure 8.5), heterochromatin is separated and arranged into distinct clumps of genetic material, the chromosomes.

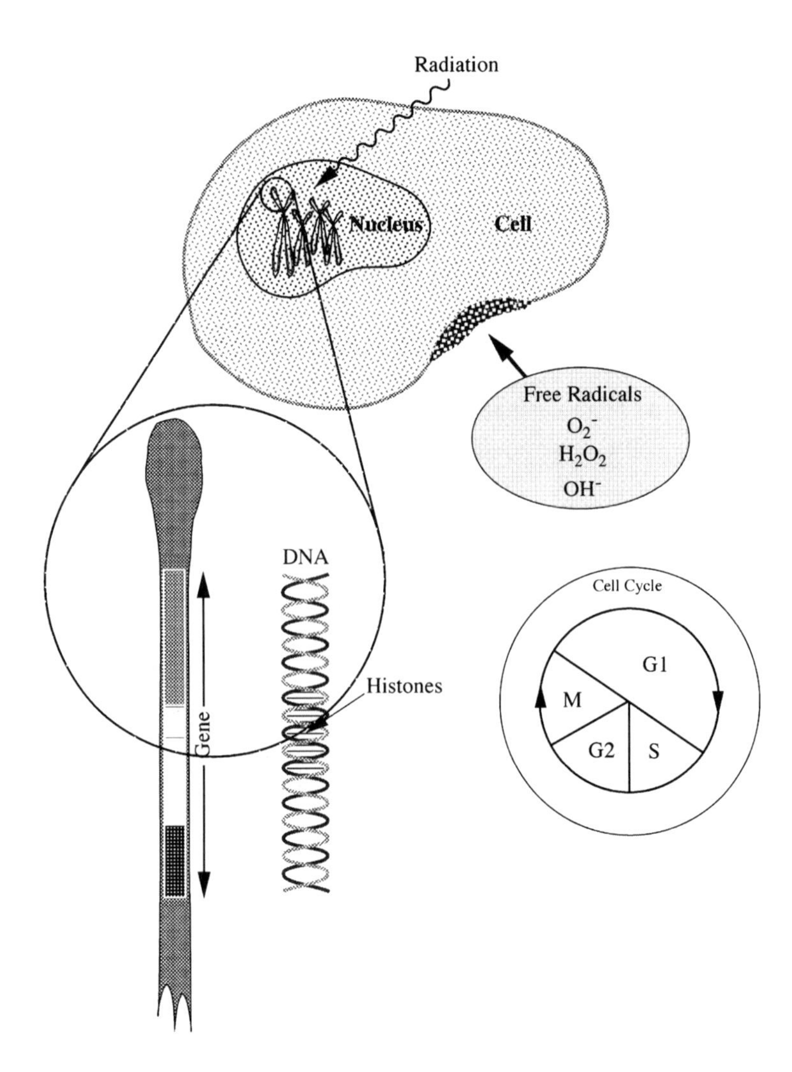

Figure 8.5 DNA, histones, and cell division. Shown are (1) the structure of DNA, histones and genetic material; (2) changes in genetic material via radiation or free radical induced damage; and (3) the cell cycle, synthesis of DNA and mitosis.

The karyotype is a fingerprint of the genetic material and is obtained by preventing termination of cell division. This allows for observation of the size and shape of the genetic material found in any cell, which is useful to determine whether contact with an implant causes changes in the genetic material, a mutation. Cellular replication can be associated with an implant in positive and negative senses. In a positive sense, wound healing is associated with cell replication and synthesis of nucleic acids and proteins. In a negative sense, cell replication in wounds after remodeling has occurred is reflective of an abnormal scarring process. Therefore, the analysis of cell replication around an implant must be done in conjunction with other information concerning the cell to be correctly interpreted.

Cell Attachment, Proliferation, and Differentiation

Cell attachment, migration, proliferation, and differentiation are phenomena associated with development and wound healing. Although it is possible to observe under what instances these processes occur without pathological complications, how these processes work is only superficially understood. Those who design medical devices can only guess at ways to improve the ability of cells and tissue to attach, migrate, proliferate, and differentiate normally around an implant or device. However, recent advances in biological mediators of these processes have provided a better understanding of how devices may be designed in the future to minimize interference with these processes. Each of these processes is examined independently in this chapter.

Cell Adhesion

Adhesion of cells to a substrate is thought to involve several steps: attachment of cells to a surface, reorganization of cytoskeletal components, formation of adhesion plaques, and deposition of an organized extracellular matrix. Adhesive interactions between cells or cells and the extracellular matrix are thought to play an essential role in several biological processes, including cellular recognition, specification and signaling, provision of positional cues during vertebrate development, cell proliferation and differentiation during wound healing, and leukocyte adherence and emigration. See Polverini (1996) and Carlos and Harlan (1994) for reviews of the role of cellular adhesion molecules involved in adhesion of leukocytes to endothelium. The adhesion of leukocytes to endothelium associated with activation of inflammation has been studied extensively. The molecules responsible for adhesion in this particular case include cellular adhesion molecules (CAMs) and consist of four families, including the integrins, the immunoglobulin superfamily, cadherins, and selectins.

Selectins consist of three members, L-, P-, and E-selectin, and are exclusively involved in the binding of leukocytes and some metastatic cells to endothelium (Figure 8.6). E-Selectin is maximally expressed on the surface of endothelial cells within three to four hours of activation by interleukin-1 (IL-1) or tumor necrosis factor-α (TNF-α). It participates in the early cascade of molecular events that lead to the adhesion of neutrophils and monocytes to the blood vessel wall during inflammation. The selectins recognize carbohydrate counterstructures expressed on glycoproteins and glycolipids.

Endothelial immunoglobulin-(Ig)-like protein is a superfamily of cell-surface receptors that are involved in recognition of antigens (C-type) or complement-binding or cell-adhesion (C2-type). Members of the C2-type (Ig) gene superfamily include CD2, CD-58 (LFA3) and CD56 (NCAM). Five members of this family expressed by endothelial cells and by leukocytes include intercellular adhesion molecule-1 (ICAM-1; CD54), ICAM-2 (CD102), vascular cell adhesion molecule (VCAM-1; CD106), platelet-endothelial cell adhesion molecule-1 (PECAM-1; CD31) and the mucosal adressin (MaAdCAM-1).

Integrins on leukocytes are transmembrane cell surface proteins that bind to cytoskeletal proteins and communicate extracellular signals. Each integrin consists of two nonlinked chains, α and β, with 8 known β subunits and 15 known α subunits that have been molecularly cloned. Integrins are arranged in subfamilies according to the β subunit. Within the integrin family, only five members have been shown to be involved in leukocyte adhesion to endothelium; the β_2 leukocyte integrins (CD11a/CD18, CD11b/CD18 and CD11c/CD18), the β_1 integrin VLA-4 ($\alpha_4\beta_1$, CD49d/CD29) and $\alpha_4\beta_7$.

The last group, the cadherins, includes V-cadherin, which is hypothesized to be involved in endothelial cell–endothelial cell interactions; however, not as much is known about this family of adhesion molecules.

Leukocyte Adhesion

Adhesion of leukocytes to vessel walls has been a subject of extensive research interest because of the leukocyte involvement in inflammation and subsequent wound healing. Initial contact of the leukocyte with the vessel wall is thought to occur as a random event brought about by local blood flow. After initial contact, some of the leukocytes are observed to roll along the vessel wall adjacent to the site of injury (Figure 8.7). Leukocyte rolling is prevalent in venules compared to arterioles, where the flow rate may be too high for enough friction to occur between the white cells and the endothelium. Adhesive forces of about

A: Selectin - Carbohydrate

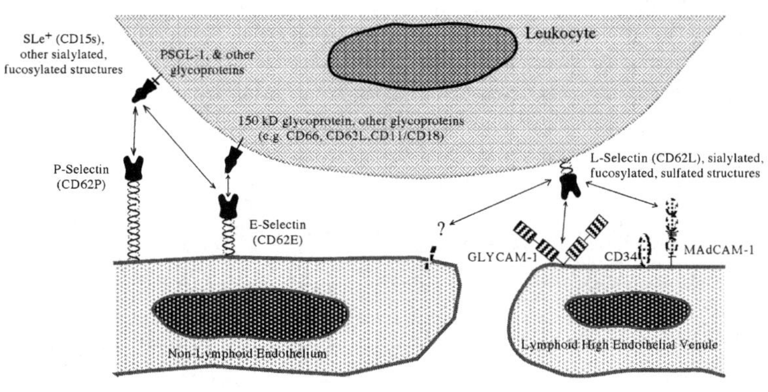

B: Integrin and Ig-Like

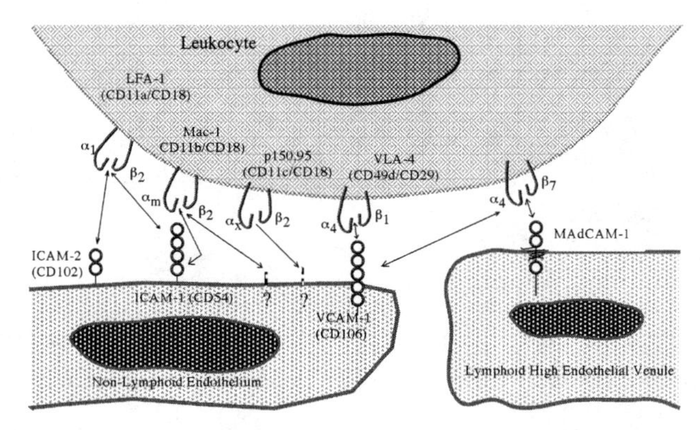

C: Other Adhesion Pathways

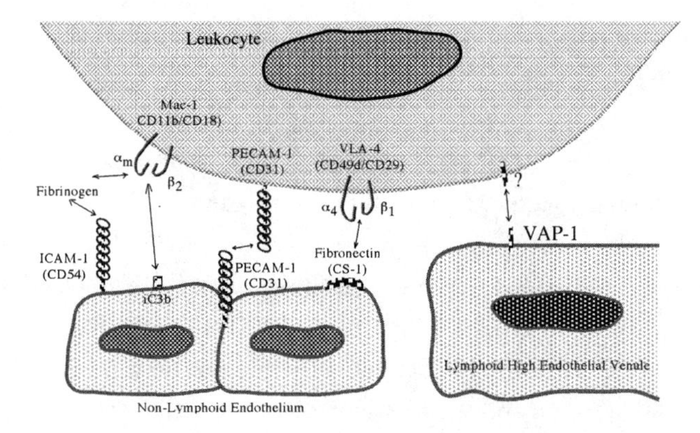

Figure 8.6 Cell adhesion molecules. Cell adhesion molecules including selectins, which bind to cell surface glycoproteins; integrins, which bind to ICAM-1, ICAM-2, and VCAM-1; and fibronectin and fibrinogen, which bind to surface adhesion molecules in nonlymphoid endothelium.

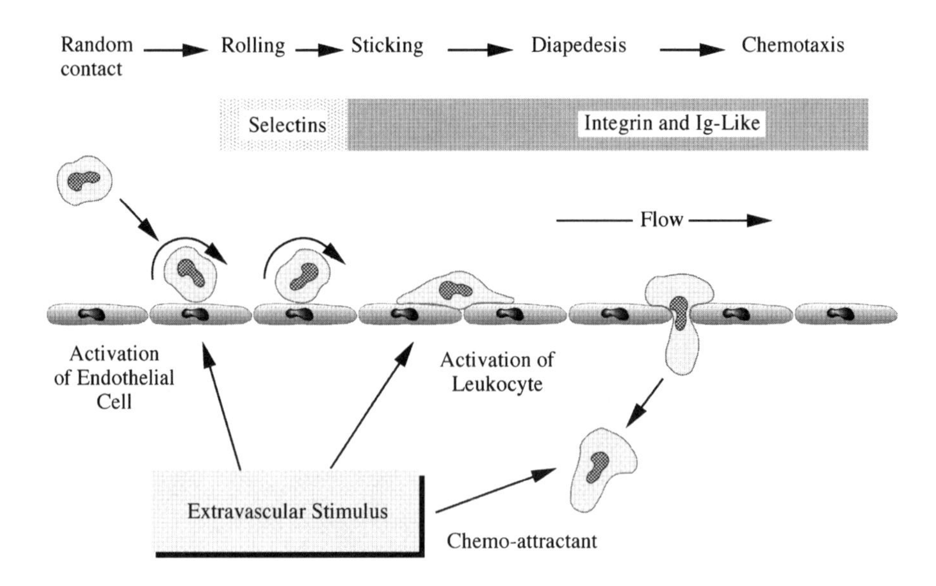

Figure 8.7 Leukocyte adhesion to vessel walls. Shown are the steps involved in rolling, sticking, diapedesis, and chemotaxis of leukocytes along and through endothelial cells.

20 dyne/cm^2 have been estimated to represent the shear forces between white cells and the wall. Selectins are believed to be involved in leukocyte cell surface interactions during rolling. It is most likely that both L-selectin and P-selectin are necessary for efficient rolling; each selectin interacts with a distinct counterstructure on the opposing cell.

In the second phase of leukocyte recruitment, activation of leukocyte integrins and expression of endothelial cell adhesion proteins occurs. Activation by molecules such as cytokines causes a conformational change in integrins, resulting in a greater ligand affinity for ligands and other subsequent events. The net result of activation in the presence of proadhesive factors is firm adhesion between integrin structures on leukocytes and endothelial Ig-like ligands.

Following tight adhesion to the vessel wall, transendothelial cell migration occurs after the cells crawl over the lumenal surface and encounter an intercellular junction and squeeze between endothelial cells or migrate across the cell membrane and through the cytoplasm. Endothelial Ig-like proteins on endothelial cells are involved in this process, as are integrins on leukocyte cell membranes. Once through the cellular layer, subendothelial cell migration involves leukocyte integrins.

Cellular Adhesion Molecules and Angiogenesis

Angiogenesis, which is the formation of new capillaries, is an important component of wound healing. Expression of several endothelial cell integrins is important in the attachment, alignment, and migration of endothelium during angiogenesis. Endothelium uses $\alpha_1\beta_1$ and $\alpha_2\beta_1$ integrins to bind to collagen types I and IV and laminin, all components of extracellular matrix. Two major receptors for the RGD (arginine-glycine-aspartic acid) sequence on fibronectin include the endothelial cell peptide-binding integrin $\alpha_5\beta_1$ as well as $\alpha_4\beta_1$. Cell surface CAMs appear to mediate interactions between endothelial cells, extracellular matrix, and cytokines, leading to angiogenesis.

The tripeptide sequence RGD is found in several adhesive proteins in the blood and extracellular matrix, including fibronectin, vitronectin, osteopontin, collagens, thrombospondin, fibrinogen, and von Willebrand factor. The RGD sequences are recognized by at least one member of the integrin family that span the membrane of a variety of cell types. On the cytoplasmic side of the membrane, the receptors connect the extracellular matrix to the cytoskeleton.

Cell Migration and Differentiation

Adhesion proteins found in extracellular matrix not only promote attachment of cells, but they also stimulate migration and differentiation. Cells move toward high concentrations of adhesive proteins, and some cells such as neurons will differentiate to form neurites when cultured on laminin.

Cell migration, also known as crawling, is in response to cell-surface stimulation by extracellular matrix molecules, such as proteins and glycoproteins or sequences such as RGD, and by lipids and small molecules when they bind to specific external cell receptors. These sensors are coupled to heterotrimeric guanosine triphosphate (GTP)-binding protein (protein G), which passes repeatedly through the membrane or other membrane-spanning proteins. Other less-specific stimulations, including electrical field exposure and mechanical forces, can also produce cell migration. Both types of stimuli initiate transmembrane signals that activate this process. Advancement of the leading edge of the cell may occur by different mechanisms. Cylindrical spikes, called filopodia, may protrude and sometimes coalesce to form extensions of the leading edge. Another mechanism is by expansion of small bubbles or blebs in the cell membrane. Ruffling is another mechanism by which cell migration occurs, and it involves lifting of the leading and trailing edges of the cell from the substrate during migration. Migration occurs via a leading lamella, which consists of thin structures free of visible organelles that extend from the organelle-rich cell body in the direction of movement. The lamellae appear to glide forward pulling the rest of the cell body. As the lamellae move forward, they maintain their forward advancement by transiently attaching to the underlying surface. This requires reversible adhesion of plasma membrane molecules to the underlying surface.

It is believed that actin filaments found in the cell cytoskeleton are responsible for the ability of the membrane–cytoskeletal complex to drag the rest of the cell behind during cell migration. Actin bundles tend to align during polymerization. They are cross-linked by actin-stabilizing proteins, one of which is filamin. Filamin is an actin-binding protein that cross-links actin filaments and results in orientation of actin neighboring filaments at high angles (Figure 8.8). Other molecules involved in stabilization of actin filaments (such as talin, vinculin, paxillin, α-actinin) link cytoskeletal actin to the β subunit of integrins on the cell surface. A dynamic cycle of actin assembly and disassembly occurs under control of transmembrane signals during the process of cell migration.

Cellular Adaptation

Events that lead to cell injury include mechanical, chemical, and electrical stimuli. Contact with foreign cells and implants may lead to cellular changes and perhaps cell death. Cell degeneration is reversible and does not always mean that cell death will follow; however, continual exposure to the insult leads to further cell degeneration and eventual death if the stimulus is not removed. Signs of reversible degeneration include cell swelling, compression of intracellular junctions, appearance of intracellular vacuoles, dilatation of the ER and Golgi apparatus, proliferation of ribosomes, and degeneration of the cell membrane.

Physical signs of irreversible degeneration or cell death include loss of intracellular attachments, contraction of the nucleolus, rupture of the nuclear membrane, evidence of autophagy, and appearance of lysosomes. Degeneration of mitochondria and disorganization and disruption of ER is accompanied by accumulation of lipids, glycogen, and protein in the cytoplasm (Figure 8.9). These changes are a result of genetic, nutritional, physical, infectious, chemical, immune, and age-related factors that are discussed later (Table 8.2).

It is necessary to consider general classes of cellular adaptation mechanisms before considering cellular responses to specific stimuli. From a morphological or structural point of view, it is difficult to distinguish in most instances

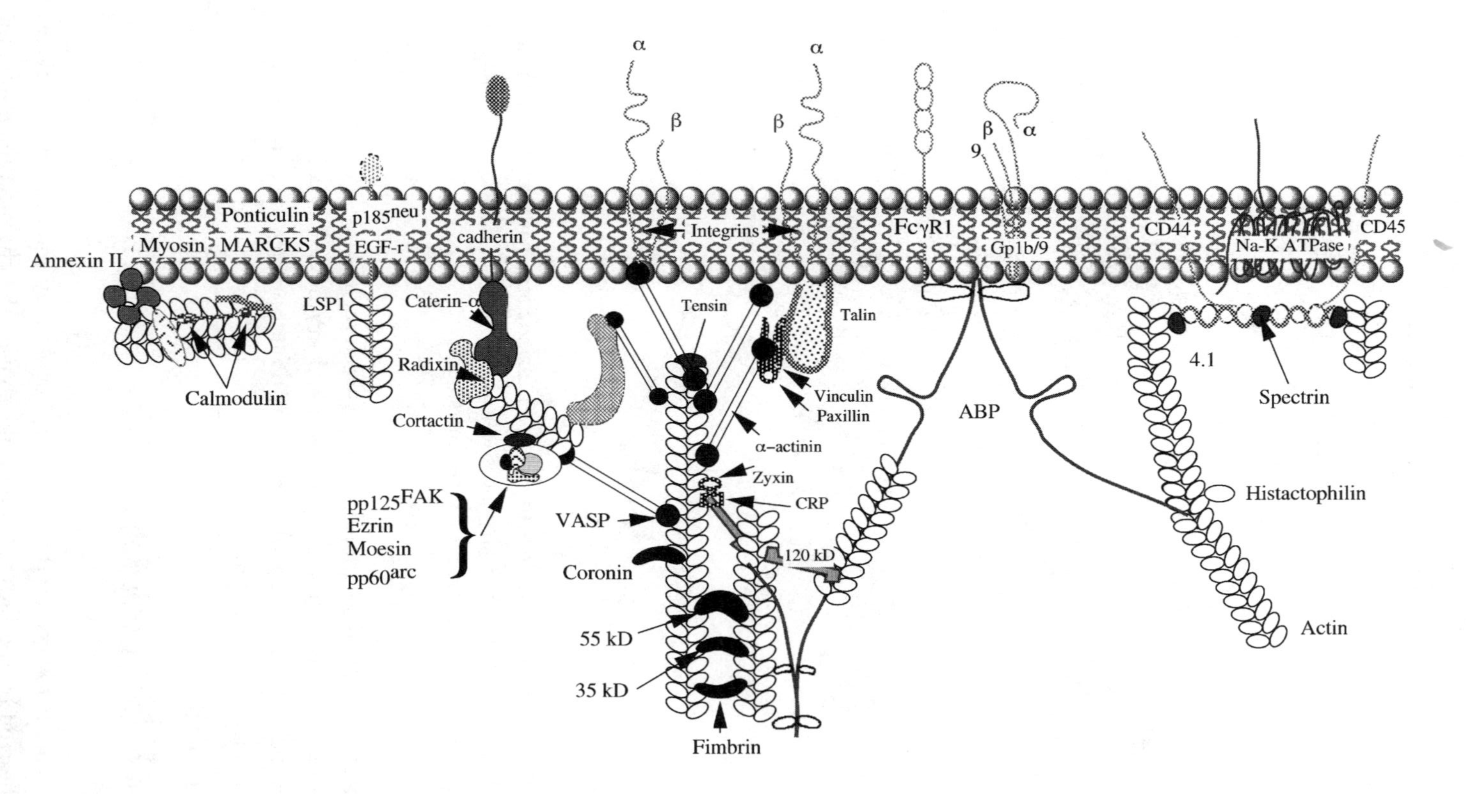

Figure 8.8 Macromolecules stabilizing cell membranes. The diagram illustrates the relationship between cell adhesion molecules (α and β integrin subunits) that stick to extracellular matrix and cell membrane and intracellular components, including actin filaments, talin, α-actinin, paxillin, and vinculin, which stabilize cell shape.

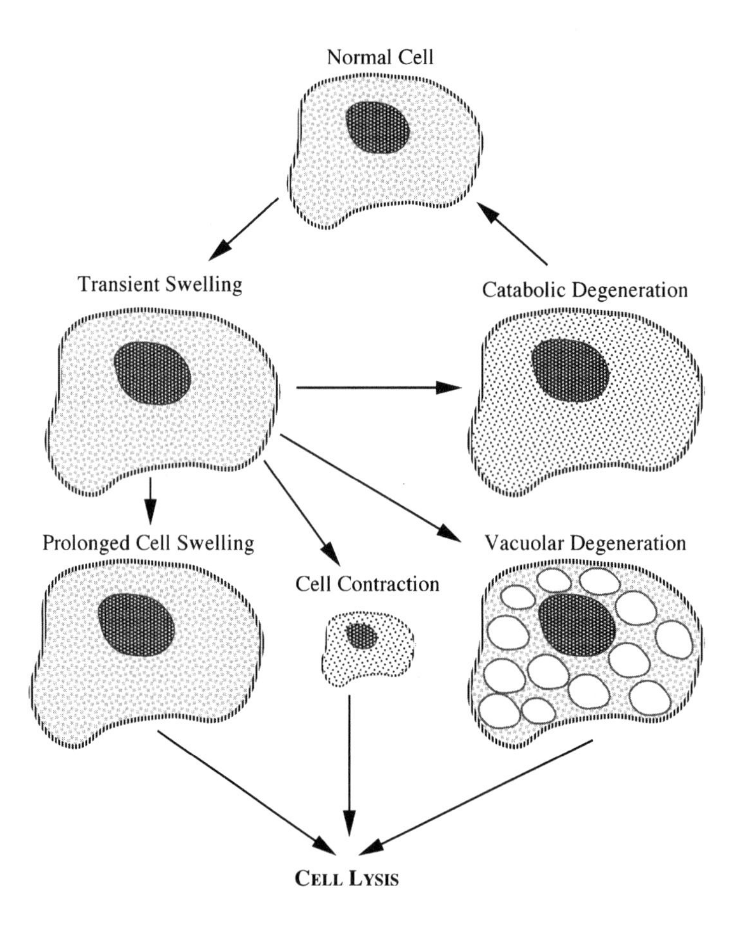

Figure 8.9 Degenerative changes to cell. Events associated with cell injury leading to death include cell swelling, degeneration of cell organelles, vacuole formation, and cell lysis via release of lysosomal enzymes.

among normal, adapted, and injured cells. Therefore, looking at the structure of a cell by light or electron microscopy is useful in studying cellular adaptation responses; however, structural changes alone do not predict the outcome of the adaptation process (Table 8.3).

Induction of Endoplasmic Reticulum

Chemicals are detoxified by the ER in the liver. As they enter the liver, enzymes are synthesized by the ER to theoretically convert the chemicals entering into

Table 8.2 Classes of stimuli leading to cell injury

Stimulus	Effect
Genetic	Dysfunctional protein
Nutritional	Delayed or inadequate wound healing
Physical	Loss or necrosis of tissue
Infectious	Tissue necrosis and delayed wound healing
Chemical	Detoxification by liver or kidney or cellular injury
Immune	Vascular leakage, antibody production
Aging	Impaired healing

Table 8.3 Cellular adaptation processes

Process	Cellular changes
Induction of endoplasmic reticulum	Dilated endoplasmic reticulum
Autophagy	Removal of worn-out organelles
Hypertrophy	Increased amount of cellular components
Hyperplasia	Increase in cell number
Metaplasia	Change in cell type
Dysplasia	Change in nuclear size and shape

harmless substances that can be excreted into the urine and feces. This is the case with some chemicals but not always. Some substances, including carbon tetrachloride, are converted into free radicals in the liver and become toxic to the cell and organism by the reactions that take place. For this reason, it is important to consider both the toxicity of breakdown products of implants as well as the metabolites of the breakdown products. A simple cell cytotoxicity study of an extract of an implant may not show that the breakdown products are toxic if the polymer requires five months to begin biodegradation. Thus, short-term biocompatibility tests based on cytotoxicity would not show that the products once metabolized by the ER are toxic to the cell. This can be revealed by analyzing whether induction of the ER at long term is associated with liver injury in animal studies.

Change in Cell Size

Hypertrophy is the name given to an increase in cell size without an increase in the number of cells that make up a tissue. Cellular hypertrophy is not necessarily pathologic in nature and must be analyzed in great detail to evaluate the source of the stimulus. For example, increased work load to the heart (right ventricle) that results from either peripheral vascular disease or from endurance exercise results in increased muscle cell size because of increased synthesis of cellular components, including enzymes, mitochondria, and ER. However, the increase in the ER in the case of endurance exercise is not induced by exogenous chemicals but by the increased workload on the heart. The outcome of cellular hypertrophy can be cell death, such as is the case when peripheral vascular resistance continues to increase, or it can be homeostasis if cell hypertrophy is not associated with any pathological condition. In this case, the outcome would be analyzed based on evaluation of the release of heart muscle proteins into the blood. Therefore, the long-term outcome cannot be accurately analyzed based only on the increased cell size. Cellular hypertrophy in the presence of an implant is likely to be associated with excessive synthesis of new proteins and other macromolecules or induction of the ER and detoxification. For this reason, it is important to understand what biochemical changes are occurring before any conclusion can be reached as to the cause of the cellular hypertrophy.

Increase in Cell Number (Hyperplasia)

When cells are stimulated mechanically or chemically, they often respond by proliferating and increasing the net cell number, which causes an increase in the volume of the tissue or organ. This process, which is called hyperplasia, can lead to normal and pathological outcomes, depending on the cell type that is proliferating. Examples of cellular hyperplasia include mechanical stimulation of the skin, resulting in proliferation of the basal epithelium and increased skin thickness, as well as abnormal proliferation of white blood cells that is associated with the onset of leukemia. In the case of epithelial cell proliferation, skin will thicken in an area that is experiencing abnormal loading without any negative consequences except that it may be less pliable. In the case of white cell proliferation, the increased number of normal leukocytes is eventually replaced with large numbers of immature blast cells, which is a sign of leukemia.

Change in Cell Types (Metaplasia)

Metaplasia is the substitution of one adult cell type for another in a tissue. This occurs under the influence of chemical, mechanical, physical, or viral signals. The most well-documented example of this change occurs in the epithelia that line the surfaces of the body. The respiratory tract is normally lined by columnar epithelia that are rectangular in cross-section. However, under the influence of chemicals found in cigarette smoke, these cells change to elongated squamous epithelial cells. The outcome of this change in the short term is not pathological; however, it is not clear over the long term whether the changes are associated with or are the predecessor of pathological conditions such as dyplasia and eventual tumor formation. For this reason, it is important to determine that cellular metaplasia around or near an implant is not associated with long-term pathological changes in tissue surrounding an implant.

Changes in Cell Characteristics (Dysplasia)

In response to chronic irritation, changes are observed in differentiated adult cells. Although these changes to cells lead to tissue and cell morphologies that are not normal, they do not necessarily mean that tissue injury and cell death will occur. The changes that are most often observed to cells include a change in cell size, shape, and loss of organization that may result in a change in the cell-staining characteristics. Dysplasia most commonly involves epithelial cells that line the entrance to the uterus. In this example, the normal epithelium is replaced with cells that have irregular nuclei and other changes that signal that some transition has occurred. Whether this is a transitional change associated with some external stimuli or it is a permanent change requires further testing. Transitional changes are associated with external stimuli, and the cell type reverses back to the original one after the stimulus is removed. However, in some cases, epithelial cell dysplasia continues. In the uterus, this is a red flag. Long-term dysplastic changes in cells ultimately require surgical intervention, which means a hysterectomy in the case of the uterus. The association between cancer of the uterus and dysplastic cell changes has been made and therefore is of concern to physicians. Any device that causes epithelial cell dysplasia when placed in contact with the lining of the uterus must be studied extensively to rule out the possibility of carcinogenicity.

Cell Injury

Cellular injury is caused by several factors that have been well documented to lead to alteration of the cell genetic material, a change in cell characteristics, and degeneration of cell organelles, leading to death. These factors are of genetic, nutritional, physical, infectious, chemical, and immune origins.

Genetic and Nutritional Factors

During replication of the genetic material found in a cell, a mutation (change in sequence of nucleic acids) can occur that leads to alterations in mRNA and the protein that is synthesized on the cell's ribosomes. The classical example of how a mutation in a single nucleic acid affects the protein that is formed is seen in sickle cell anemia. In sickle cell anemia, a mutation occurs that results in substitution of the amino acid valine for a glutamic acid. As a result of this change, the hemoglobin molecule cannot maintain its 3-D structure inside of red cells at low oxygen tensions, and the hemoglobin and cell collapses into a sickle shape. This change causes red cells to get stuck in the capillaries, resulting in loss of the oxygen-transporting ability of the blood. This change results in anemia and resultant kidney, heart, and lung problems as well as chronic skin ulcers.

Genetic mutations can also occur as a result of cell and tissue contact with the chemicals and radiation that are used to process medical devices. It is well known that solvents used to clean parts, such as trichloroethylene, have been shown to cause mutations leading to leukemias in humans. This has led to the decreased use of chlorinated solvents and other chemicals to clean manufactured products and process implants. However, because some solvents are still used in manufacturing processes, there is the possibility of an adverse outcome after the use of these products.

Nutritional factors, such as excesses or deficiencies, can lead to implant failure or disease states. Vitamin C is required for wound healing and cross-linking of the extracellular matrix. Absence of vitamin C leads to impaired wound healing and implant failure because most implants are stabilized by a fibrous capsule. Other factors are necessary to support normal wound healing, so normal nutritional health is essential for the successful outcome of any procedure involving an implantable device.

Physical Factors

Physical factors, including thermal contact, excessive pressure, radiation exposure, and electrical contact, play a role in causing cellular adaptation and injury responses. Destruction of the skin or even internal injury is caused by thermal, electrical, and radiation exposure. Contact of the epidermis, dermis, and subcutaneous tissues with surfaces that are only 10 °C above body temperature can induce cell injury by accelerating cellular metabolism and by inducing vascular injury. The duration of temperature elevation is important in determining the extent of injury, which is reflected by the degree of redness of the injured area. In thermal injuries, the degree of redness reflects the extent of vascular dilatation and release of vascular components into the extracellular matrix.

Exposure to low temperatures results in tissue injury by a different mechanism. Low temperatures cause vasoconstriction of blood vessels and decreased flow, which can cause cell death because of decreased oxygen supply. However, with freeze injuries, no damage occurs to the extracellular matrix, and for this reason redness and increased local circulation are absent.

Excessive pressure produced by balloon inflation in a catheter or movement of an implant can produce mechanical injury to cells. Patients undergoing prolonged bed rest who have excessive pressure on their skin from bony areas of the body may experience skin and muscle tissue necrosis.

Radiation injury caused from cosmic rays, ultraviolet and visible light, microwaves, or radio waves can cause cell injury, depending on the frequency and duration. Ionizing radiation leads to ejection of an electron from the target cells or cells. This leads to the creation of positively and negatively charged ions. Electromagnetic waves with sufficient energy result in ionization of atoms; these include X-rays, γ-rays, and cosmic photons. Ionizing radiation involves two subclasses: oscillating electromagnetic waves, including X-rays and γ-rays, and high-energy subatomic particles, such as an α-particle. Ionizing electromagnetic waves penetrate deeply into tissues because of their low masses and transfer energy to electrons in cellular molecules. Charged particle radiation penetrates only a few centimeters but transfers energy to tissue because of the high particle mass. The unit of absorbed ionizing radiation is the rad. One rad is the amount of radiation that deposits 10^{-2} joules of energy per kilogram of tissue. For X-rays and γ-rays, the unit of exposure dose is the roentgen. A dose of one roentgen to living matter results in energy absorption of about one rad. A dose of 500 rad is sufficient to destroy a cell by causing polymer chain cleavage in DNA, cell membrane proteins, and phospholipids.

Cells that divide at a high rate, such as epithelium in skin or the gastrointestinal tract, bone marrow cells, and lymphocytes in the lymph nodes, are very sensitive to radiation. Differentiated mature cells that have a low mitotic activity, including muscle and brain cells, require high doses of radiation to be altered.

Many devices are sterilized terminally, at the end of the manufacturing process after they are packaged, by exposure to a dose of about 2.5 mega (million) rads using a ^{60}Co source for γ-rays. Exposure to γ-rays kills bacteria, fungi, and viruses. However, it is important to analyze how the exposure to γ-rays or any sterilization process affects the chemical structure of an implant.

Infectious Agents

Exposure to infectious agents, such as bacteria, viruses, and fungi, are potential complications of surgery. Biomaterials processed under nonsterile conditions are contaminated and must be sterilized before implantation or evaluation of biocompatibility in cell culture models. Infectious agents cause tissue inflammation and cellular hyperplasia and hypertrophy and must be deactivated.

Viruses must gain entrance into a cell before becoming infectious. Viruses contain either RNA or DNA in single- or double-stranded forms within a glycoprotein envelope. The envelope is coated on its inner layer with a protein layer called the capsid. Viruses have diameters of 100 to 300 nm and have all the enzymes necessary to replicate themselves. Upon entrance into the cell, they direct the host cell to replicate the viral membrane capsid and nuclei acid.

In contrast, bacteria and fungi produce products that are infectious when present extracellularly. Endotoxins that are secreted by bacteria cause elevated temperature and infection. Even if only bacterial endotoxins and not live bacteria are present in the implant, it may fail biocompatibility tests. Bacterial infection of terminally sterilized products is also observed. This is associated with the implantation of devices into and through wounds contaminated with bacteria on the surface. During and after most surgeries, prophylactic antibodies are given to prevent infection. If infection is not resolved by antibiotic therapy, then the implant must be removed.

Chemicals

Chemicals found in our environment or used to process materials can produce cell injury and must be removed before implantation of a medical device (Figure

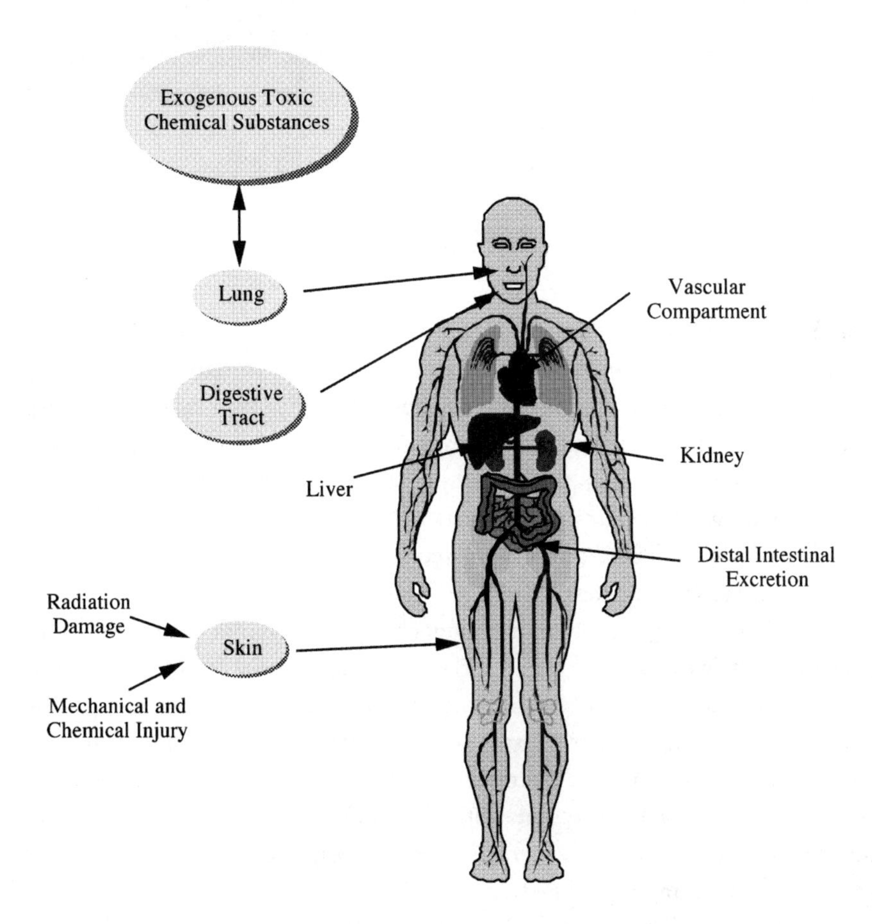

Figure 8.10 Chemical, mechanical, and radiation-induced injury. Injury can be induced by environmental chemicals entering the body through the air or by ingestion and then subsequently being absorbed into the blood, liver, kidney, and intestines. Mechanical and radiation damage occurs by direct contact with the skin.

8.10). In addition, all chemical treatment procedures used to manufacture an implant must be designed to avoid exposing device workers to harmful environments. In most cases, chemicals enter into the body through the lungs, skin, and gastrointestinal tract. Every step in device manufacturing must be evaluated as to its possible harmful effects because any particles created or solvents used in manufacturing can potentially be ingested. Ingested particles or solvents can be eliminated by excretion into the urine and feces or can be detoxified by cells in the liver and kidneys. Cellular components, such as the lysosomes, can detoxify a toxic chemical or can modify a chemical into a toxic form. In many cases, any plasticizers or solvents that are used to process nonpolar polymers can be harmful to cells and tissues and are not recommended for use.

Inflammation and Immune Reactions

During the response to injury resulting in implantation of a device, inflammatory cells are observed to surround an implant and attempt to remove it by phagocytosis. If the device cannot be removed, it is surrounded by a capsule to separate it from the host. The reactions to a foreign material or host if left unresolved will cause tissue destruction that can be irreversible. Release of factors from dying cells, including interleukins, cause further tissue destruction or cell death in an unwanted fashion. Therefore, as is discussed in Chapter 9, the inflammatory process can cause tissue destruction.

When a foreign protein or polymer is introduced into the body, immune cells immediately recognize that a surface is foreign or not "self." Cells respond by either secreting or synthesizing antibodies. In addition, activation of other helper cells can augment the response to produce higher levels of antibodies. Antibody–antigen complexes formed can lead to injury by triggering hypersensitivity through reactions of IgE with mast cells and release of mediators of inflammation or by precipitation of complexes with IgG in joints and tissues, causing pain and impeding function.

Although the primary biological function of inflammation and immune processes is to protect the host against invasion of pathogens and other foreign materials, it sometimes leads to tissue destruction. To minimize adverse effects of biological systems in the presence of implants, we must always recognize the signs of adverse reactions and the possible causes. A small percentage of patients will react to any foreign material used as an implant. It is not unusual for these patients to have serious allergic reactions to device implantation. All implants must be removed from these patients if a marked inflammatory reaction is noticed.

Summary

All implants have pathological complications associated with their use. Responses include adaptation in cell size and type as well as induction of inflammation and immune responses. The outcome of all these events can be normal or pathological, depending on the cell changes involved and the level of stimulation. In most cases, the outcome can only be assessed by studying the effects of different stimuli in model systems to determine if device implantation causes transitory or permanent changes. Although transitory changes are tolerable, permanent changes suggest that a device may have long-term adverse consequences for the host. The risk–benefit balance associated with use of a device must always be weighed based on understanding the pathobiological changes that take place as well as the short- and long-term outcomes. Depending on the health of a patient and the likely consequences of no intervention, a device may be used to prolong a life even when there are significant risk factors.

Suggested Reading

Carlos T.M. and Harlan J.M., Leukocyte-Endothelial Adhesion Molecules, Blood 84, 2068, 1994.

Polverini P.J., Cellular Adhesion Molecules: Newly Identified Mediators of Angiogenesis, Am. J. Pathol. 148, 1023, 1996.

Rosen S.D. and Bertozzi C.R., The Selectins and Their Ligands, Curr. Opin. Cell Biol. 6, 663, 1994.

Ruoslahti E. and Pierschbacher M.D., New Perspectives in Cell Adhesion: RGD and Integrins, Science 238, 491, 1987.

Stossel T.P., On the Crawling of Animal Cells, Science 260, 1086, 1993.

9 Wound Healing

Introduction

The ability of humans to maintain their normal physiological functions involves many complex interwoven biological pathways. Part of this homeostasis involves normal turnover of tissues found throughout the body. In addition, it involves the ability of mammalian tissues to repair themselves after injury. Without the ability of tissues to repair themselves, we would be unable to withstand the trauma of daily life or the invasion of bacteria, fungi, and viruses. In addition, the ability to heal after chemical, mechanical, electrical, and biological trauma makes life as we know it possible. In this chapter, we will discuss how healing occurs and what processes are involved. As we discuss below, healing is a multistep process that involves biological components found in blood and elements that make up extracellular matrix. These elements are components of systems that prevent excessive bleeding, remove exogenous debris, promote new tissue deposition, and allow resumption of normal physiological processes. The scope of this chapter is to discuss the relationship between these components and systems that promote homeostasis.

One of the most challenging areas of study medical scientists have dealt with in the twentieth century is why some tissues heal by regeneration and others heal by the formation of aligned collagenous (scar) tissue. In mammals, injury to most tissues, whether external or internal, results in blood clotting, fibrosis, and functional repair; however, most tissues do not regain their original structure and function. When an injury to the skin occurs that results in loss of only the epidermal or superficial layer, the epidermis will regenerate by cell migration from the wound edges. However, if both the epidermis and the underlying layer of dermis are injured, then dermis that is lost will be replaced with scar tissue, and the epidermis will migrate over scar tissue. On a superficial level, we understand that the presence of an intact basement membrane over the dermis supports epidermal regeneration; we do not understand the exact information contained in the basement membrane that signals basal epithelium to divide and migrate. Therefore, much of our understanding of wound healing is descriptive.

Mechanical, chemical, biological, or electrical trauma to tissues leads to activation of a series of processes that limit bleeding (blood coagulation), lead to clot removal (fibrinolysis), target foreign bacteria for removal (complement system), and stimulate dilatation of the vasculature (kinin system). These systems limit injury and initiate the healing process; however, if left unchecked they can also lead to further tissue injury. It is important to determine how these biological systems and inflammatory and immune responses affect wound healing. This chapter discusses how each of these systems contributes to healing of dermal tissue, because in many cases, scar tissue, much like that seen in dermal healing, is observed to be deposited throughout the body during wound healing.

Biological Cascades Involved in Healing

Blood contains a number of soluble components that are involved in the biological cascades. In addition, it contains cells involved in the wound-healing process. The protein components are involved in osmotic pressure regulation (albumin), blood clotting (Figure 9.1), fibrinolysis (Figure 9.2), bradykinin formation (Figure 9.3), complement activation (Figure 9.4), and specific antibody recognition (Figure 9.5). In addition, cellular components called growth factors

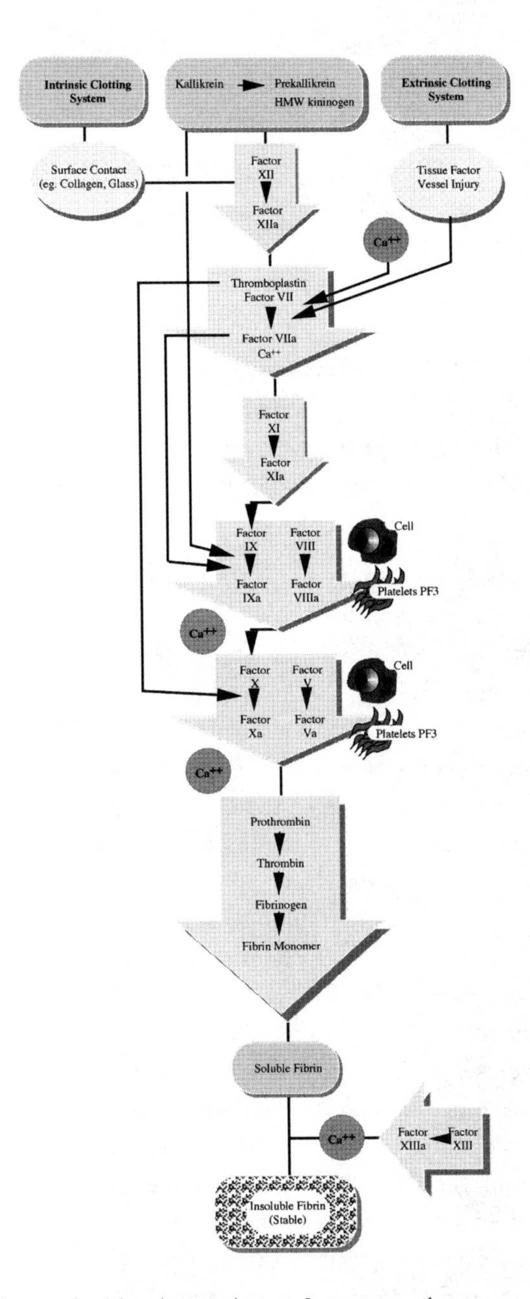

Figure 9.1 Pathways for blood coagulation. Intrinsic and extrinsic coagulation pathways are triggered by contact with collagen, activation of the kinin system, or via tissue injury. Activation of either the intrinsic or extrinsic pathways leads to formation of a fibrin clot that is cross-linked by activated factor XIII. Fibrin clot formation initiates fibrin clot degradation via activation of fibrinolysis.

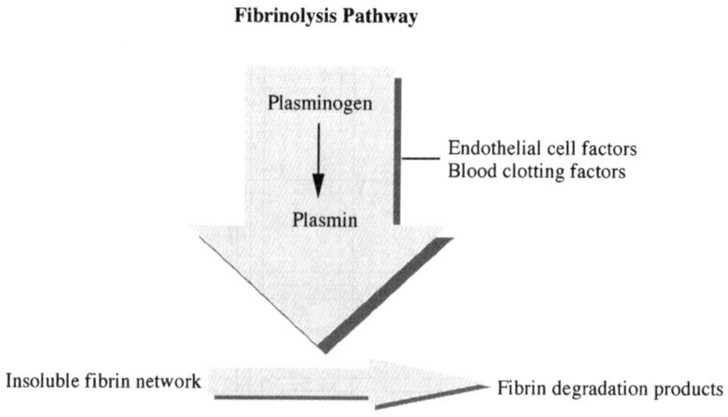

Figure 9.2 Activation of clot lysis. The diagram illustrates activation of fibrinolysis via plasmin generation and subsequent release of fibrin degradation products.

release macromolecular components that promote tissue synthesis by cells (Figure 9.6).

Temporally, the immediate response to tissue injury is activation of inflammation and wound healing. This leads to migration of inflammatory cells to the wound site, release of cytokines and growth factors that promote cell proliferation and protein synthesis, and activation of kinin, complement, blood clotting, and fibrinolytic cascades (Silver and Parsons, 1991). Unfortunately, if not controlled, this sequence of events can lead to deposition of fibrous scar tissue, preventing proliferation of progenitor and other cell types that are needed to regenerate new tissue with the morphology and properties of the original tissue.

Blood contains a series of proteins and blood cells that play important roles in wound healing. Soluble blood proteins (see Figure 9.1), including Hageman factor (factor XII) and fibrinogen (factor II), play a critical role in allowing blood vessels to be self-sealing and in preventing loss of blood and subsequent death through the process of blood clotting. Hageman factor is activated by exposed fragments of collagen fibers that are found on the surface of torn blood vessels. Activated Hageman factor initiates the intrinsic blood coagulation cascade, leading to the formation of an insoluble fibrin network that binds proteins and cells together. The same process can also occur by direct cell injury and release of tissue factors that directly activate factor VII and short circuit insoluble fibrin network formation via the extrinsic clotting system. In addition to protein clotting, during blood coagulation platelets adhere to cut segments of the vessel wall and release agents such as ADP, causing platelet aggregation. The purpose of blood clotting and platelet aggregation is to prevent uncontrolled bleeding

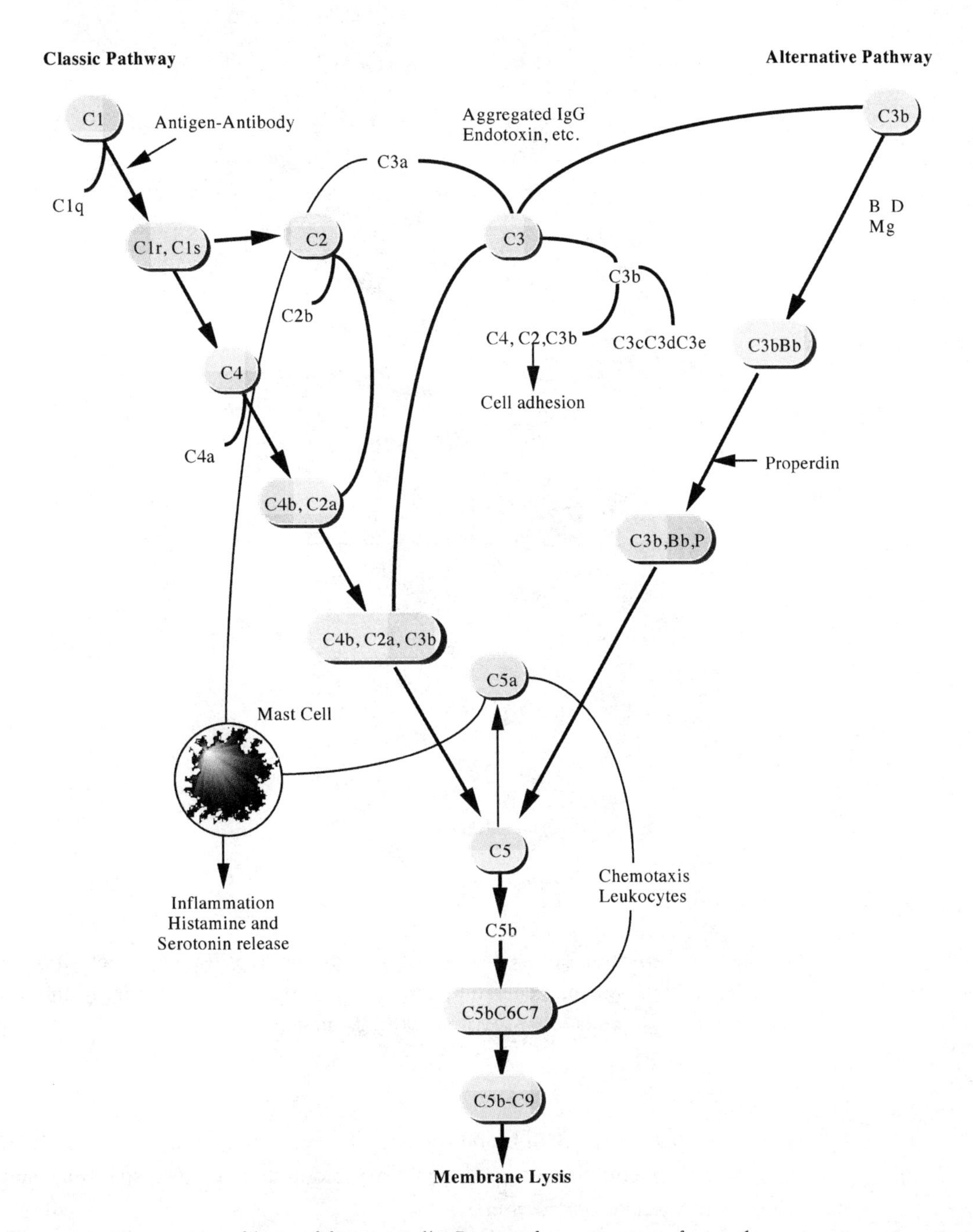

Figure 9.3 Activation of lysis of foreign cells. Process for activation of complement system via antigen–antibody complex, the classic pathway, and via aggregated antibody or bacterial endotoxins, the alternate pathway. Formation of a C5bC6C7C8 complex via either pathway leads to membrane lysis of bacteria or foreign cells.

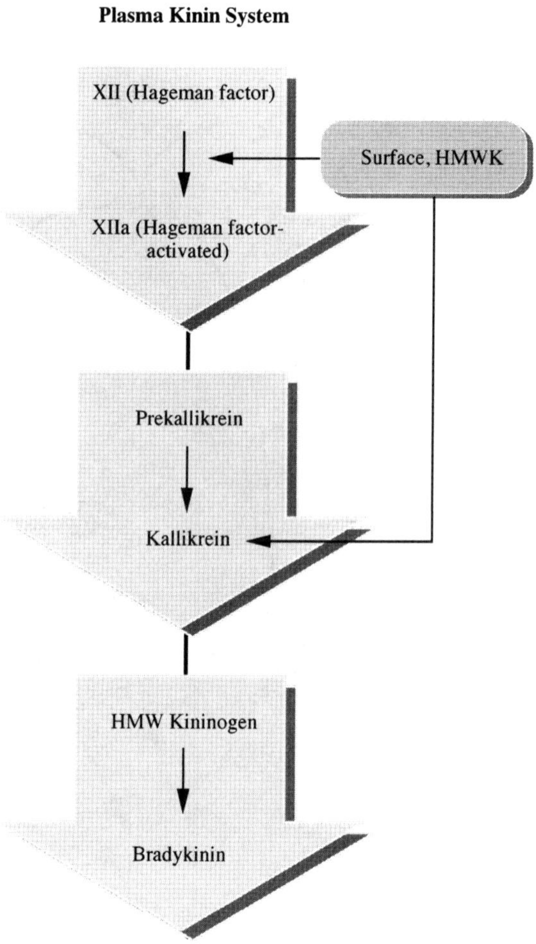

Figure 9.4 Activation of vasodilation. Shown are the steps leading to activation of the plasma kinin system, generating bradykinin, a vasoactive agent. Bradykinin release prolongs vasodilation associated with inflammation.

due to traumatic rupture of blood vessels. However, once a blood clot is formed it must subsequently be removed for normal blood flow to be reestablished and to prevent vascular obstruction.

Removal of blood clots occurs by an enzymatic process called fibrinolysis that lyses the fibrin network. This limits the obstruction of blood flow through a vessel. Fibrinolysis (Figure 9.2) occurs when plasminogen, an enzyme precursor found in blood plasma, interacts with the cross-linked fibrin network in the pres-

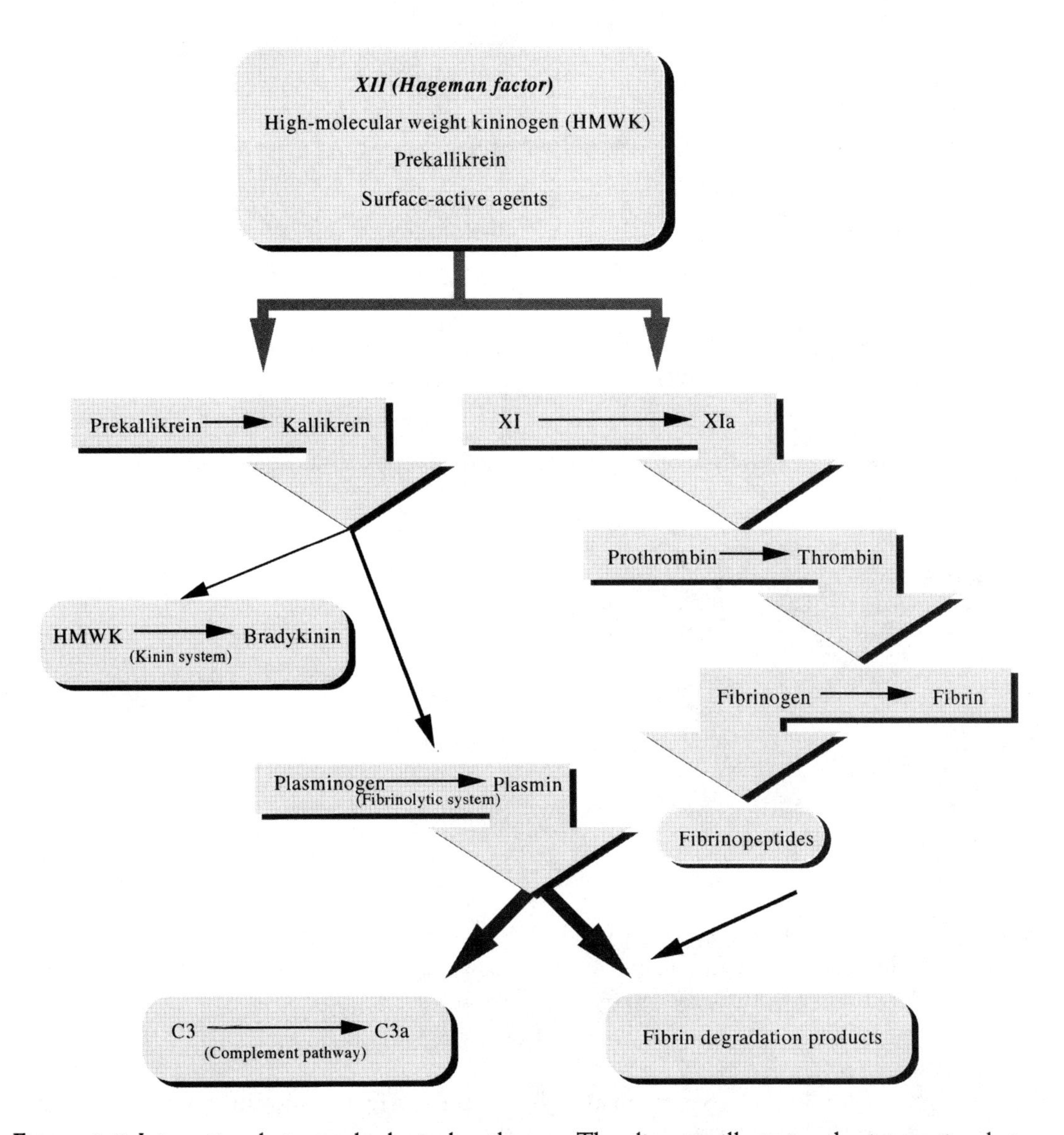

Figure 9.5 Interaction between biological pathways. The diagram illustrates the interaction between blood clotting, kinin, complement, and fibrinolysis pathways.

ence of cellular and plasma factors. This leads to formation of plasmin, degradation of the fibrin network, and release of fibrin degradation products.

Another plasma factor cleaved by activated Hageman factor is called prekallikrein. This factor is converted to kallikrein during blood clotting, which

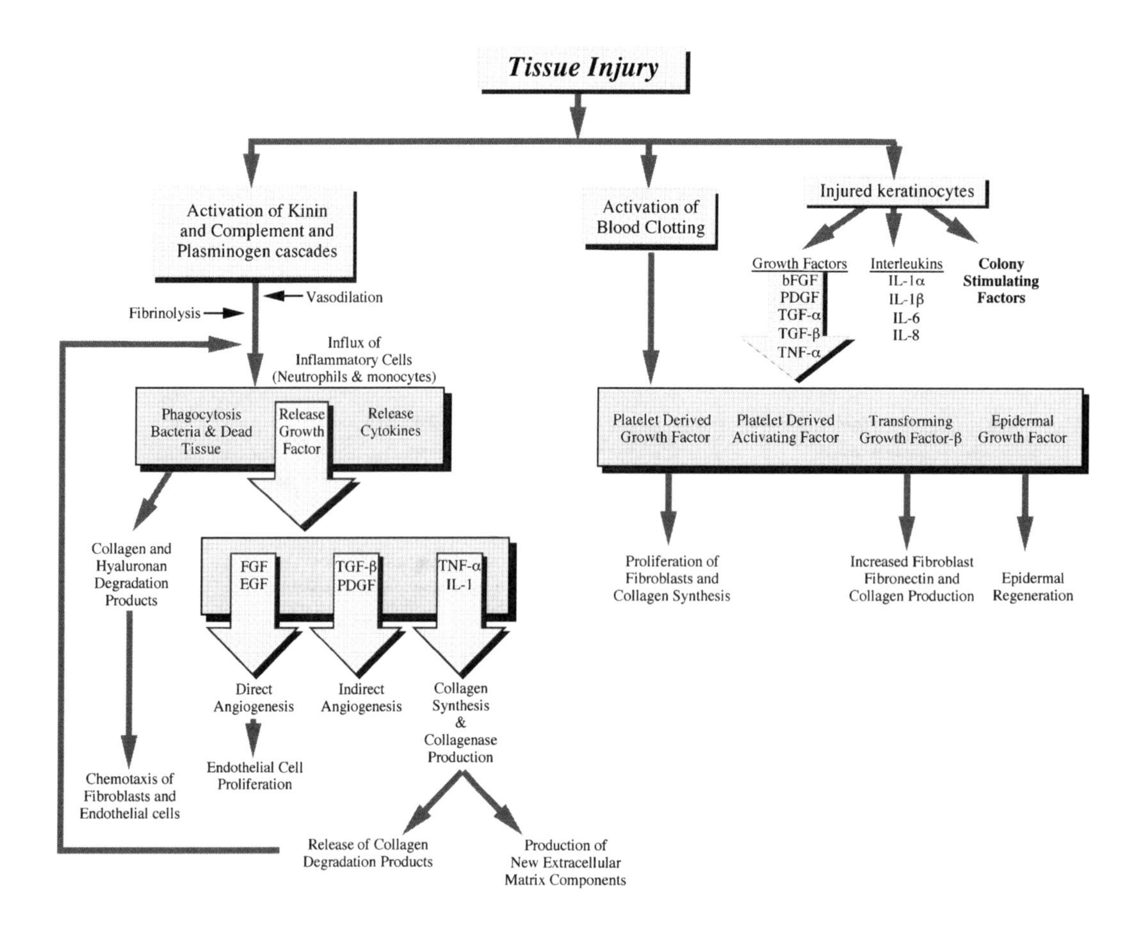

Figure 9.6 Relationship between tissue injury and release of cytokines and growth factors. The illustration shows injury and activation of kinin, plasminogen, and complement cascades that lead to inflammation, release of cytokines, and growth factors, triggering scar tissue deposition.

leads to the conversion of high molecular weight kininogen to bradykinin (Figure 9.3). Bradykinin is vasoactive and functions to stimulate vascular permeability.

The complement system is another plasma-derived biological system that is triggered during contact with foreign materials such as bacteria and implants; it involves about 11 soluble serum proteins that can lyse bacteria in the presence of an antibody response to foreign cell surface macromolecules. Fragments of complement-activated proteins induce an inflammatory response, so excessive complement activation is undesirable. There are two pathways by which membrane lysis can occur. The classic pathway is triggered by an antibody–antigen complex on the foreign cell surface that binds complement subunit component

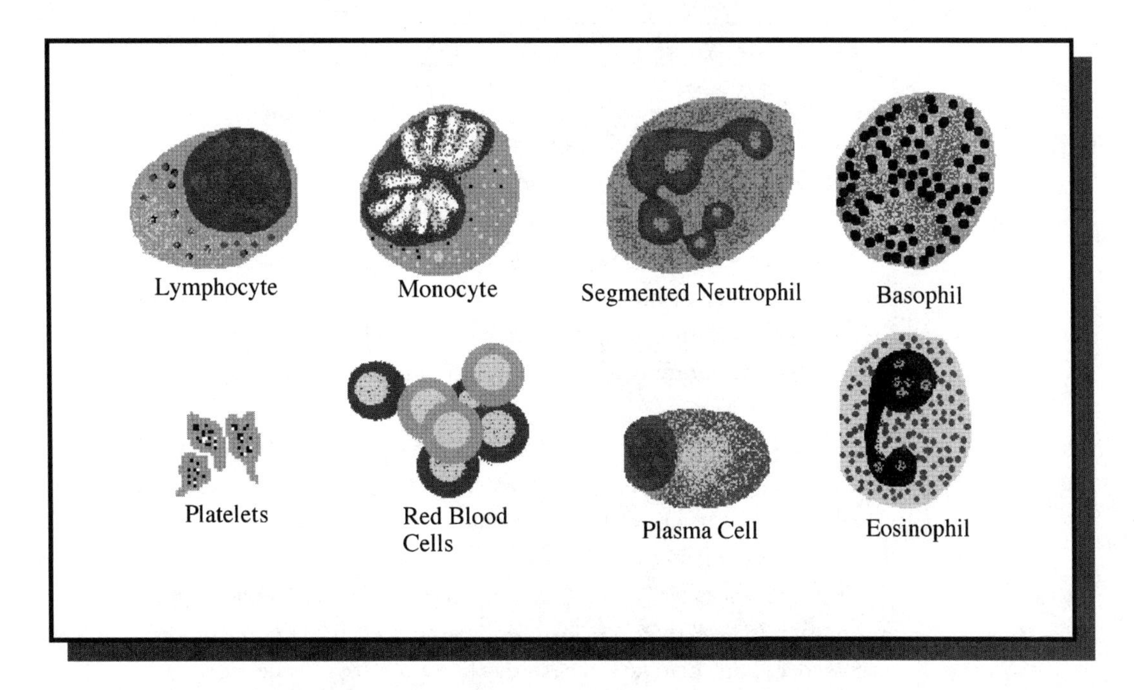

Figure 9.7 Blood cell morphology. Lymphocytes are 6 to 20 μm in diameter and are characterized by a pale cytoplasmic ring and a dense dark staining nucleus. Depending on the size of the cell, it is characterized as small or large. Monocytes are 15 to 20 μm in diameter and have a kidney-shaped nucleus with folds over nuclear lobes. Neutrophils are about 10 to 14 μm in diameter and have a multilobed nucleus connected by strands of chromatin. Plasma cells make antibodies and are observed to have an eccentric nucleus and a dark-staining cytoplasm. Basophils have a multilobed nucleus and a cytoplasm filled with dark-purplish-staining granules. Eosinophils are characterized by the presence of numerous light-red-staining granules. Platelets and red blood cells lack nuclei and are recognized by their contents of densely staining granules and hemoglobin, respectively.

C1q, which activates the C1r subunit (Figure 9.4). This subsequently leads to release of fragments C3a and C5a, which mediate inflammation and tissue swelling by causing histamine release from mast cells and by attracting white blood cells to the area of injury. Membrane lysis occurs as a result of formation of a complex involving fragments C5bC6 and C7, which in the presence of C9 form a full-thickness defect in the bacterial cell membrane, leading to cell death. The alternate pathway involves conversion of C3 into fragments that in the presence of properdin forms active C3b. At this point, the classic and alternate pathways join. Figure 9.5 shows the relationship between blood clotting, kinin, complement, and fibrinolysis systems.

Cells Involved in Wound Healing

The cellular components involved in wound healing include red blood cells, platelets, white blood cells, and cells of the extracellular matrix, including fibroblasts and endothelial cells (Figure 9.6). Red cells are enucleated cells surrounded by a deformable membrane in the form of a disk; they are filled with the protein hemoglobin and comprise about 45% of the blood volume, with about 5×10^6 cells per milliliter of blood (Figure 9.7). Red cells circulate for about 120 days before they are removed by the reticuloendothelial system. They are trapped in fibrin networks during coagulation of blood and result in clots with a characteristic red color.

Leukocytes (see Figure 9.7) are white blood cells that are found at a concentration of 4,000 to 10,000 per milliliter of blood. There are several subclassifications of leukocytes. Neutrophils, also known as neutrophilic leukocytes or polymorphonuclear leukocytes, make up 40 to 75% of the white blood cell fraction; lymphocytes, 20 to 45%; monocytes, 2 to 10%; eosinophils, 1 to 6%; and basophils, less than 1%. The only other cell type found in blood and not discussed yet are platelets. These disklike cells lack a nucleus and stick to cut blood vessels. They are present at a concentration of 150,000 to 400,000 per milliliter of blood and play an active role in plugging leaky vessels in conjunction with fibrin.

Both neutrophils and monocytes are phagocytic cells that are able to digest foreign and dead cellular and noncellular material. These cell types find their way to traumatized tissue during the initial phases of wound healing and provide a vehicle for removal of dead tissue before wound tissue can be deposited. Neutrophils have cell nuclei with two to five lobes and contain granules in the cytoplasm. These granules, called lysosomes, are actually membrane-lined vacuoles containing hydrolytic enzymes. Lysosomes are used to break down bacteria, denatured proteins, and other foreign or worn-out materials (Figure 9. 8). Phagocytosis is the process that leads to break down of foreign materials. It involves incorporation of foreign material into the cell (endocytosis), formation of a membrane-lined vacuole containing foreign materials (phagosome), and fusion of the phagosome with a lysosome (secondary lysosome). After digestion, the material can remain within the vacuole inside the cell (residual body) or it can be released from the cell by exocytosis. Neutrophils are found circulating in peripheral blood as well as along the margins of small blood vessels. Within 12 hours of injury, they migrate into tissues to the sites of injury where they undergo phagocytosis for a period of four to five days. If the injury persists for more than five days, then neutrophils are replaced by monocytes at the wound site.

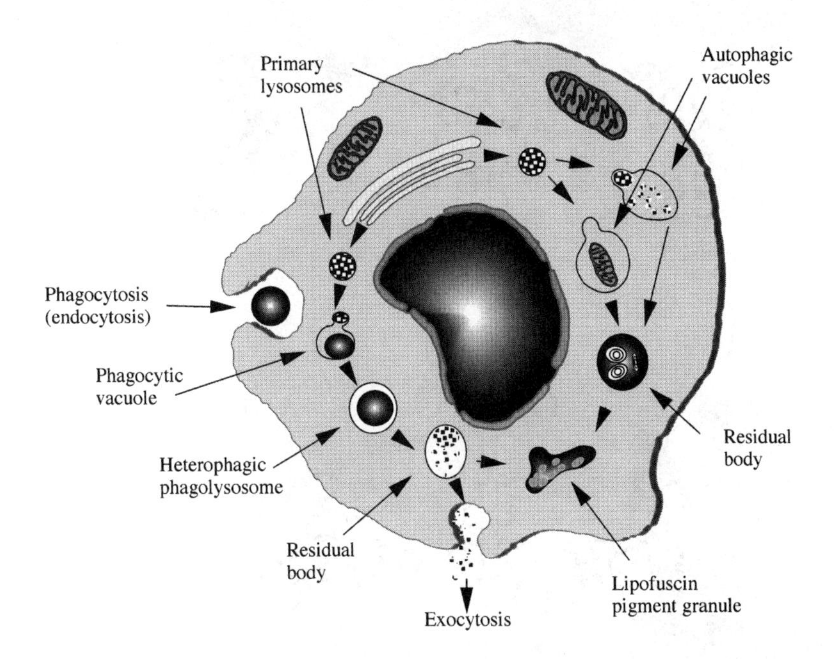

Figure 9.8 Representation of phagocytosis. Illustration of lysosomal breakdown of phagocytosed material. The drawing shows autophagy of internal cellular components as well as lysis of extracellular material.

Monocytes are larger than other leukocytes and have an average diameter of about 15 to 20 μm. In blood smears, they have an almost kidney-shaped nucleus and fine pink cytoplasmic granules. Monocytes are released into the circulation from the bone marrow, where they mature and migrate in a day or two into the liver, spleen, lymph nodes, and lungs. In these tissues, they comprise the macrophages of the reticuloendothelial system that ingest particles and worn-out materials. In connective tissue and sites of injury, macrophages remove foreign and worn-out materials by the process of phagocytosis.

Lymphocytes are a group of white blood cells that are associated with immunological responses to foreign materials. They have diameters from 6 to 20 μm and possess large nuclei; they are subclassified as B and T cells. B cells are transformed on contact with foreign antigens into plasma cells that make antibodies (Figure 9.9). These antibodies bind to foreign cells and promote cell death, or they form antibody–antigen complexes that can be removed by the reticuloendothelial system. T cells participate in cell-mediated immunity by helping or suppressing antibody production or by promoting cell lysis.

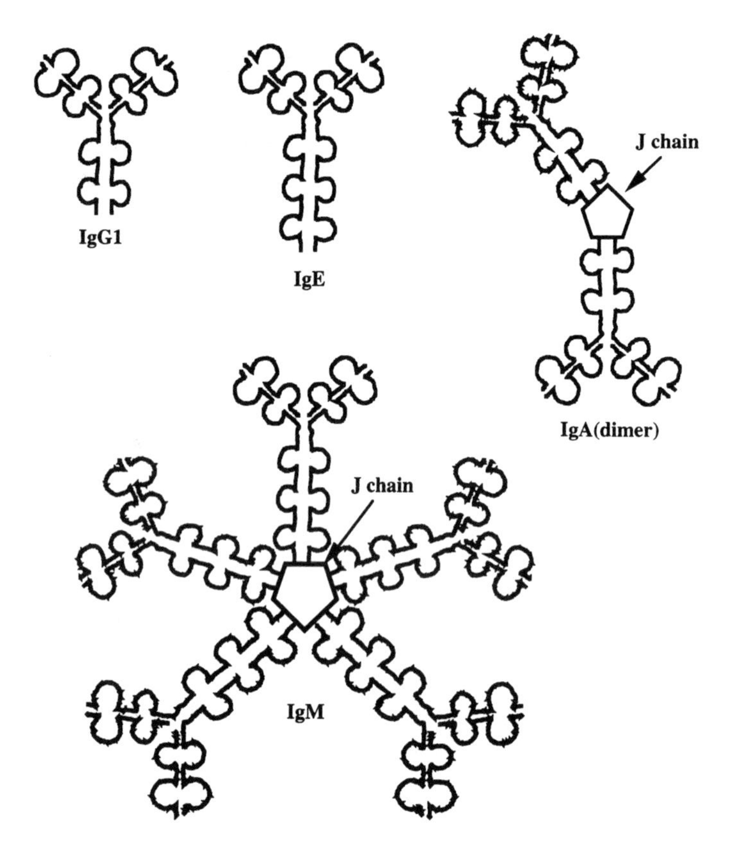

Figure 9.9 Schematic diagrams of various antibody (Ig) isotypes. IgG and IgE are present in the blood as monomers, whereas secreted forms of IgA and IgM are dimers and pentamers, respectively; the latter are stabilized by the J chain.

Inflammatory and Immunological Aspects of Wound Healing

Injury to cells and tissues as well as activation of the kinin, complement, and fibrinolytic systems causes vasodilation, an influx of inflammatory cells, release of growth factors, release of proteolytic enzymes, degradation of dead tissue, and deposition of new connective tissue (Figure 9.6). As discussed earlier, vasodilation is a consequence of release of the vasoactive complement fragments C3a and C5a and bradykinin. It results in vascular permeability and the ability of white blood cells to permeate through blood vessel walls into the wound area. Once inflammatory cells are present in the vicinity of the wound, denatured collagen and fibrin degradation products help promote migration of the white blood cells, specifically neutrophils, monocytes, and lymphocytes to the site of cellular injury. Neutrophils and monocytes release growth factors and cytokines and assist with the digestion of foreign and necrotic (dead) tissue. Lymphocytes, including T and B cells, are involved in producing antibodies and cell-mediated lysis of foreign material; these immune responses are humoral (B-cell dependent) and cell-mediated (T-cell mediated) responses.

Immunological Aspects of Wound Healing

During evolution, plants and animals have developed a variety of mechanisms to defend themselves from bacteria, viruses, and other nonhost elements. These defenses include phagocytosis and cellular and humoral responses. The normal defense against foreign matter consists of natural and acquired immunity. Natural immunity involves no prior exposure to the foreign material and is not enhanced by exposure. This type of immunity involves the inflammatory process, including neutrophils and macrophages. Acquired immunity is specific and requires exposure to a foreign material, an antigen, and is magnified by a subsequent exposure to the antigen. Unfortunately, both natural and acquired immunity cause tissue destruction as a negative consequence of the wound-healing process. Immunological aspects of healing are important in the removal of foreign cells and dead tissue.

Lymphocytes are important in wound healing because they have the capacity to recognize and react with specific foreign molecules, and they are the primary effectors of antigen-specific immune responses. All lymphocytes are derived from stem cells and subsequently differentiate into B (bone marrow derived) or T (thymus derived) cells. A third type of lymphocyte lacks the characteristics of T or B cells; they are called null cells and include natural killer cells (NK). Cytolytic T cells are another subset of T cells that kill target cells expressing specific antigens on the cell membrane. T cells are characterized at different stages of development by specific cell-surface markers; such cells include $CD4^+$, T-helper cells, and $CD8^+$, T-suppressor cells. Antibody-producing cells in the presence of T-helper cells up regulate antibody production to foreign cells (Figure 9.10), while T-suppressor cells down regulate the production of antibodies by plasma cells. In general, $CD4^+$ T cells secrete proinflammatory lymphokines such as interleukin 1 (IL-1) and recognize antigens on foreign cells containing class II major histocompatibility complex (MHC) markers, which are discussed later.

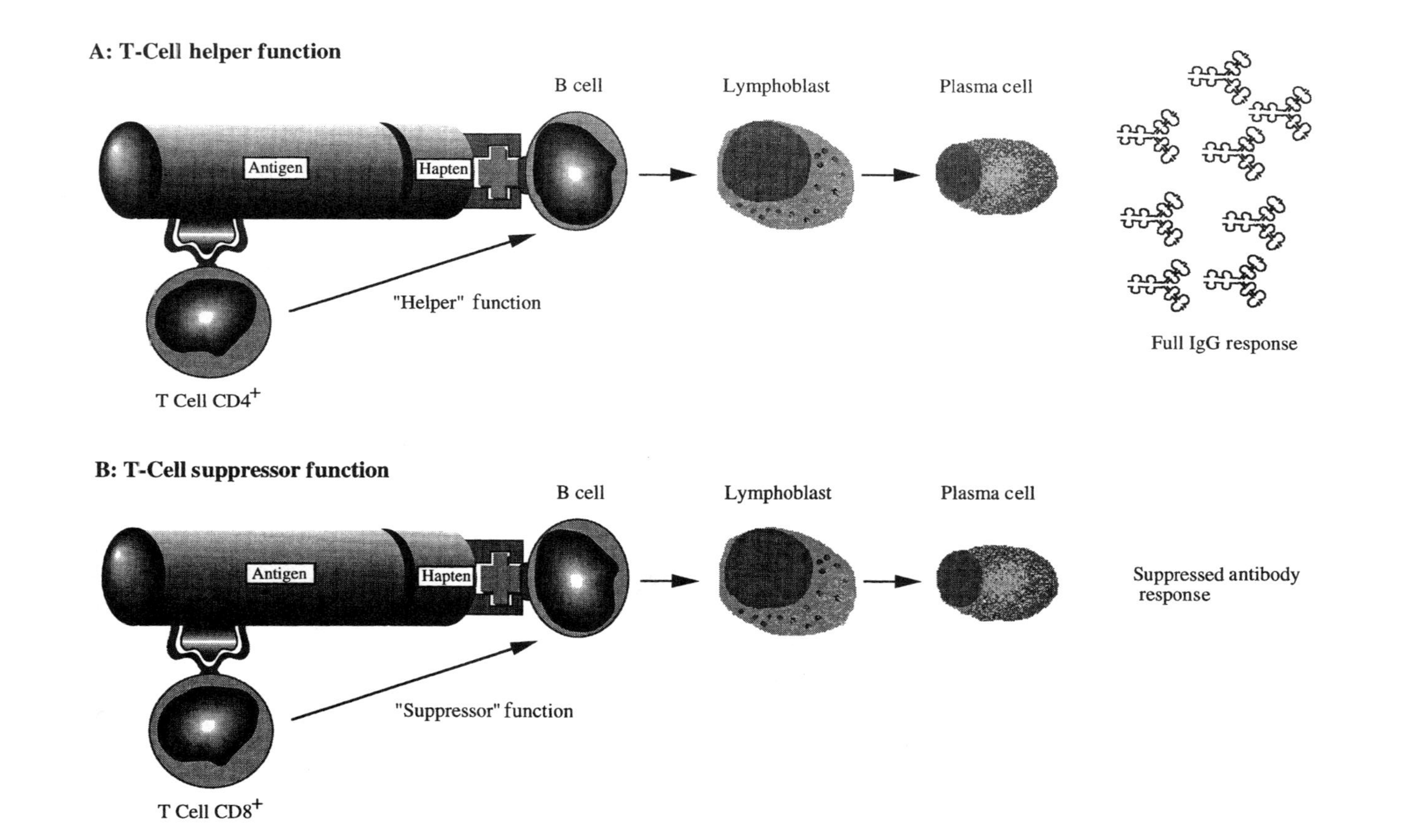

Figure 9.10 Effect of helper and suppressor T cells on plasma cell function. The diagram illustrates the roles of $CD4^+$ helper cells in promoting an antibody response to a foreign antigen and $CD8^+$ suppressor cells in limiting the response to foreign antigens. Helper cells promote the antibody response to foreign class II MHC antigens, thereby improving opsonization and phagocytosis by macrophages through complement binding. Suppressor cells down regulate antibody production and opsonization; however, they react with class I MHC markers and stimulate cytotoxic T lymphocytes.

In comparison, $CD8^+$ T cells secrete molecules that exert suppressor and cytotoxic functions, and these cells recognize class I MHC markers on foreign cells. Most cytolytic T cells are $CD8^+$ and recognize class I MHC products expressed on the surface of foreign cells. Cytolytic T cells react with class I MHC markers and release a cytotoxin that leads to target cell lysis (Figure 9.11). In comparison, natural killer (NK) cells are a subset of lymphocytes found in blood and lymphoid tissue (lymph nodes and spleen) that lack specific receptors for antigen recognition on the surface of foreign cells. NK cells possess the ability to kill certain tumor cells or cells infected by viruses and are not antigen dependent (Figure 9.12). Lymphocytes are important components of wound healing in response to foreign cells, as is the case when tissue is transplanted or during bacterial invasion. They are involved in removal of foreign cellular material that is present in wounds.

Major Histocompatibility Complex

All cells contain membrane-bound protein surface markers that identify the origin of the cell. Because these markers were first discovered on leukocytes, they were called human leukocyte antigens (HLA) and were found to be coded for by a series of genes on chromosome 6 in the human. The MHC markers are the main targets that are involved in the rejection of transplanted organs and in the recognition of foreign cells during wound healing. MHC genes code for three major classes of molecules, designated as class I, II, and III (Figure 9.13). The class III region codes for molecules in the complement system and is involved in the process of nonspecific lysis of foreign cells. The MHC markers are important in determining whether a host cell will respond to a foreign or implanted cell.

Class I antigens found on the surface of cells are coded for by genes in the A, B, and C regions of the MHC. The class I products are all similar in structure and are expressed on cells of nearly all tissues. These products consist of two chains, a 44,000-MW transmembrane glycoprotein called α_2-microglobulin (heavy chain), which is a product of the MHC, and a non-MHC light chain (MW 12,000), which is called β_2-microglobulin (see Figure 2.35A). There are many forms of the heavy chain, and tissue matching for transplantation is done by matching the subclasses of class I A, B, and C antigens using antibodies specific to each subclass. The subclasses of class I antigens are expressed co-dominantly (there are two copies of every gene—one is inherited from each parent—products of both genes are expressed when genes are co-dominant), and

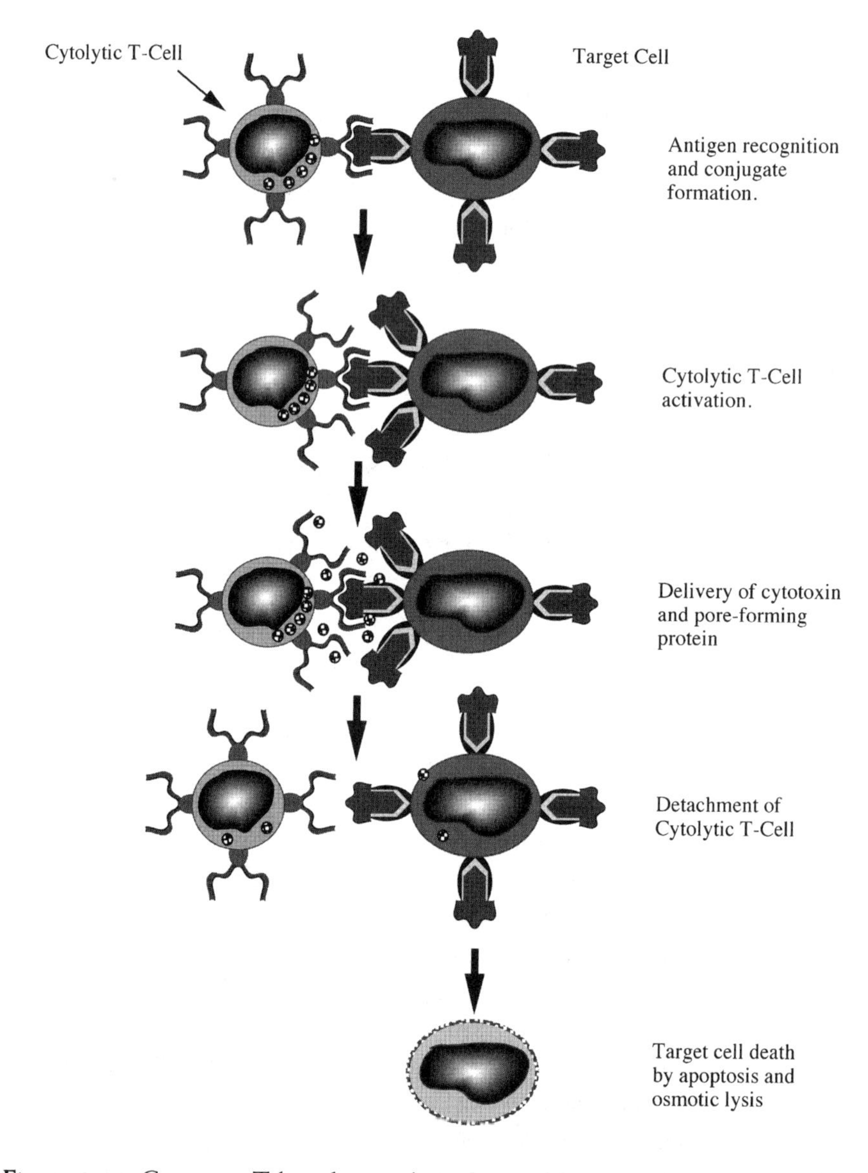

Figure 9.11 Cytotoxic T-lymphocyte lysis. Lysis of foreign cells occurs via $CD8^+$ cell recognition of foreign class I MHC markers and subsequent lysis of foreign cells.

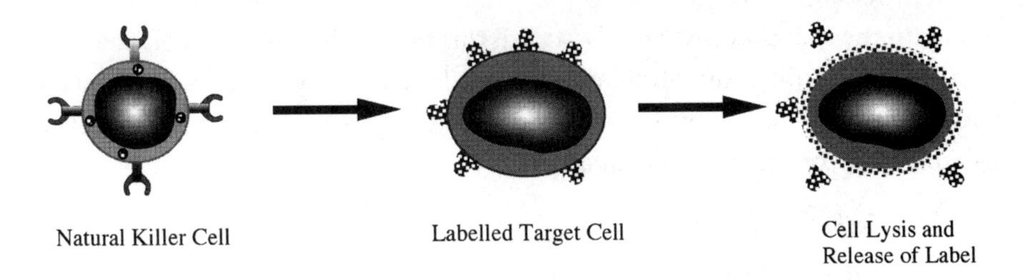

Figure 9.12 Natural killer cell cytotoxicity. Natural killer cells can cause direct lysis of target cells in the absence of MHC molecules, releasing target label.

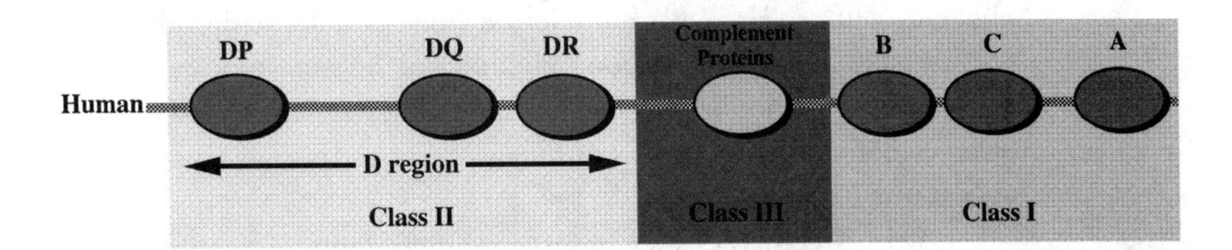

Figure 9.13 Location of MHC markers on chromosome 6. The schematic map of the MHC on human chromosome 6 shows the sites that control synthesis of class I and II markers found on cell surfaces. Each class II region consists of DQ, DR, and DP genes that code for three different cell surface products. These products are recognized by T-helper cells and lead to increased antibody synthesis in the presence of foreign cells. The class I region contains genes for A, B, and C that code for other cell surface molecules that are recognized by suppressor T cells. Both class I and II MHC products should be identical for successful transplantation of cells between donor and recipient.

cells in all tissues express class I antigens derived from both genes inherited by both parents. These markers are present on all foreign cells and are recognized by cytotoxic T cells during graft rejection and during killing of virus-infected cells.

Class II histocompatibility antigens are coded for by at least three loci in the D region: DP, DQ, and DR (Figure 9.13). These regions code for molecules that are primarily found on macrophages and B cells. Class II molecules consist of two noncovalently linked glycoprotein chains with MWs of 34,000 (β chain) and 29,000 (α chain), both of which span the entire cell membrane (see Figure 2.35B). These chains are expressed co-dominantly, and cells contain markers inherited from both parents. The class II markers are important for interactions between immune cells, particularly in presentation of antigens to T cells.

When host and recipient cell class I and II MHC products do not match, this sets into motion a series of reactions that lead to the cutting off of the blood supply to the foreign cells. Simply stated, this means that tissues implanted into hosts that have different MHC markers are killed by $CD8^+$ cells or stimulate release of lymphokines by $CD4^+$ cells. In either event, the implanted cell is killed directly or induces inflammation.

Cell-Derived Mediators

Several mediators act as effectors of inflammation and immunity during wound healing. These mediators include products that derive from cell membrane phospholipids and products that are released from granules within the cell. These mediators stimulate vasodilation, migration of inflammatory cells toward the wound area, and smooth muscle contraction. Release of these mediators prolongs inflammation and phagocytosis of foreign material; this prevents completion of wound healing.

Several cell types release vasoactive mediators as a result of injury, including platelets, mast cells, basophils, neutrophils, endothelial cells, monocytes, macrophages, and injured cells such as keratinocytes (see Figure 9.6). These cells release preformed mediators such as histamine, serotonin, lysosomal enzymes, or metabolites of phospholipids from granules within the cell. Histamine and serotonin (Table 9.1) are directly vasoactive, whereas phospholipid metabolites are indirectly vasoactive by attracting other cells involved in inflammation. Injury

Table 9.1 Cellular mediators of inflammation and immunity

Cell	Agent	Function
Endothelial	Plasminogen activators and inhibitors	Activate and deactivate fibrinolysis
Macrophage (monocyte)	Prostaglandin metabolites, leukotrienes, and platelet-activating factors	Activate neutrophils, monocytes, endothelial cells, vascular smooth muscle cells, vasodilation, and vascular permeability
Mast cell, basophil	Histamine and leukotrienes C_4, D_4, E_4, and SRS	Smooth muscle contraction, increase vascular permeability

Note: SRS = slow-reacting substance.

or activation to cell membrane phospholipids generates arachidonic acid metabolites and prostaglandins that act as vasodilators, including prostaglandins I_2 and E_2 (PGI_2 and PGE_2), and vasoconstrictors (PGD_2 and thromboxane A_2). These are products of the cyclooxygenase pathway and are inhibited by aspirin and indomethacin.

Arachidonic acid metabolites of cell membrane phospholipids, including eicosanoids and leukotrienes, stimulate smooth muscle contraction and enhance vascular permeability via another mechanism, the lipogenase pathway. Neutrophils secrete leukotriene B_4, which is chemotactic to neutrophils, monocytes, and macrophages. Mast cells, basophils, and macrophages secrete leukotrienes C4, D4, and E4, also known as slow-reacting substances of anaphylaxis (SRS) that stimulate contraction of smooth muscle and enhance vascular permeability.

Cytokines are low molecular weight substances that are involved in mediating natural immunity, regulating lymphocyte function, activating inflammatory cells, and stimulating hematopoiesis (formation of blood cells). During natural immunity, cytokines are released from mononuclear phagocytes (monocytes and macrophages) that protect against viral infection and initiate inflammatory reactions that protect against foreign bacteria. These cytokines include type I interferons (IFNs), tumor necrosis factor (TNF), and interleukins 1, 6, and 8 (IL-1, IL-6, and IL-8). Interferons are a group of proteins that are secreted from a variety of cells; they inhibit viral replication and cell proliferation, increase the lytic capability of natural killer cells, and increase the expression of class I MHC molecules, thereby boosting the effector phase of cytolytic T-cell mediated killing of foreign cells (Table 9.2).

Tumor necrosis factor is primarily released from mononuclear phagocytes; it causes vascular endothelial cells to become adhesive for leukocytes (neutrophils, monocytes, and leukocytes), allowing accumulation of neutrophils at sites of inflammation. It also activates neutrophils, eosinophils, and mononuclear phagocytes to kill bacteria and stimulates other phagocytes to produce IL-1, IL-6, TNF, and IL-8. It stimulates antibody production by B cells and augments cytotoxic T-lymphocyte mediated lysis of cells.

Interleukin-1 functions to enhance the proliferation of $CD4^+$ T cells and growth and differentiation of B cells. It also stimulates mononuclear phagocytes and vascular endothelium to produce IL-1 and IL-6. Interleukin-6 causes hepatocytes to synthesize fibrinogen that is required for the acute response to injury. It also acts as the principal growth factor for activated B cells. Interleukin-8 serves as the principal secondary mediator of inflammation.

Lymphokines are cytokines that are elaborated by antigen-activated T lymphocytes most commonly possessing the $CD4^+$ and $CD8^+$ surface markers;

Table 9.2 Cytokines involved in wound healing

Cytokine	Cell source	Target cell	Function
Type I interferon	Mononuclear phagocyte (fibroblasts)	All natural killer cells	Antiviral, antiproliferative
Tumor necrosis factor	Mononuclear phagocyte, T cell	Neutrophil, endothelial cell	Activator, inflammation, blood clotting
Interleukin-1	Mononuclear phagocyte	T cell, B cell, endothelial cell	Co-stimulate, activation of inflammation and blood clotting
Interleukin-6	Mononuclear phagocyte endothelial cell, T cell	T cell, B cell	Co-stimulator
Interleukin-8	Mononuclear phagocyte, endothelial cell, fibroblast, T cell, platelet	Leukocyte	Chemotaxis and activation

they help both cell-mediated and humoral immune responses. Lymphokines that act on other lymphocytes include IL-2, IL-4, and transforming growth factor-β (TGF-β). Interleukin-2 is synthesized by activated $CD4^+$ cells and plays a role in growth and activation of T and NK cells; it also promotes antibody synthesis by B cells (Table 9.3). Interleukin-4 is a growth and differentiation factor for B lymphocytes and may be necessary for both proliferation and antibody secretion by B cells. It is a growth factor for $CD4^+$ cells and mast cells and activates macrophages. Transforming growth factor-β inhibits T-cell proliferation and maturation of cytolytic T lymphocytes. It promotes the synthesis of extracellular matrix and of cellular receptors for matrix proteins.

Lymphokines that act primarily on other inflammatory cells include interferon-γ (IFN-γ), lymphotoxin, IL-5, and macrophage inflammatory protein (MIP). Interferon-γ is an activator of mononuclear phagocytes that kill bacteria and tumor cells. It also increases class I and II MHC molecule expression and

Table 9.3 Lymphokines involved in wound healing

Lymphokine	Cell source	Target cell	Primary effect
IL-2	T cell	T cell	Growth cytokine production
		NK cell	Growth activation
		B cell	Growth antibody synthesis
IL-4	$CD4^+$ cell	T cell	Growth
		B cell	Activation and growth
TGF-β	T cell, mononuclear phagocyte	T cell	Inhibit activation and proliferation
		Mononuclear phagocyte	Inhibit activation
IFN	T cell, NK cell	Mononuclear phagocyte	Activation
		Endothelial cell	Activation
		NK cell	Activation
		All	Increase class I and II MHC molecules
Lymphotoxin	T cell	Neutrophil	Activation
		Endothelial cell	Activation
TL-5	T cell	B cell	Growth and activation
MIP	T cell	Mononuclear phagocyte	Conversion from motile to immotile state

Note: IFN = interferon; IL = interleukin; MHC = major histocompatibility complex; MIP = macrophage inflammatory protein; NK = natural killer; TGF = transforming growth factor; TL = T lymphocyte.

causes differentiation of B and T cells. It activates neutrophils, NK cells, and vascular endothelium. Interleukin-5 stimulates growth of antigen activated B cells and stimulates growth and differentiation of eosinophils. Migration inhibition factor inhibits macrophage motility.

Cell-derived mediators of inflammation cause vasodilation and increase blood flow into the wound area, attract and recruit inflammatory cells, and kill and ingest foreign and mismatched cells in preparation for deposition of new tissue.

Cell Adhesion Molecules Involved in Wound Healing

Cell adhesion is important in many phases of wound healing, including adhesion of white blood cells to vessel walls and adhesion of extracellular matrix cells to the underlying tissue. There are several classes of adhesion molecules that are involved in these processes. Integrins are cell-surface molecules involved in adhesion of cells to each other and to extracellular glycoproteins. In addition, glycoproteins, such as fibronectin, are involved in adhesion to extracellular matrix components, such as collagen fibers. Both of these types of adhesion processes are important in wound healing.

Integrins are a family of cell-surface molecules expressed by most cells. They are composed of two subunits, α and β. There are two principal types of integrins: those that play a role in cell–extracellular matrix interactions and those that play a role in cell–cell adhesion. They are expressed on the surface of leukocytes and extracellular matrix cells to enhance binding to the site of injury. The very late activation antigens (VLA) or β_1 integrins are expressed on many cell types, including fibroblasts, leukocytes, platelets, endothelial cells, and epithelial cells; they function in binding to extracellular matrix components, including collagen, laminin, fibronectin, or vitronectin. They also participate to a lesser degree in cell–cell interactions.

The β_2 integrin family, found primarily on leukocytes, comprises three members: $\alpha_1\beta_2$ (LFA-1), $\alpha_m\beta_2$ (Mac-1), and $\alpha_x\beta_2$. Leukocyte-function-associated antigen 1 (LFA-1) is expressed on all leukocytes and mediates leukocyte binding to cells expressing intercellular adhesion molecules-1, -2, or -3 (ICAM-1, ICAM-2, ICAM-3) expressed on activated vascular endothelium. Interactions between LFA-1 and ICAM-1 coordinate with MHC complex class II antigen binding to mediate antigen presentation to $CD4^+$ T cells, leading to T-cell activation.

The β_3 integrins include integrins expressed on platelets that are involved in adhesion to fibrinogen (and other blood proteins) and to other cells found in extracellular matrix. Other integrin families have been isolated and are less well characterized.

A number of cell adhesion molecules are involved in wound healing, including fibronectin and laminin, lectin–cell adhesion molecules (Lec-CAMs), endothelial cell leukocyte adherence molecule (ELAM-1), and leukocyte adhesion molecule (LAM-1). Others are expressed by endothelial cells and leukocytes, including GMP-140 expressed on the surface of endothelial cells and platelets after stimulation by histamine, thrombin, or cytokines. This molecule mediates recruitment of inflammatory cells. The ELAM-1 is synthesized by cytokine-activated endothelial cells and enhances recruitment of leukocytes. ICAM-1 is expressed by cytokine-stimulated endothelial cells and leukocytes and binds to receptors on neutrophils and macrophages. The ICAM-2 is expressed by endothelial cells and binds to a receptor on neutrophils.

Cell adhesion molecules are key to providing adherence of white blood cells to vessel walls before movement of these cells through the vessel wall into the wound area. They are also important in adhesion of cells to extracellular matrix that is laid down during wound healing.

Growth Factors Involved in Wound Healing

Growth factors are a class of polypeptide chains that are produced by monocytes and connective tissue cells. These factors either up regulate or down regulate the proliferation of connective tissue cells, such as fibroblasts, as well as regulate the synthesis of connective tissue macromolecules (Figure 9. 6). Several growth factors are involved in wound healing, including those elaborated by keratinocytes: basic fibroblast growth factor (bFGF), platelet-derived growth factor (PDGF), transforming growth factor α (TGF-α), transforming growth factor β (TGF-β), and those released by other cells, including epidermal growth factor (EGF), acidic fibroblast growth factor (aFGF), and insulin-like growth factors (IGFs). Both IL-1 and TNF-α have effects on cell proliferation and connective tissue synthesis.

During wound healing, cell proliferation and synthesis of new connective tissue macromolecules occur simultaneously by cells in the wound. The elaboration of specific growth factors and the concentration of each factor are key to the promotion of wound healing and the limitation of deposition of excessive scar tissue. In reality, there is considerable overlap in the wound-healing capability of these molecules. They were originally named for the first target cells that they were shown to stimulate.

Epidermal growth factor primarily stimulates the growth of epithelial cells and plays a role in nerve regeneration, tendon healing, and blood vessel formation (Sporn and Roberts, 1986; Van Brunt and Klausner, 1988) Both FGF-α and FGF-β stimulate endothelial cells and mesenchymal cells. The PDGF produces a chemotactic response in fibroblasts and smooth muscle cells, attracts inflammatory cells, and causes collagen synthesis. Transforming growth factor-β stimulates connective tissue synthesis and activates macrophages to induce angiogenesis. Insulin-like growth factors regulate the synthetic activity of cells. Macrophages secrete PDGF, TGF-α, TGF-β, and bFGF, which serve as major growth stimulators during wound healing.

Growth factors are elaborated in wounds by macrophages, platelets, and connective tissue cells to stimulate cell replication and synthesis of new tissue. They also play a role in preventing excess collagen deposition and scarring.

General Wound-Healing Process: Inflammatory Phase

Cellular injury, as shown in Figure 9.6, sets into motion a cascade of events, including influx of inflammatory cells, release of cytokines, production of collagenases and other enzymes to remove dead cellular and noncellular materials, phagocytosis of bacteria and dead tissue, and activation of blood clotting, kinin, complement, and plasminogen cascades. This leads to the deposition of new granulation tissue that is subsequently remodeled into a scar. Injured keratinocytes in the skin release growth factors, cytokines, and colony-stimulating factors that in combination with PDGF, EGF, and TGF-β cause increased proliferation of fibroblasts, increased collagen synthesis, and epidermal regeneration at the surface of skin. All these events overlap in time; wound healing and the influx and activation of inflammatory cells are events that occur simultaneously. Figure 9.14 illustrates the factors that are involved in vasodilation and subsequent influx of inflammatory cells into the wound area. Vasodilation increases the amount of blood flow into the wound area, allowing more white blood cells to adhere to the inner lining of the vessel wall and subsequently migrate through the wall into the injured area.

At the microscopic level, the changes that are first seen after trauma to tissue include vasodilation caused by release of histamine from mast cells, prostaglandins from all injured cells, and serotonin from platelets. These agents cause a loss of intercellular contacts between endothelial cells that line the blood vessels, resulting in leakage of fluid and migration of cells into the wound area. Vasodilation is further mediated by contact of mast cells, basophils, and platelets with collagen and thrombin and by activation of complement components C3a and C5a. Other stimulators of vasodilation include degradation products of fibrin networks, contact of factor XII with collagen (activates the kinin system), and release of leukotrienes and lymphokines. Release of lymphokines and leukotrienes activates lymphocytes and continues to up regulate vascular permeability. The migration of phagocytic cells into the wound area is critical to removal of dead tissue that precedes new tissue deposition.

Removal of dead cells, clotted blood proteins, and denatured tissue components is accomplished via phagocytosis. During phagocytosis, dead cellular

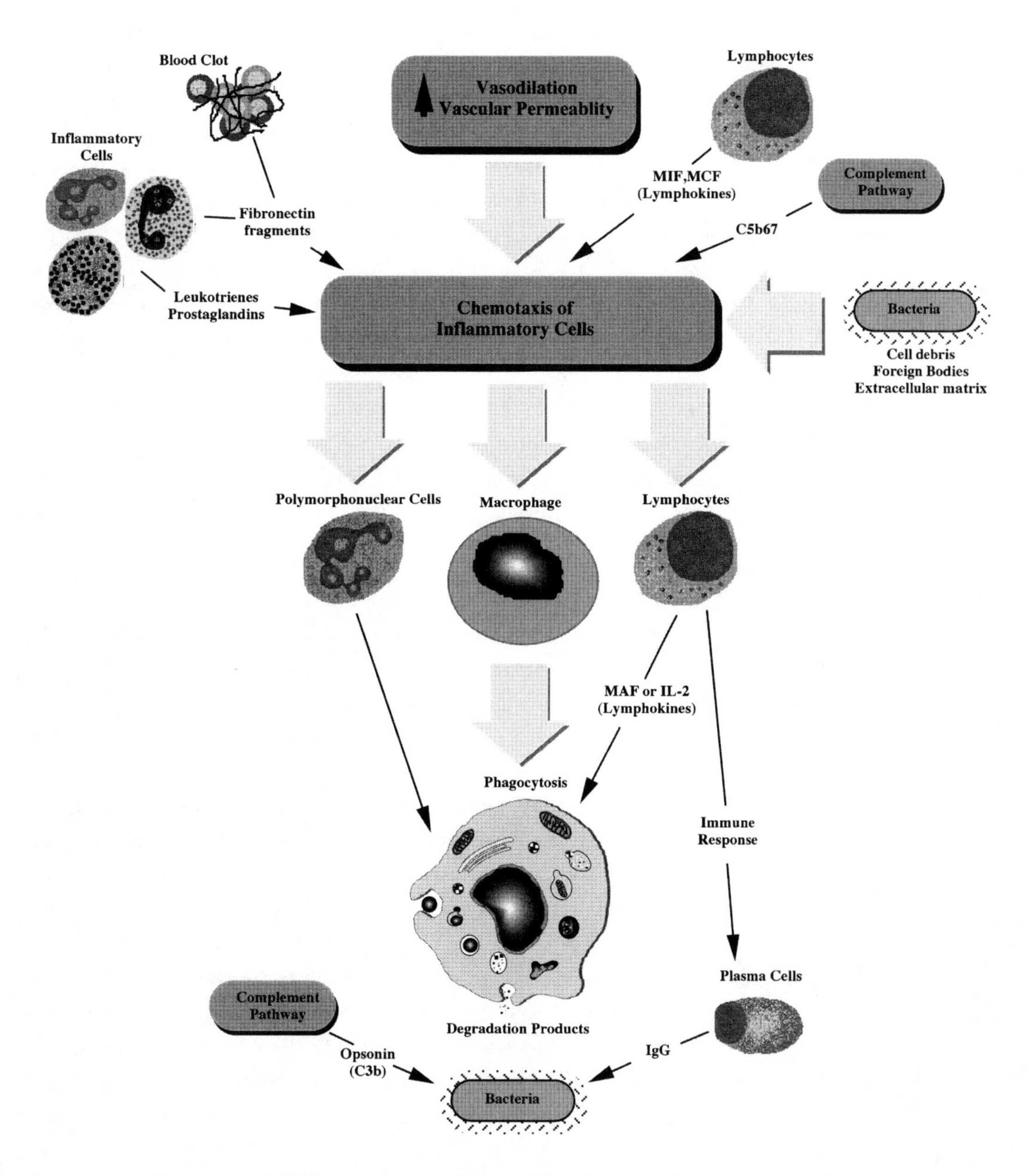

Figure 9.14 Effectors of chemotactic stimulators. The drawing illustrates the release of chemotactic stimulators, including leukotrienes, prostaglandins, lymphokines, and complement, and the resultant effects, including phagocytosis, release of antibodies, and destruction of foreign cells.

material and denatured protein are engulfed first by neutrophils during acute inflammation and subsequently by macrophages during chronic inflammation. Granulation tissue is being laid down by endothelial cells and fibroblasts at the same time tissue debris is being removed. The balance between inflammation and removal of dead tissue and the formation of new tissue is regulated by the rate at which new inflammatory cells are recruited into the wound area. Once the chemotaxis of inflammatory cells is decreased by removal of the stimulus, which is either cellular trauma, release of lymphokines, activation of the complement pathway, activation of blood coagulation, release of leukotrienes, and vascular vasodilation, then the proliferative phase of wound healing can begin.

The involvement of the immune system during inflammation occurs when foreign cells and tissues present at the wound site by virtue of the mismatch of the class I and class II cellular markers. These foreign cells exhibit surface markers that trigger either proliferation of plasma cells and antibody production by T-helper cells or direct cell lysis by cytolytic T cells. The macrophage plays a role in triggering antibody production and T-cell proliferation.

Proliferative Phase of Wound Healing

The inflammatory phase limits bleeding and attracts white blood cells to the wound area to clean up damaged tissue, bacteria, and dead cells. Once inflammation is triggered, this also triggers proliferation of fibroblasts and endothelial cells that are involved in laying down new connective tissue at the site of injury. The proliferative phase involves migration of fibroblasts, endothelial cells, and epithelial cells into the wound area, followed by synthesis and deposition of new tissue as a consequence of release of growth factors (Figure 9.15). Chemical mediators, such as platelet-activating factor, cause platelet aggregation and release of PDGF, which in turn stimulates fibroblast and endothelial cell migration.

Fibroblasts initially synthesize and deposit extracellular matrix in a random fashion to form granulation tissue. Granulation tissue consists of fibroblasts surrounded by extracellular matrix and endothelial cells involved in capillary formation. Synthesis and deposition of granulation tissue is enhanced by FGF, EGF, ascorbic acid, oxygen, glycosaminoglycans, proteoglycans, and glycoproteins. In contrast, IL-1 and prostaglandins inhibit synthesis of extracellular matrix.

Biochemically, the composition of extracellular matrix changes during wound healing. Initially, the collagen type III/type I ratio is high and decreases with healing time. The concentration of hyaluronan is initially high in dermal

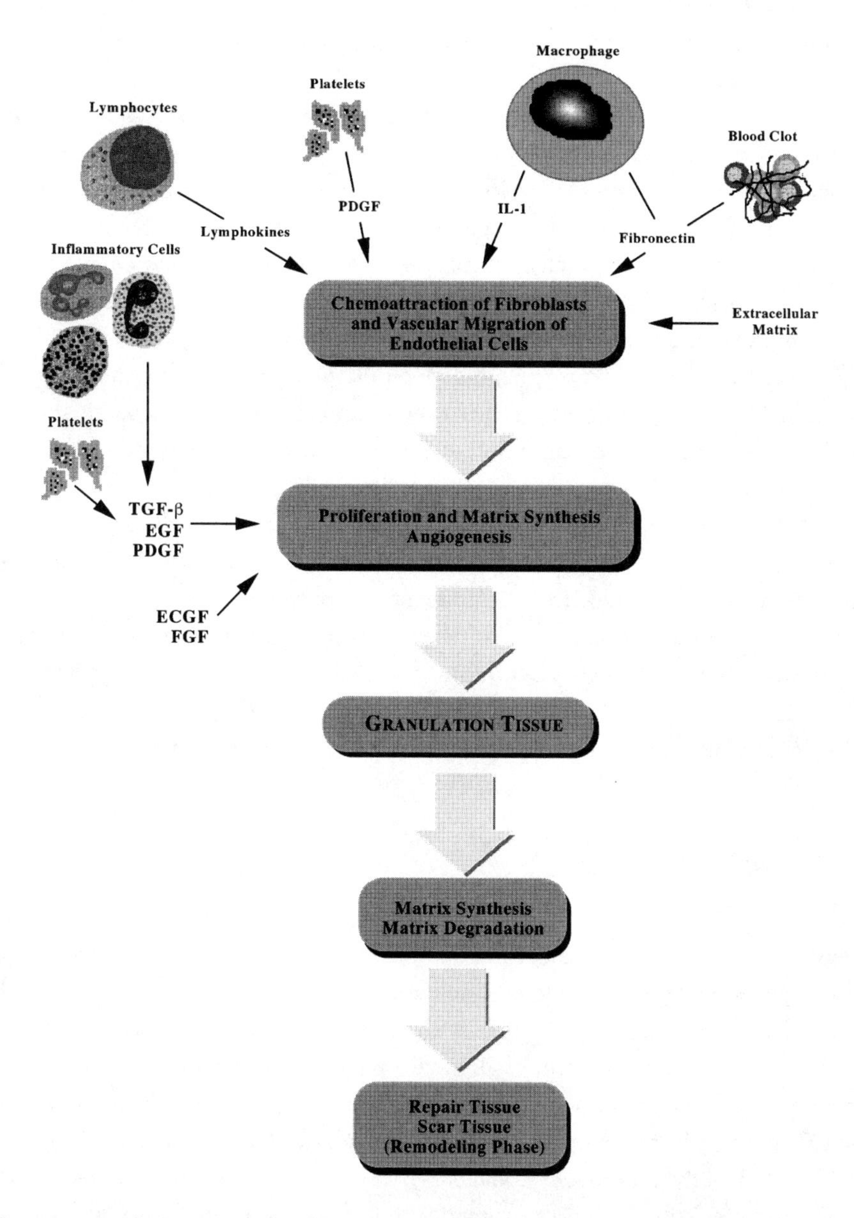

Figure 9.15 Interaction of inflammatory and extracellular matrix cells. Shown are the relationships among release of inflammatory mediators and growth factors, attraction of endothelial cells and fibroblasts into the wound area, synthesis and degradation of granulation tissue, and remodeling and deposition of scar tissue. Prolonged release of inflammatory mediators increases the amount of granulation tissue deposited and results in abnormal scarring, whereas rapid resolution of inflammation promotes repair and minimizes scarring.

wound tissue and is subsequently replaced with chondroitin and dermatan sulfates.

Proliferating and migrating fibroblasts contain large actin bundles within the cytoskeleton that are present in the cell cytoplasm near the cell membrane. These bundles are parallel to the long axis of the cell, and they appear to connect the cell membrane through integrins and heparan sulfate proteoglycan to extracellular matrix receptors, such as fibronectin. These fibroblasts have been referred to as myofibroblasts and may be involved in contraction of the wound by movement of the edges during healing. Angiogenesis and the formation of new blood vessels is triggered by macrophage-derived factors, low oxygen concentration, lactic acid accumulation, vasoactive agents, and heparin and heparan sulfate. Replication and migration of endothelial cells results in capillary bed formation from the wound edges toward the wound center.

Reepithelialization in skin is another event that frequently occurs during the proliferative phase. This event corresponds to the movement of epithelial cells from the free edges to the center of the defect. Release of chemotactic stimuli, including fibronectin and other matrix components, leads to movement of epithelial cells over newly synthesized basement membrane components. Epithelialization requires deposition of granulation tissue that contains thin collagen fibers within the wound and requires the defect contour to be smooth.

Remodeling Phase

Remodeling is an event that occurs immediately after granulation tissue is laid down in the wound. The role of granulation tissue is to provide material to fill the defect. After the defect is filled with randomly organized collagen fibers, it is then remodeled in the form of aligned collagen fibers, which is also known as normal scar. Scar formation requires degradation of granulation tissue by release of collagenases from inflammatory and other cells found in the extracellular matrix as well as by subsequent collagen synthesis and deposition. Deposition of large-diameter collagen fibrils, primarily type I, are associated with an increase in wound tensile strength. A direct relationship has been observed between accumulation of collagen within the wound and the development of mechanical strength during the initial period of wound healing. Even after six months, the strength of dermal wounds normally is less than 80% that of normal skin. In deep wound sites, excessive accumulation of fibrous collagen can be found in the form of whirls and nodules; this type of healing is referred to as hypertrophic scar tissue. This tissue is characterized as having a reduced extensibility and appears to have increased rigidity. Hypertrophic scar tissue occurs as a consequence of excessive granulation tissue deposition. The strength of scar tissue never reaches more than 80% of the original skin strength, which is believed to be due to the decreased diameter of collagen fibrils and fibers in repair tissue compared to that of the original tissue.

Capsule Formation Around Implants

At the interface surrounding implants, the trauma and cell necrosis associated with insertion of the implant (see Figure 9.6) lead to the release of cytokines and growth factors and initiate inflammation and healing responses. The inflammation associated with implant placement causes an influx of inflammatory cells. If left unresolved, this results in implant loosening, migration, degradation, and eventual failure. A small inflammatory response is observed around all implants, leading to granulation tissue formation, remodeling, and capsule formation. However, if the implant releases low molecular weight components, it moves or causes pressure on the surrounding tissue, then the inflammatory process becomes chronic, and the capsule deposited around the implant grows in size.

A capsule forms around all implants that are stable. Its size depends on the porosity of the surface and the biodegradation rate of the implant. Implants with pores greater than about 50 mm in diameter are infiltrated during the proliferative phase by cells. In this case, only a small capsule is formed and becomes incorporated into the implant surface. If the implant is biodegradable and the biodegradation occurs within about 10 weeks of implantation, no capsule is formed. If the implant is permanent and does not biodegrade, then a small capsule remains throughout the life of the implant. Solid implants are associated with larger capsules, especially if the surface of the implant is unstable. In the extreme case, capsules can grow in thickness until they begin to collapse or cause contraction of the implant. This is discussed further in the section on pathobiological responses to silicone implants.

Cartilage Wound Healing

Unlike skin and extracellular matrix throughout the body, cartilage is a specialized tissue with specialized wound-healing properties. This is a consequence of its special role in dissipating loads in joints (hyaline cartilage), transferring loads between tendons and bone (fibrocartilage), and providing flexible support to external structures (elastic cartilage). The specialized wound-healing response is dictated by chondroblasts and chondrocytes, special cells that elaborate the extracellular matrix composed of proteoglycans and collagen fibrils; it is the amount, type, and organization of these components that differentiates the various types of cartilage. The wound-healing response sometimes alters the structure and properties of the repair tissue compared to the normal tissue because each cartilage type has a specific architecture that results in a particular set of mechanical properties. It therefore becomes important to elucidate the structure of each cartilage type before the wound-healing behavior can be understood.

Because cartilage is composed primarily of type II collagen and proteoglycans that give it unique mechanical properties, any wound-healing response that does not lead to replacement of the appropriate collagen types and proteoglycans will result in abnormal function of the tissue. As discussed earlier with reference to Figure 9.6, the wound-healing response in general leads to the deposition of type I collagen, either in the form of normal extracellular matrix fibers or as aligned fibrils in the form of a scar. The deposition of large-diameter type I collagen fibrils in cartilage during wound healing would alter the structure of cartilage and negatively affect the physiology of the tissue. Therefore, it is important to understand under what conditions cartilage will regenerate by deposition of type II collagen and proteoglycans as opposed to healing by deposition of scar tissue.

Articular cartilage is difficult to regenerate. Loss of articular cartilage from the joint surface associated with osteoarthritis currently is corrected by total joint replacement. It is far more commonly observed that articular cartilage is repaired by deposition of fibrocartilage than by hyaline cartilage. Therefore, it is important to determine if it is possible to prevent fibrocartilage replacement of hyaline cartilage.

Articular cartilage repair has been studied extensively because of the difficulty of repairing worn-out cartilage on the knee joint. The literature is unclear concerning the conditions that support cartilage regeneration. Some workers have reported that chondrocytes found in the femoral condyle of immature animals would repair a defect by proliferation and hyaline cartilage matrix synthesis (Calandruccio and Gilmer, 1962). Another report suggests that partial-thickness defects that do not reach the subchondral bone do not heal after 66 weeks; however, full-thickness defects after 16 weeks are completely filled with immature cartilage (DePalma et al., 1966). A third report concluded that articular cartilage defects heal by deposition of fibrous tissue and fibrocartilage (Campbell,

1969), and a fourth indicated that full-thickness defects less than 9 mm in diameter healed after three months (Convery et al., 1972). In the study by Convery and associates, full-thickness wounds greater than 9 mm in diameter did not heal.

Three factors appear important in determining whether cartilage repair occurs: the depth of the defect, the maturity of the cartilage, and the position of the defect on the surface. Even under optimum conditions, however, repair of cartilage defects with hyaline cartilage is difficult because of the slow rate at which chondrocytes proliferate in comparison with the rate of proliferation of cells that produce scar tissue. In most cases, the repair is achieved by a mixture of fibrocartilage and fibrous tissue. The amount of fibrocartilage can be maximized by increasing the depth of the defect through the subchondral bone. The mechanism of fibrocartilage formation appears to be mediated via the proliferation and differentiation of mesenchymal cells of the marrow, and not by chondrocytes of residual cartilage (Shapiro, 1993).

Fibrocartilage regenerates more readily than does articular cartilage, which might be expected based on the previously cited studies. Robinson (1993) reported that mandibular condylar cartilage regeneration required six months to complete after creation of a full-thickness defect. This supports the observation that fibrocartilage is found after wounding of cartilage surfaces in general.

Possible Ways to Improve Hyaline Cartilage Regeneration

Because of the poor regeneration capacity of articular cartilage, several different approaches have been used to improve hyaline cartilage healing. Biologic grafts have been used in the wound area, including perichondrium, osteochondral, meniscus, epiphyseal growth plate, periosteum, and mandibular cartilage (Silver and Glasgold, 1995). Positive results have been reported after use of perichondral grafts and vascularized perichondrium.

Another approach has been to isolate autogenous or allogeneic chondrocytes from cartilage and then expand these cells in culture in the absence or presence of substrates. Reports of successful cartilage regeneration using cell-culture-expanded autogenous chondrocytes or progenitor cells can be found in the literature (Grande et al., 1987; Nakahara, 1991).

The final approach has been to use exogenously administered growth factors to stimulate cartilage regeneration. The major growth factors affecting cartilage are somatomedins or insulin-like growth factors (IGFs). Other growth factors, including FGF, PDGF, cartilage-derived growth factor (CDGF), and EGF, are peptides similar to the somatomedins and have been shown to affect cartilage metabolism.

Tendon and Ligament Healing

The mechanism of tendon and ligament healing is similar to that of other soft tissues; these tissues heal by inflammation, granulation tissue formation, and remodeling. In a cut tendon, blood initially fills the defect. Blood-derived activation of clotting, complement, and other cascades initiates inflammation and migration of inflammatory and connective tissue cells into the wound. Fibroblasts within the tendon, called tenocytes, migrate into the area of injury, and inflammatory cells from other connective tissue layers that hold the tendon together (epitenon and endotenon) migrate to the site of injury. All structures surrounding the tendon, including the synovial sheath, subcutaneous tissue, fascia, and bone periosteum, provide fibroblasts and endothelium for healing. These cells migrate to the wound area and begin to proliferate. Collagen synthesis begins about one week after injury, and subsequently capillaries begin to form. Type I collagen is the predominant type of collagen in granulation tissue.

During the remodeling phase that occurs one to two months after injury, the strength of the tendon begins to increase as type I collagen is laid down and cross-linking occurs. Mobilization of the limb is associated with increased proteoglycan content and mechanical strength. The strength of the repaired tendon never reaches that of the original tendon, which is believed to reflect the decreased diameter and amount of collagen fibrils in the repair tissue.

Peripheral Nerve Repair

Transection of a peripheral nerve produces a discontinuity to the axon and to the basement membranes that line the outer membrane of nerve tubes. After transection, the axons located in the stump furthest removed from innervation (the distal stump) are phagocytosed by Schwann cells and macrophages. Schwann cells are wrapped around the axon and make the myelin sheath. In the distal stump, Schwann cells proliferate, and the basement membrane tubes remain intact. Axonal sprouting occurs from the proximal nerve stump, across the site of injury, and into the distal stump. Misdirection of the sprouting axons results in failure to reestablish contact with the original target organs and thus impairs functional recovery. Deposition of fibrous tissue during granulating and remodeling phases of wound healing inhibits axon sprouting and reconnection of the nerve stumps across the wound area.

Hard Tissue Repair

Bone heals via two mechanisms that regenerate hard tissue: endochondral and membranous bone formation. Endochondral bone formation involves the formation of a cartilage analogue that subsequently mineralizes. Membranous bone formation involves mesenchymal cells derived from the periosteum that subsequently differentiate into osteoblasts at the wound site. As with other tissues, blood present within the wound area causes activation of the biological cascades, causing chemotaxis of inflammatory cells and the initiation of inflammation.

Endochondral bone formation occurs at the epiphyseal plate at the ends of long bones and is responsible for long bone growth. This type of regeneration involves formation of a cartilage analogue that is gradually transformed into calcified tissue, similar to what happens at the interface between cartilage and bone on the joint surface. The bone that is laid down is characterized by layers of parallel lamellar bone.

In contrast, membranous bone formation does not require a cartilage intermediate and is directly formed by mesenchymal cells that differentiate into osteoblasts that lay down osteoid directly. Membranous bones include the calvarium, facial bones, clavicle, mandible, and bone immediately subperiosteal. Hard tissue formed in this manner is called osteonic bone.

Bone regeneration is actually a combination of membranous and endochondral processes. The stages of repair include induction, inflammation, soft callus formation, callus calcification, and remodeling. The inductive phase involves recruitment of viable cells from the periosteum or endosteum to migrate into the wound to begin to make new bone. Other cells, including fibroblasts, endothelial cells, and muscle cells, may differentiate into cartilage and bone-forming cells in the presence of factors in the wound.

Osteoinductive proteins that promote differentiation of cells into bone-forming cells are found in the wound area. These factors, including bone morphogenetic protein (BMP) and TGF-β, have been isolated and tested for their bone-inducing capabilities.

Inflammation begins immediately upon wounding and can last for a period of months, depending on whether the stimulus is removed. In the presence of an implant, the inflammatory phase can go on for years. Soft callus formation involves deposition of a combination of osteoid from beneath the periosteum through membranous bone formation and formation of cartilage by endochondral bone formation. Once the soft callus fills the wound, it begins to mineralize, and by three to four weeks, it has sufficient rigidity to prevent gross motion of the fracture.

During the hard callus phase, the soft callus is gradually converted into bone, and through macrophage and osteoclast activity, all the dead tissue is removed. Callus laid down during the mineralization process is converted into fiber and woven bone during remodeling, which takes three to four months in humans. The final phase is remodeling of fiber and woven bone into lamellar bone through osteoclast and osteoblast activity.

Cardiovascular Wound Healing

Wound healing of vessels in the cardiovascular system is a very complex process. Vascular channels that undergo thrombosis to the point that the vessel lumen is occluded must be opened surgically or via the use of enzymes so that inflammation and wound healing do not result in scarring and permanent occlusion of the vessel lumen. Much of our understanding of cardiovascular wounds comes from studying the wound-healing response to porous vascular grafts.

Porous vascular grafts have been observed to heal in fashion suitable to make a three-layered arterial wall that consists of an external fibrous capsule, a middle layer of the porous prosthesis, and an inner layer of fibrous capsule (Wesolow, 1982). Healing is initiated by the accumulation of blood in the interstices of the graft, leading to blood coagulation and fibrin deposition to produce a blood-tight graft. Fibrin deposition attracts inflammatory and connective tissue cells into the graft, leading to phagocytosis of fibrin and deposition of new collagen during the early phases of wound healing. Remodeling of the collagenous tissue leads to scar tissue deposition and contraction of the graft. Scar tissue contraction can also lead to perigraft bleeding and hematoma or aneurysm. Degeneration of the inner fibrous capsule can lead to graft occlusion or sloughing of the inner capsule. These complications lead to new healing processes that must be resolved for wound healing to be complete.

The formation of the neointima in animal aortas treated with textile prostheses occurs by the ingrowth of smooth muscle cells and endothelium from the proximal and distal aortic stumps over the inner fibrin layer. Inner layers of cells derive from cells entering the prosthesis body, including fibroblasts, myofibroblasts, and spindle-shaped smooth muscle cells.

Summary

Healing is a complex set of events that begins by triggering blood clotting and activation of the fibrinolysis, complement, and kinin cascades. Once these cascades are activated and inflammatory cells are recruited into the wound area, these cells release cytokines that attract other inflammatory cells into the wound and induce phagocytosis of necrotic material. Growth factors released by platelets and inflammatory cells induce migration of connective tissue cells into the wound and proliferation of these cells. Connective tissue is next synthesized for repair of the extracellular matrix. Macrophages and neutrophils within the wound digest dead cells and denatured collagen, making space for deposition of new material. Type III collagen is initially deposited as thin fibrils. During remodeling, thick type I collagen fibrils are deposited that are associated with an increased tensile strength. The healing response is completed when all necrotic tissue is removed and the new extracellular matrix fills the wound.

Suggested Reading

Abbas A.K., Lichtman A.H., and Pober J.S., Cellular and Molecular Immunology, W.B. Saunders, Philadelphia, chapters 2, 3, 5, 7, 11, and 13, 1991.

Barlow Y. and Willoughby J., Pathophysiology of Soft Tissue Repair, Br. Med. Bull. 48, 698, 1992.

Beutler B. and van Huffel C., Unraveling Function in the TNF Ligand and Receptor Families, Science 264, 667, 1994.

Calandruccio R.A. and Gilmer W.S., Proliferation, Regeneration, and Repair of Articular Cartilage of Immature Animals, J. Bone Joint Surg. 44A, 431, 1962.

Campbell C.J., The Healing of Cartilage Defects. Clin. Orthop. 64, 45, 1969.

Clark J.M., The Organization of Collagen Fibrils in the Superficial Zones of Articular Cartilage, J. Anat. 171, 117, 1990.

Convery F.R., Akeson W.H., and Keown G.H., The Repair of Large Osteochondral Defects: An Experimental Study in Horses, Clin. Orthop. 82, 253, 1972.

DePalma A.F., McKeever C.D., and Subin D.K., Process of Repair of Articular Cartilage Demonstrated by Histology and Autoradiography With Tritiated Thymidine, Clin. Orthop. 48, 229, 1966.

Elner S.G. and Elner V.M., The Integrin Superfamily and the Eye, Invest. Ophthalmol. Vis. Sci. 37, 696, 1996.

Kehrl J.H., Alvarez-Mon M., Delsing G.A., and Fauci A.S., Lymphotoxin Is an Important T Cell-Derived Growth Factor for Human B Cells, Science 238, 1144, 1987.

Grande D.A., Singh I.J., and Pugh J., Healing of Experimentally Produced Lesions in Articular Cartilage Following Chondrocyte Transplantation, Anat. Rec. 218, 142, 1987.

Horton W.A., Morphology of Connective Tissue: Cartilage, in Connective Tissue and Its Heritable Disorders, Wiley-Liss, New York, p. 73, 1993.

Henney C.S., The Interleukins as Lymphocyte Growth Factors, Transplant. Proc. 21, 22, 1989.

Jerusalem C., Hess F., and Werner H., The Formation of a Neo-Intima in Textile Prostheses Implanted in the Aorta of Rats and Dogs, Cell Tissue Res. 248, 505, 1987.

Nakahara H., Goldberg V.M., and Caplan A.I., Culture-Expanded Human Periosteal-Derived Cells Exhibit Osteochondral Potential In Vivo, J. Orthop. Res. 9, 465, 1991.

Pierce G.F., Macrophages: Important Physiological and Pathologic Sources of Polypeptide Growth Factors, Am. J. Respir. Cell Mol. Biol. 2, 233, 1990.

Robinson P.D., Articular Cartilage of the Temporomandibular Joint: Can It Regenerate? Ann. R. Coll. Surg. Engl. 75, 2316, 1993.

Ruoslahti E., Integrins, J. Clin. Invest. 87,1, 1991.

Samuelsson B., Dahlen S.-E., Lindgren J.A., Rouzer C.A., and Serhan C.N., Leukotrienes and Lipoxins: Structures, Biosynthesis, and Biological Effects, Science 237,1171, 1987.

Shapiro F., Koide S., and Glimcher M.J., Cell Origin and Differentiation in the Repair of Full-Thickness Defects of Articular Cartilage, J. Bone Joint Surg. 75A, 532, 1993.

Silver F.H and Glasgold A.I., Cartilage Wound Healing: An Overview, Otolaryngol. Clin. North Am. 28, 847, 1995.

Silver F.H. and Maas C.S., Biology of Synthetic Facial Implant Materials, Facial Plast. Surg. Clin. North Am. 2, 241, 1994.

Silver F.H. and Parsons J.R., Repair of Skin, Bone and Cartilage, in Applications of Biomaterials in Facial Plastic Surgery, edited by A.I. Glasgold and F.H Silver, CRC Press, Boca Raton, FL, p. 65, 1991.

Sporn M.B. and Roberts A.B., Peptide Growth Factors and Inflammation, Tissue Repair, and Cancer, J. Clin. Invest. 78, 329, 1986.

Tiggs M.A., Casey L., and Kosland M.E., Mechanism of Interleukin-2 Signaling: Mediation of Different Outcomes by a Single Receptor and Transduction Pathway, Science 243, 781, 1989.

Van Brunt J. and Klausner A., Growth Factors Speed Wound Healing, Bio/ Technology 6, 25, 1988.

Weiss C., Rosenberg L., and Helfet A.J., An Ultrastructural Study of Normal Young Adult Human Articular Cartilage, J. Bone Joint Surg. 50A: 663, 1968.

Wesolow A., The Healing of Arterial Prostheses: The State of the Art, Thorac. Cardiovasc. Surg. 30, 196, 1982.

10

Pathobiological Responses to Implants

Introduction

The introduction of implants in the 1950s revolutionized the field of medicine. Vascular grafts, heart valves, orthopedic implants, contact lenses, wound dressings, and other implants have extended the life spans and improved quality of life. However, the introduction of this new technology is also associated with pathobiological responses that in some cases can be life-threatening. Many of the adverse responses to implants are associated with interactions between components of biological pathways and inflammatory mediators with the implant surface or materials emanating from implants. These responses include complement depletion and increased predisposition to infection, local inflammation and foreign body response to implant materials, symptoms of systemic connective-tissue-like disorders with silicone implants, phagocytosis of wear debris by macrophages and migration to local lymph nodes, immediate hypersensitivity to implant materials, nonphysiological calcification leading to implant failure, extrusion of plastic implants from beneath the surface of tissues, and intimal hyperplasia associated with vascular stents.

The number of complications associated with the use of implants is small, and few epidemiological studies have been published identifying the risks. These studies are usually retrospective reviews of adverse effects after the appearance of a number of clinical reports. This has made the design of the next generation of devices more complicated because evaluation of the outcome associated with the previous generation of devices sometimes takes several decades to become public. The end result of this problem is that our knowledge of the pathobiological responses to implants in general is limited except in the few cases where a negative outcome has been associated with major health concerns. In this chapter, some of the pathobiological responses that have been observed with implants are discussed.

Pathobiological Complications with Silicone Implants

Silicone materials are used for facial, breast, and finger implants. In addition, they have been used in the form of tubes and shunts for relieving excess pressure in different areas of the body. The successful outcome of procedures with most silicone implants led to the wide use of this material in medicine. In the 1990s, potential side effects believed to be associated with the use of silicone-gel-filled implants were brought to the public's attention. Some of the local side effects, including capsular contraction leading to pain and distorted breast shape, were already well known; however, reports of systemic symptoms normally associated with connective tissues disorders, including fatigue, muscle weakness, joint problems, and loss of memory, began to concern the public.

Chemistry of Silicone Implants

Silicone-containing polymers are derived by reaction of elemental silicon with methyl chloride. This reaction forms dimethyl chlorosilane, which when hydrolyzed in water generates poly(dimethylsiloxane), commonly known as silicone (Figure 10.1). This generates a high molecular weight gum that is mixed with

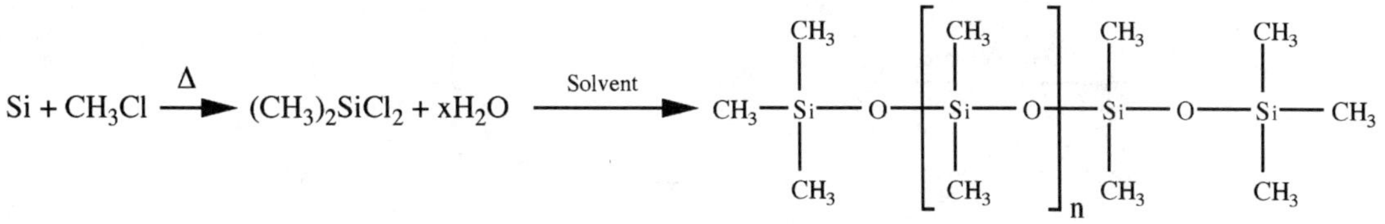

$$\mathrm{Si} + \mathrm{CH_3Cl} \xrightarrow{\Delta} (\mathrm{CH_3})_2\mathrm{SiCl_2} + x\mathrm{H_2O} \xrightarrow{\text{Solvent}} \mathrm{CH_3{-}Si(CH_3)_2{-}O{-}[Si(CH_3)_2{-}O]_n{-}Si(CH_3)_2{-}O{-}Si(CH_3)_2{-}CH_3}$$

Figure 10.1 Synthesis of silicone polymers. The diagram shows synthesis of poly(dimethylsiloxane) from elemental silicon and methyl chloride at elevated temperature, producing dimethylchlorosilane. Poly(dimethylsiloxane) prepolymer is prepared by reacting dimethylchorosilane with water that is later repolymerized, copolymerized, or blended with other siloxanes to form the final material.

silica particles (SiO_2) and cross-linked to form a stable polymer (Frisch, 1983). Medical-grade silicone fluid is poly(dimethylsiloxane) with trimethylsiloxy groups blocking the ends of the polymer chains. It is a clear, colorless, odorless, tasteless fluid with a MW more than 800. Different viscosity grades are available and are made by varying the length of the polymer chain. Medical-grade silicone rubber is made from poly(dimethylsiloxane) chains with MWs more than 500,000. These chains are mixed with fumed silica particles to reinforce the product. Cross-linking is achieved by introduction of vinyl derivative chains of vinylmethylsiloxane or by incorporation of peroxides to cross-link via a free radical mechanism. Polymer, reinforcing particles, and cross-linking agents are mixed using a compounding mill consisting of two metal rollers that squeeze the material into a sheet. Materials after rolling can be screened to remove large particles. The final stock is then processed into final form by compression molding, transfer molding, extrusion, calendering, dispersion coating, and hand lay-up (Figure 10.2). Each of these processes can contribute impurities that end up in the final product.

Local Reactions to Silicone Implants

Injectable silicone has been used to augment breast tissue in women since the 1960s (Orentreich and Orentreich, 1991). Although originally manufactured in the United States by Dow Corning Corporation in an electrical grade, it was subsequently manufactured into food and medical grades for other applications. Medical-grade 360 silicone fluid, as it was known in the trade, and solid silicone were used in sutures, disposable hypodermic syringes and needles, testicular implants, uterine rings, scleral buckling materials, coverings for cardiac pacemaker

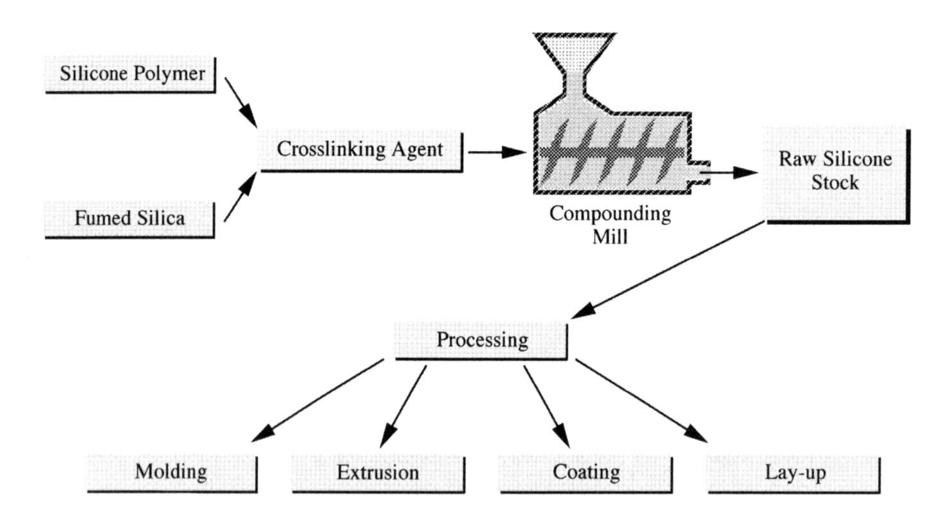

Figure 10.2 Steps used to prepare medical-grade silicone. Silicone polymer, polymerized as described in Figure 10.1., is processed by adding fused silica and crosslinking agents during compounding on a mill to form raw silicone polymer stock. The stock polymer is then processed into final form by molding, extrusion, coating, or laying-up.

leads, and renal dialysis tubing. It was also used for coating containers for medicinals and blood products and in protective skin coatings. In 1965, Dow Corning developed a more highly purified form of this material to be used for soft tissue augmentation; however, this material was never approved for use by the Food and Drug Administration (FDA). Injectable silicone oil was introduced to augment women's breast tissue without FDA approval.

Injection of silicone into soft tissue and interpretation of the local responses is complicated by a number of parameters. The first is the compounding of silicone oil with other materials, including olive oil, which sometimes led to complications that were disastrous. Many of these complications were viewed as resulting from use of unpure materials or from the technique used for injecting the material. Some physicians still believe that silicone fluid does not induce abnormal reactions but that the impurities in the material used to dilute the silicone or the injection method cause the tissue response (Orentreich and Orentreich, 1991). In any event, it is accepted that one local problem associated with injection of large amounts of silicone is the migration of the material along tissue cleavage planes, which in some locations is promoted by muscular contraction. The nonsilicone component of mixtures was hypothesized to cause an inflammatory reaction around the silicone to trap it in the injection site.

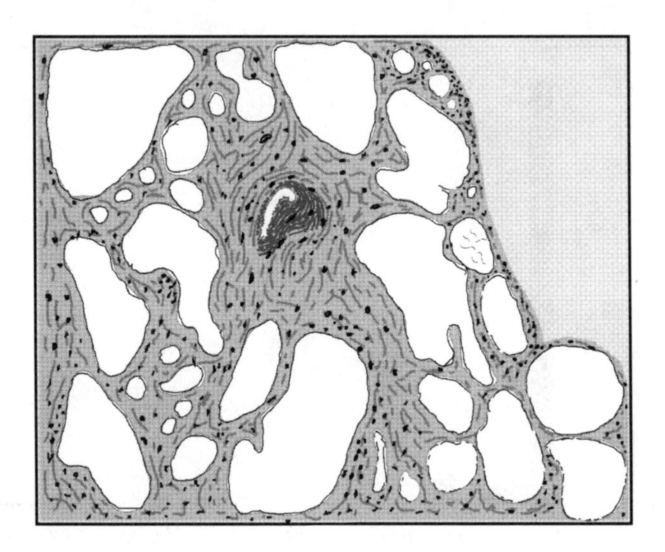

Figure 10.3 Vacuoles seen in tissues surrounding silicone-gel-filled implants. Illustration of empty vacuoles (white areas) that are seen surrounding gel-filled implants. Typically, the regions surrounding breast implants contain aligned collagen fibers in the form of scar tissue, fibroblasts, some fat tissue and glandular structures, macrophages, and giant cells. This diagram shows that much of the space between vacuoles contains aligned collagen fibers (gray areas) that are laid down as a result of a chronic inflammatory response as well as nuclei of fibroblasts and mononuclear inflammatory cells (black dots).

In addition to local problems with injection of silicone oil, implantation of medical-grade silicone products in the form of solid implants is associated with granuloma formation as part of a nonspecific immunological response. Local reactions include capsule formation around the implant, coating with a layer of macrophages, and giant cell and granuloma formation at the site of implantation. Some of this inflammation may be the result of silicone-induced fat cell necrosis that mediates chronic inflammation and fibrosis (Silver et al., 1995).

Shanklin and Smalley (1996) have summarized the pathological responses that have been documented in the literature with silicone. They report that the histopathology of lesions arising in capsular tissues surrounding and adjacent to silicone implants consist of macrophages, sometimes in granulomas; lymphocytes, often in surrounding blood vessels; and fibrous tissue. Silicone is found in vacuoles that appear as empty spaces after the tissue has been processed for histology (Figure 10.3). The vacuoles range in size (Figure 10.4). In some areas, macrophages contain silicone deposits too small to see under the light micro-

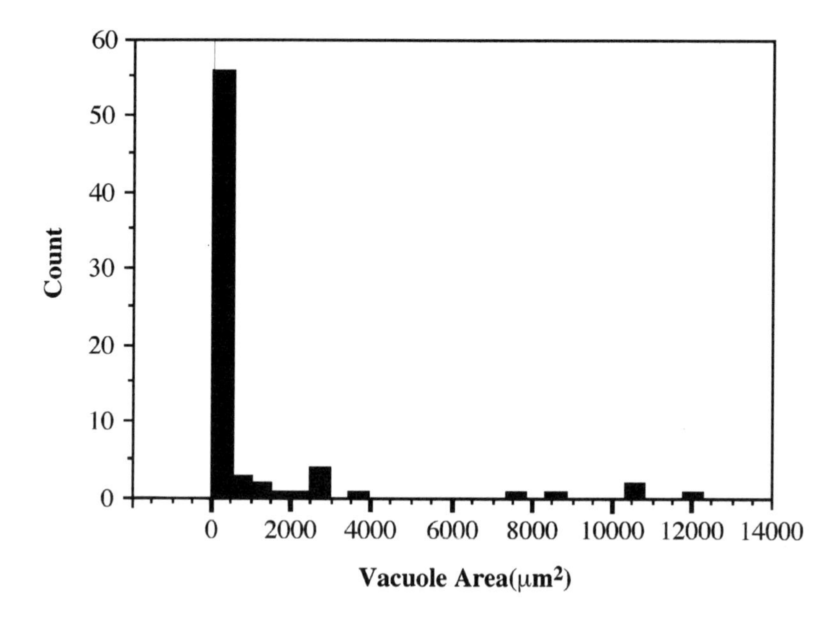

Figure 10.4 Area distribution of vacuoles seen surrounding silicone-gel-filled implants. The diagram illustrates the number of vacuoles seen as a function of cross-sectional area, illustrating that most of the vacuoles seen fall in the 0 to 500 μm^2 area range. These vacuoles may be derived from dead fat cells because they are also observed in fat cell necrosis, which does not involve the presence of an implant.

scope; however, they exhibit a fine granular material in the cytoplasm and appear swollen (Figure 10.5). Other pathological responses include T cells wrapped around small venules (lymphocyte vasculitis) (Figure 10.6), and granulomas with intense infiltration of T lymphocytes (Figure 10.7) after implant rupture.

Complications associated with silicone breast implants include asymmetry, fibrous capsule formation, calcification of the fibrous capsule, capsular contraction, excessive hardness, extrusion of the implant, implant deflation, implant displacement, infection, numbness, postoperative lactation, sensitization, interference with cancer detection, and a possible association with systemic connective tissue disorders (Table 10.1). Symptoms similar to those observed in systemic connective tissue disorders, such as systemic sclerosis (also referred to as scleroderma), systemic lupus erythematosus (SLE or lupus), rheumatoid arthritis, polymyositis-dermatomyositis, Sjögren syndrome, and connective tissue disease (a mixture of the symptoms of the other diseases) (Table 10.2), have been observed.

The observation of connective tissue disease in patients receiving silicone implants has been reported; however, the association of these diseases with the

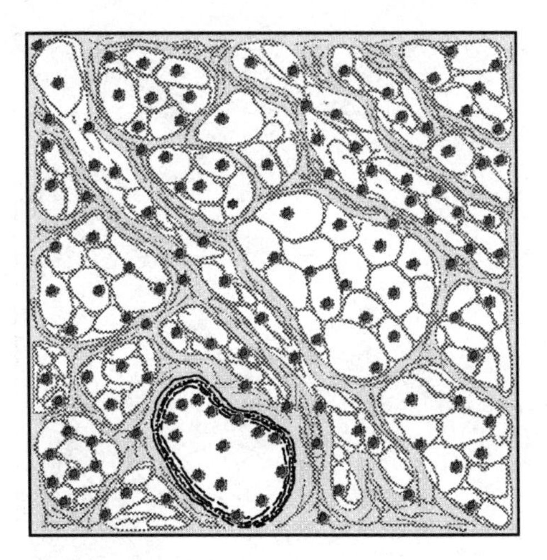

Figure 10.5 Macrophage response to silicone-gel-filled breast implants. This is a schematic illustration of a response showing macrophages (dark nuclei) surrounded by cytoplasm (light areas). The cytoplasm is larger than that seen with normal monocytes. The cytoplasm appears grainy or foamy, hence the name foamy macrophages. Foamy macrophages appear in clusters that sometimes are seen near multinucleated giant cells. The foamy appearance is related to the presence of intracellular granules.

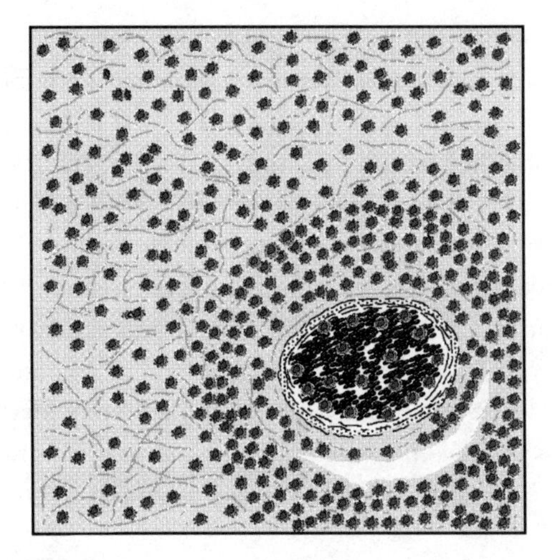

Figure 10.6 Vasculitis associated with silicone-gel-filled implants. The mononuclear infiltrate (dark cell nuclei), in this case T cells, is found wrapped around small venules (circular element at lower right in diagram) found in capsular tissues, muscles, and fat surrounding the implant. The venule is filled with red blood cells, which is normal, but it is surrounded by a thick coating of mononuclear cells that thicken the vessel wall.

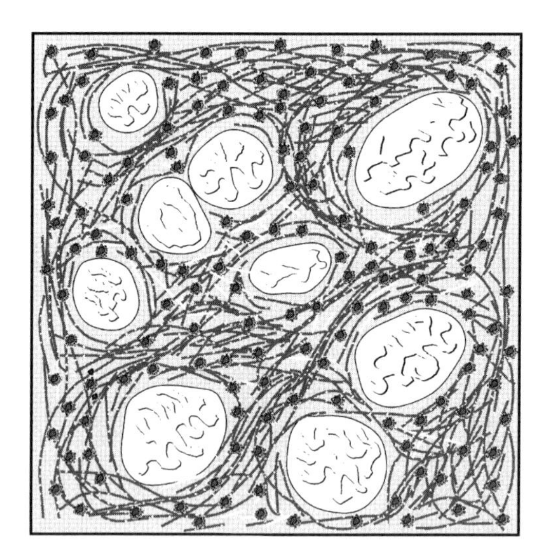

Figure 10.7 Vacuoles containing fragments of silicone implants. Upon rupture of silicone-gel-filled implants, particles are also seen within vesicles. These particles are refractile under the light microscope because they appear almost shiny against a clear background.

presence of an implant is still as subject of intense debate. There are reports of development of systemic sclerosis after augmentation mammoplasty or alloplastic chin implantation. However, most women with breast implants had normal tests of common immunologic tests. In one study, 35% of women with silicone breast implants had high levels of antibodies to types I and II collagens (Vasey et al., 1992). However, in an epidemiological study published in 1994, there were no statistically significant data to associate connective tissue diseases with breast implantation (Gabriel et al., 1994). These data suggest that silicone implants may heighten the response of patients predisposed to immune disease but they alone are unlikely to be responsible for causing autoimmune diseases. If this is the case, care must be taken to determine which patients may be at risk for developing immune diseases and limit their exposure to silicone implants.

Inflammation Induced by Wear Particles

Hip or knee replacement by implants restores mobility and stability in people with degenerative joint disease, femoral neck fracture, and osteonecrosis (Figure 10.8). Total joint implants are made of a metallic component composed of tita-

Table 10.1 Complications associated with silicone-gel-filled breast implants

Complication	Clinical problem
Asymmetry	Noticeable prosthesis
Fibrous capsule formation	Hardening of breast tissue
Fibrous capsule calcification	Hardening of breast tissue
Implant displacement	Noticeable prosthesis
Implant deflation	Surgical removal
Infection	Implant removal
Numbness	Discomfort
Postoperative lactation	Embarrassment
Sensitization	Discomfort
Association with systemic disorders of connective tissue	Morbidity

Table 10.2 Connective tissue disorders associated with silicone-gel-filled breast implants

Disorder	Clinical symptoms
Systemic lupus erythematosus	Butterfly rash, red raised patches, photosensitivity, oral ulcers, arthritis, serositis, renal disorder, hematological disorders, and antinuclear antibodies
Polymyositis-dermatomyositis	Proximal muscle weakness, muscular inflammation, elevated muscle enzymes in serum, electromyographical abnormalities, and skin rash
Rheumatoid arthritis	Characteristic radiographic findings of cartilage destruction, turbid synovial fluid with decreased viscosity, and rheumatoid factor
Sjögren syndrome	Dry eyes and dry mouth
Systemic sclerosis	Raynaud phenomenon, collagenization of skin, atrophy of hands, and muscular weakness

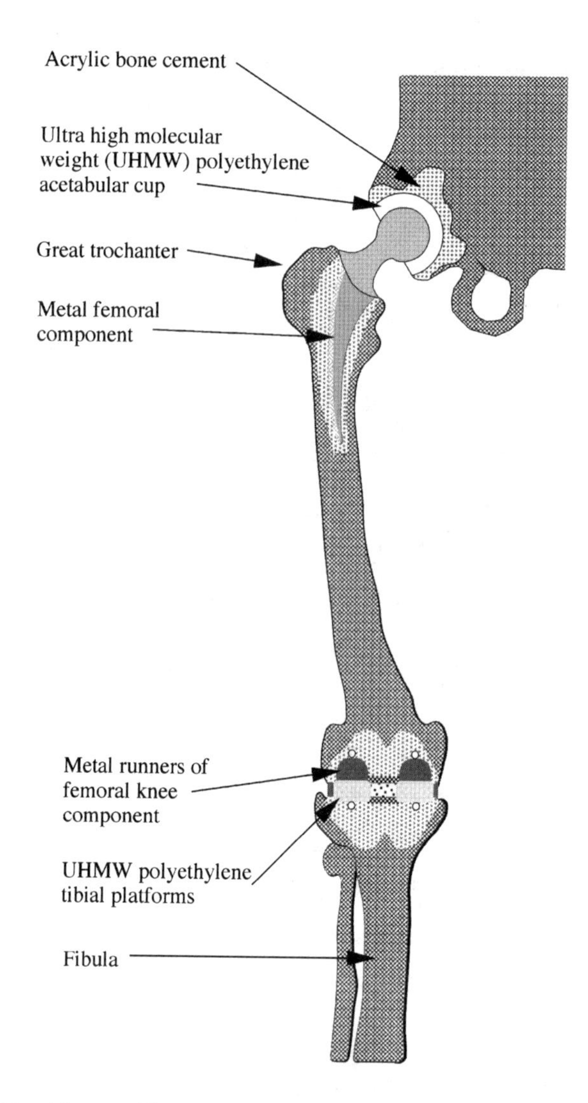

Figure 10.8 Total hip and knee surface replacement implants. Components of hip and knee implants consist of acrylic bone cement, ultra high molecular weight polyethylene, and metal components.

nium, stainless steel, or cobalt-chromium and a polymeric articular surface, usually composed of ultra high molecular weight polyethylene. In addition, certain types of prostheses are held in place by cement composed of poly(methyl methacrylate). Implant failure occurs by loosening of the cement and subsequent motion of the implant in contact with the cortical bone. The etiology of implant loosening involves wear particles formed by abrasion resulting from friction and micromotion between the implant and the neighboring host tissue. Wear particles have been found in tissues surrounding implants and are associated with the local inflammatory response that leads to formation of fibrous tissue or bone tissue resorption and subsequent prosthetic loosening.

Migration of metal particles released from total joint replacements and phagocytosed by macrophages has been reported to reach local lymph nodes and distant tissues. Macrophages containing cobalt and chromium or titanium alloys and polyethylene particles have been identified and associated with particle migration to lymph nodes and subsequent lymph node enlargement (Figure 10.9). However, in one study, lymph node enlargement appeared to be associated with polyethylene particles (Basle et al., 1996). In addition, poly(methyl methacrylate) cement particles are implicated in aseptic loosening by a chronic inflammatory reaction (Goldring et al., 1983). The mechanisms thought to underlie implant loosening are believed to involve particle-mediated macrophage activation and release of inflammatory mediators that cause bone resorption. Fibrous membranes observed at the bone–cement interface of failed cemented total joint arthroplasties contain macrophages associated with intracellular and extracellular poly(methyl methacrylate) particles (Goldring et al. 1986). The macrophage response to particulates appears to result in bone resorption by a number of different possible mechanisms. Activated macrophages release inflammatory mediators, including cytokines and collagenases, that act to cause bone destruction. Small particles capable of being phagocytosed (<12 μm) are a critical factor in bone resorption, and particle load also appears to be a factor (Gonzalez et al. 1996). Neither small nor large particles appear to be directly cytotoxic to macrophages, suggesting that prolonged inflammation and fibrosis are associated with phagocytosis of implant particles. One study found a higher frequency of large particles in failed knee joints compared with failed hip joints reflected the perceived higher rate of delamination and fragmentation of tibial and patellar polyethylene implants compared to acetabular polyethylene implants (Hirakawa et al., 1996).

Complement Activation by Biomaterial Surfaces

The complement system is an important component of the host's defense mechanisms against infection and mediates an immunological response to injury. It is one of the major components of blood that is activated by contact of blood with

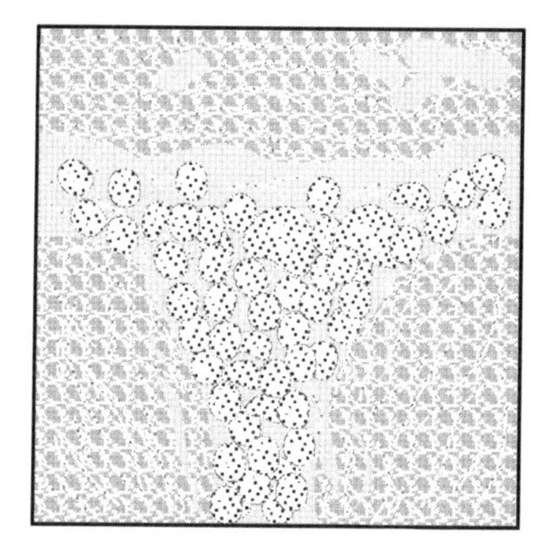

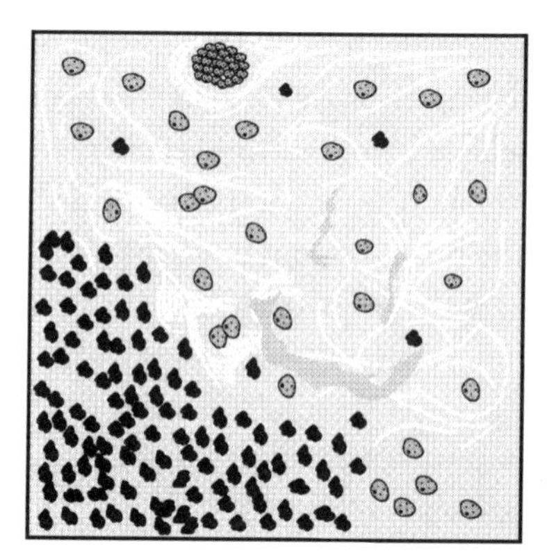

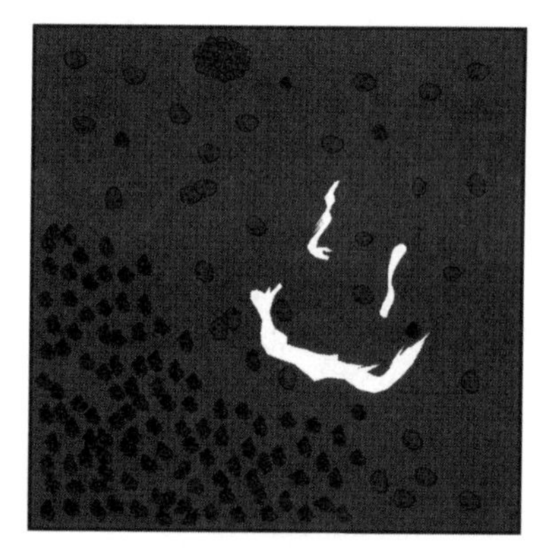

Figure 10.9 Wear particles released from total joint implants. Small metallic particles within macrophages in tissue surround an implant (top). Particles phagocytosed by inflammatory cells are in the micrometer size range and appear as dots within the cell. Middle and lower diagrams show the relationship between inflammatory cells (middle) and large polyethylene particles (lower). Large polyethylene particles are not ingested by phagocytic cells; however, they are surrounded by these cells.

a surface. Complement pays a role in the body's defense against "non-self," and complement activation leads to attachment of specific complement components to bacteria, immune complexes, and up regulates cellular receptors for uptake of these opsonized moieties. Complement activation also leads to the lysis of cells via the membrane-attack complex.

The clinical use of catheters, oxygenators, prosthetic heart valves, hemodialysis membranes, and blood oxygenators is known to be associated with complications such as thrombosis, thromboembolism, pulmonary dysfunction, and infection. Some of these pathological conditions are believed to be the direct result of complement and leukocyte activation (Gemmell et al., 1996).

The activation of complement during cardiopulmonary bypass appears to result from the contact of blood with the mesh liner, commonly made from nylon, in bubble oxygenators (Hakim, 1993) (Figure 10.10). Polyester or polypropylene substitution for nylon decreases the amount of complement activation. The means of complement activation is via the alternate pathway, similar to the activation caused by passage of blood through other extracorporeal circuits.

Activation of complement has been observed after hemodialysis using cellulosic membranes (Figure 10.11). Although the original cuprophan membranes were very active in C3a generation, newer materials were developed that are much improved. However, cellulose membranes are still used clinically because of their low cost and because after treatment with formaldehyde their ability to activate complement on reuse is markedly diminished. Adverse symptoms have been reported in the literature as a result of activation of complement by cuprophane membrane dialyzers (Hakim et al., 1984). The symptoms include pruritus, shortness of breath, wheezing, back pain, hypotension, fatigue, and anorexia.

Complement activation and repetitive exposure to dialysis membranes have led to a decrease in predialysis neutrophil count, possibly reflecting the neutropenia that occurs during dialysis. A deterioration of the functional determinants of neutrophils from patients who have undergone hemodialysis has been found and correlates with the higher infection rates for patients treated with cellulosic membranes (Vanholder et al., 1991).

Long-term catheters are usually comprised of silicone polymers and are used for treatment of malignancy, systemic infection, and parenteral nutrition. A major complication with these procedures involves infection that begins in the subcutaneous space or in the catheter lumen. High levels of complement activation by silicone are used to account for the greater inflammation seen around silicone catheters and the high rate of infection (Marosk et al., 1996).

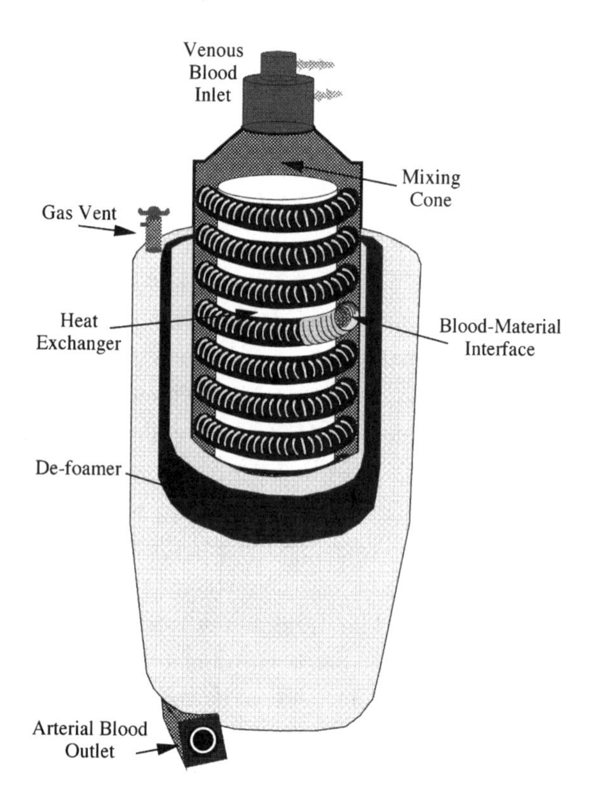

Figure 10.10 Bubble type oxygenator. Venous blood enters the top of the device and flows through a membrane that is wrapped around a core housing. The core housing has spaces between the wraps for mass transfer. Gas that contains oxygen is flushed over the outside of the membrane within the core of the device and passes into the blood by diffusion. Oxygenated blood passes out of the membrane and is circulated back to the arterial side of the circulation.

Restenosis After Vascular Stenting

Coronary artery narrowing is a consequence of atherosclerosis and deposition of plaque, lipid, collagen, and mineral. The consequence of narrowing the coronary arteries is angina and the inability to perform exercise. Medical treatment for this condition includes the use of percutaneous coronary transluminal angioplasty (PTCA) and coronary bypass in extreme conditions where PTCA is unlikely to work. In PTCA, an expandable balloon is introduced into the coronary arteries through the femoral or other superficial artery. The balloon is positioned in the coronary artery and expanded to crack the plaque that narrows the lumen (Figure 10.12). The use of PTCA for treatment of atherosclerotic lesions in the coronary

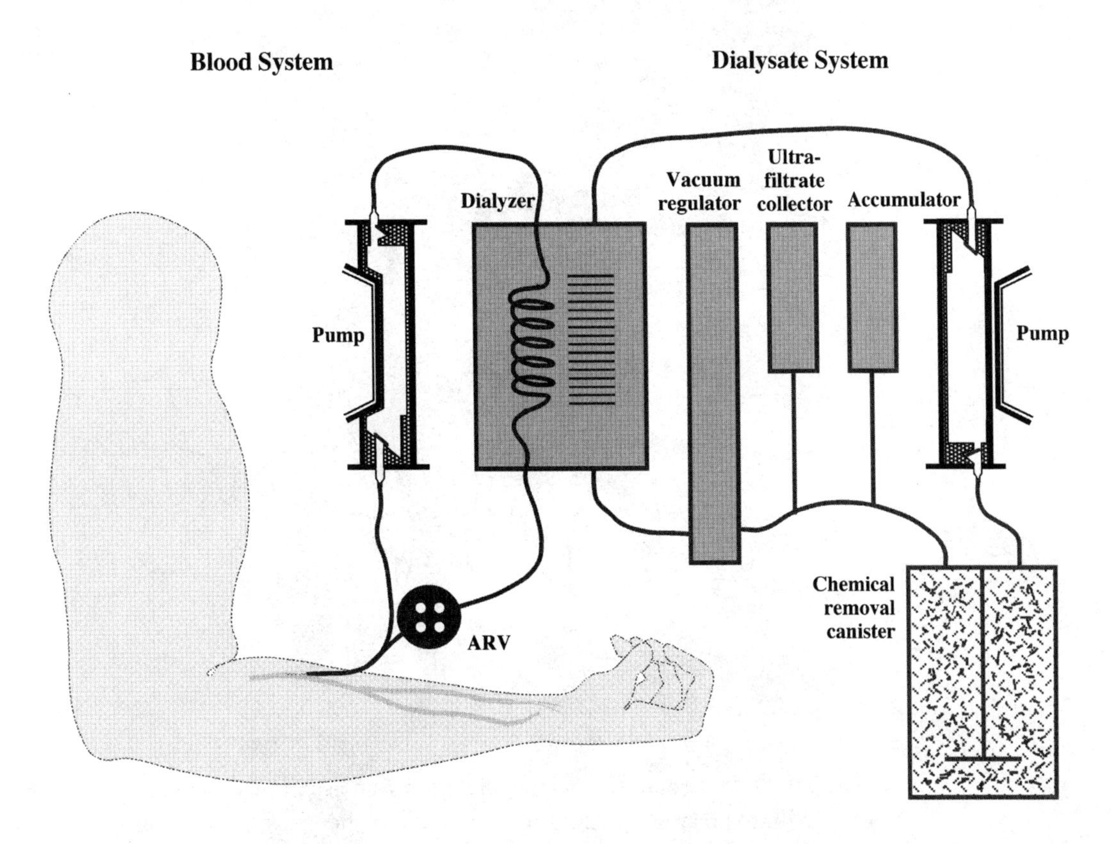

Figure 10.11 Blood dialysis using a membrane dialyzer. The diagram illustrates how kidney dialysis occurs. Blood is removed from the patient's arm and is passed through a membrane that is bathed in a solution low in solutes. Solutes such as urea flow from the patient's blood into the dialysis solution by diffusion. Dialysed blood is returned back to the patient, and urea is removed from the dialysis solution by the chemical removal canister.

arteries has increased in frequency in recent years. The success rate for this procedure is high, and the incidence of immediate or short-term occlusion of the site of angioplasty is 2 to 5% (Myler et al., 1987). Restenosis or narrowing of the lumen, caused by thrombus and proliferation of vascular smooth muscle cells several months after the procedure, can occur and has an incidence of 20 to 40% (Gruntzig et al., 1987).

Metallic supports referred to as endovascular coronary stents have been developed to increase luminal diameter beyond that achieved with balloon angioplasty and to limit restenosis. These devices are inserted through an artery and expanded in the area where the coronary artery was ballooned. The pathobiological complications associated with the use of endovascular stents are

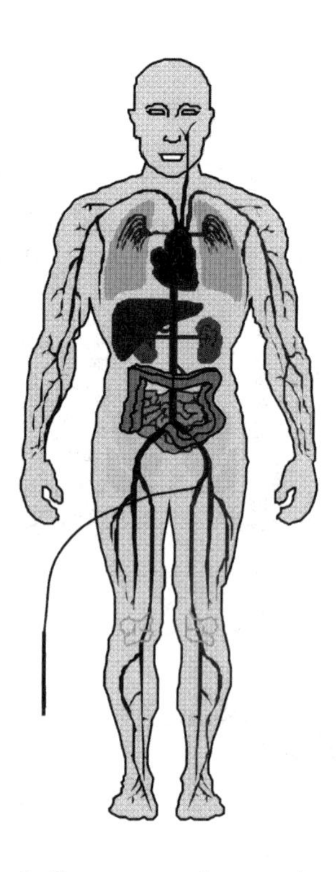

Figure 10.12 Procedure for balloon angioplasty and stent placement. A guide wire is inserted into an artery, such as the femoral, and then pushed until it is in the site of stenosis (narrowing). A balloon is inserted over the guide wire and fed until it is also in the site of stenosis. The balloon is inflated, and the plaque is plastically deformed. A stent is introduced over the guide wire and expanded. The stent is used to prevent intimal hyperplasia and restenosis.

thrombosis and restenosis. Overexpansion of the artery lumen during stent expansion stimulates smooth muscle proliferation and neointimal hyperplasia.

Vascular stents were first investigated by Dotter in 1969. Subsequent studies led to the development of self-expanding spring-loaded stents and balloon-expandable stents (Schatz, 1988) (Figure 10.13). Although many of the spring-loaded stents were compressed during delivery, a second mechanism of delivery of self-expanding stents involved thermal memory affects. Metals such as nitinol change their configuration upon exposure to heat. Stents made of nitinol start out with small diameters at room temperature and then are expanded to their final diameters when their temperature approaches that of the body. Palmatz and co-workers introduced the concept of a balloon-expandable stent. Stents of this

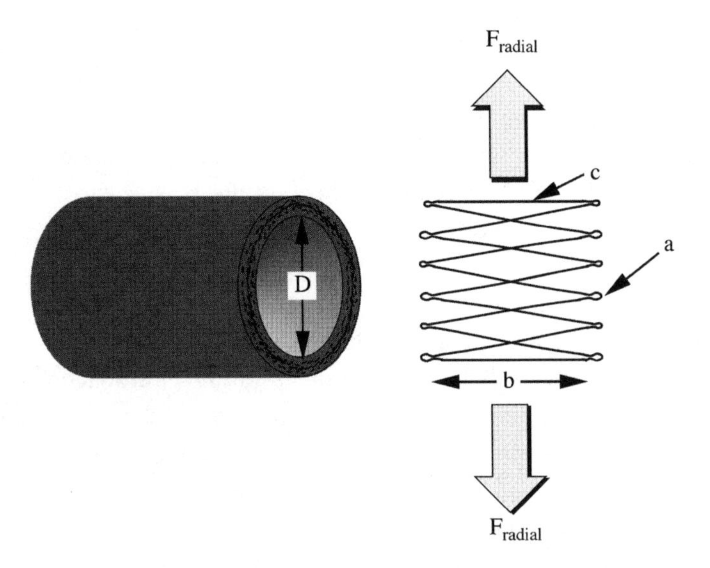

Figure 10.13 Self-expanding stent. Some of the factors that contribute to the generation of a force that keeps the vessel from undergoing stenosis include the number of bends or points (a); the length of legs (b), the gauge of wire (c), and the diameter of the blood vessel (d). The spring is stretched lengthwise to collapse it during insertion, and it is then allowed to contract lengthwise when it is in place, which results in a radial force. The radial force holds it in place, preventing it from migrating and preventing intimal hyperplasia and restenosis.

type are plastically expanded by expansion of a balloon within the lumen of the stent.

Events associated with restenosis of vascular stents include surface thrombogenicity of the stent metals, subsequent thrombosis and wound healing, and stent flexibility (Palmatz, 1992). The immediate biological response to a stent occurs at the surface. The texture of the surface is important in causing thrombosis because rough surfaces induce higher degrees of blood clotting and platelet aggregation. Fibrin is quickly deposited on the stent surface in areas where the stent wire contacts the vessel surface. In areas not covered by the wire, the tissue protruding between the wire is not covered by fibrin. These areas not covered by thrombus between the wire of the stent serve as sites for migration of normal endothelial cells over the thrombus-covered areas. Use of antithrombotic therapies decrease the amount of blood clots that are attached to the stent. In a few days to weeks, the thrombus is replaced by the deposition of fibromuscular tissue around the struts of the stents; subsequently it gradually increases in thickness.

Expansion and overexpansion of the vessel lumen during stenting acts as a potent stimulus for smooth muscle proliferation, leading to restenosis; however, some investigators believe that surface material and geometric configuration of stents may be more important than placement diameter in determining intimal hyperplasia and thrombosis because monocytes are believed to be important modulators of stent-induced intimal thickening (Rogers and Edelman, 1995). However, it is not possible to separate the effects of thrombosis and pressure on the trauma and healing associated with PTCA.

Formation of a neointima is a normal wound healing response triggered by damage to the vessel wall. Cracking of the plaque and introduction of foreign stenting material causes vessel wall injury. Exposure of subendothelial components causes thrombus formation and release of platelet-derived growth factor (PDGF). PDGF is a potent stimulator of smooth muscle cells, causing migration and proliferation. Subsequent deposition of collagen and extracellular matrix macromolecules causes lumenal narrowing and restenosis. Factors that prevent proliferation of smooth cells, such as exposure to low doses of radioactivity, limit intimal hyperplasia.

Thrombosis and Small-Diameter Vascular Grafts

Lumenal narrowing in small-diameter (<6 mm) vascular grafts fabricated from synthetic polymers is much higher than that of large-diameter (>6 mm) grafts and is thought to result from activation of blood clotting and platelet aggregation that lead to graft occlusion. The major events associated with thrombosis that result when a vascular graft is placed in the arterial system include protein adsorption onto the surface, protein activation, platelet adhesion and activation, and thrombus attachment and growth (Figure 10.14) (Reynolds et al., 1993). Protein adsorption can occur in a hydrophilic fashion, which is typically reversible, and in a hydrophobic fashion, which is largely irreversible. Once an initial monolayer of protein is present, additional layers of protein adhere to the surface that are affected by the flow rate and surface charge.

Surface activation of factor XII (Hageman factor) is a result of the surface properties of the material and is followed by platelet adhesion because of diffusion and collision with the surface. Platelet activation and aggregation depend on platelet concentration, flow rate, and time. The amount of thrombus deposition is proportional to the surface area of the exposed device and may be increased by deposition of fibrinogen into cracks on the surface.

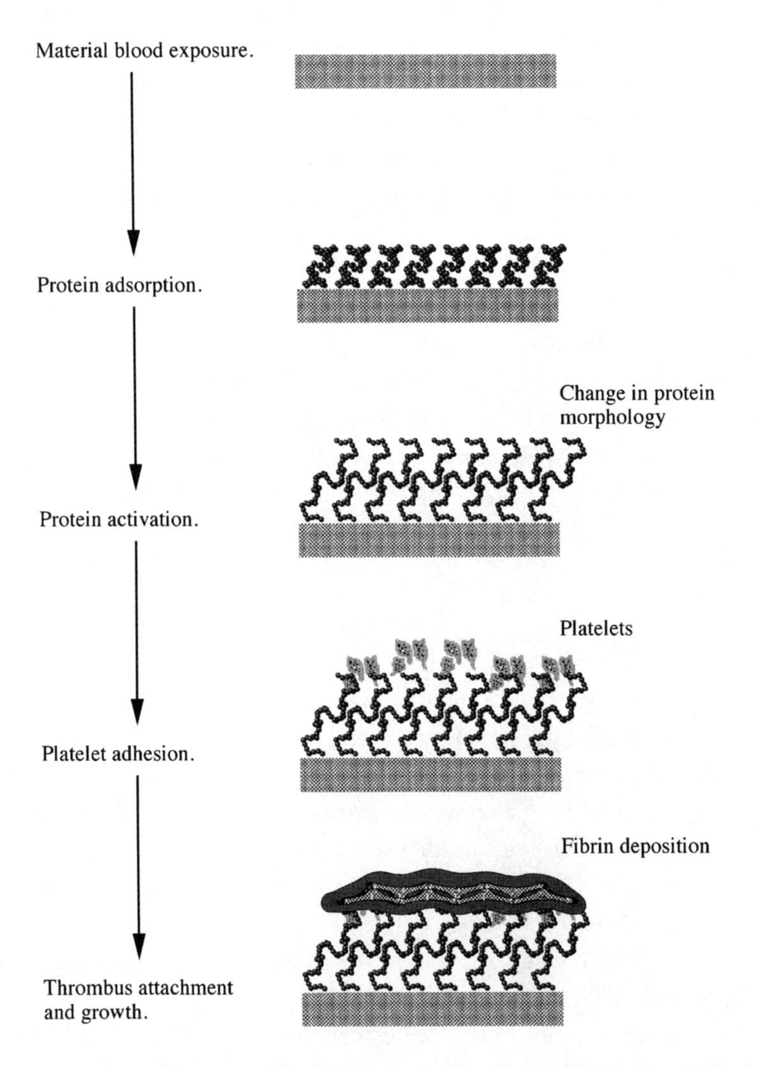

Figure 10.14 Events associated with thrombosis of vascular grafts. The diagram illustrates the adsorption of proteins from the blood to a material surface, followed by activation, platelet binding, and fibrin deposition. Although some surfaces can be pacified by binding albumin or other inert macromolecule, in general this sequence of events leads to thrombus attachment and growth.

Patency rates of the available small-diameter vascular grafts made using synthetic polymeric materials are much lower than those obtained using autogenous grafts in above- and below-the-knee applications (Veith et al., 1986). Failure of these grafts has been attributed to compliance mismatches between the graft and the host tissue and thrombogenicity of synthetic grafts. Compliance mismatches at the anastomotic sites are a result of the increased stiffness of synthetic grafts over that of host arteries. Finite element analysis of the artery/poly(tetrafluoroethylene) anastomosis, in which diameters of the artery and graft were matched at 0 mm Hg, revealed the development of tensile and shear stresses in the graft near the anastomosis at 100 mm Hg (Matsumoto et al., 1994). This mismatch caused a zone of hypercompliance in the host vessel wall near the anastomosis sites.

Reduction in thrombogenicity of synthetic polymeric surfaces has been accomplished by coating the surface with endothelial cells (ECs). This has been attempted by "seeding" the polymer surface with low numbers of ECs at surgery and allowing gradual growth to confluence in vivo or by "sodding" ECs at high densities to produce a cell-coated graft at implantation. Cell-seeded grafts are shown to retain their cell coating after implantation and remain patent on 1-mm-diameter (Ahlswede and Williams, 1994) and 4-mm-diameter grafts.

Ectopic Mineralization of Implants

The deposition of hydroxyapatite into mineralizing cartilage at the ends of long bones normally occurs during endochondral bone formation and by direct deposition into extracellular matrix during membranous bone formation. This process is referred to as mineralization or calcification because calcium as well as phosphate and hydroxyl ions are deposited. However, calcification also occurs in association with a wide variety of cardiovascular and noncardiovascular medical devices, and it is the leading cause of failure of bioprosthetic heart valves (Schoen et al., 1988). The calcification of tissues such as heart valves that are normally nonmineralized is called ectopic.

During normal mineralization, osteoblasts derived from undifferentiated primitive mesenchymal or differentiated chondrocytes produce a collagenous matrix called osteoid, and they indirectly control mineralization through release of matrix vesicles containing apatitic mineral. In avian tendons, localization of apatite within matrix vesicles is first seen, which is followed by apatite in the extracellular matrix and then in neighboring collagen fibrils. The mineralization

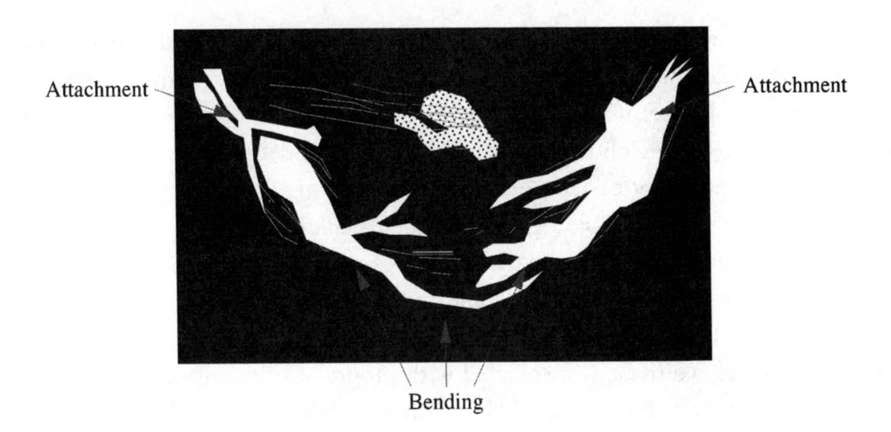

Figure 10.15 Calcification of bioprosthetic heart valves. Illustration of the points of calcification of bioprostheses that occur at the attachment site to the cordae tendineae and at the site of maximum bending.

of bone organic matrix is based predominantly on the molecular organization of collagen type I into a quarter-staggered array because mineral crystal formation commonly reflects the periodic structure of collagen fibrils (Arsenault, 1988). In addition to the role of type I collagen, control of mineralization of cartilage involves proteoglycans and the C-propeptides of collagen types II and X (Boskey, 1989). The breakdown of proteoglycans or changes in their structural role in cartilage or bone increase mineralization and ectopic calcification (Buckwalter, 1987).

Calcification is the principal cause of clinical failure of heart valve bioprostheses fabricated from glutaraldehyde-pretreated porcine aortic valves, and it is also an important complication of valves made of glutaraldehyde-treated bovine pericardium. Failure rates for porcine aortic valves are as high as 50% at 12 to 15 years, with more than 75% of failures resulting from calcification-induced stenosis or tearing (Levy et al., 1991). Clinical and experimental studies suggest that dynamic mechanical stress and strain promote calcification, especially in areas of leaflet flexion (Figure 10.15). The earliest mineral deposits appear to be localized to transplanted connective tissue cells and not to extracellular collagen. As the implantation period increases, cell-associated deposits increase in size and number and appear to cut into collagen fibrils.

One theory suggests that the earliest events in calcification of bioprosthetic valves involve cellular devitalization caused by exposure to glutaraldehyde and disruption of cellular calcium regulation, resulting in an influx of calcium ions into the cell. This subsequently causes a reaction with membrane-bound organelles high in phosphorus, such as mitochondria, forming apatite. Alkaline phosphatase, which normally hydrolyzes the phosphoester bond in phospholipid-releasing phosphate, is present in glutaraldehyde pretreated bovine pericardial tissue (Levy et al., 1991), and blockage of alkaline phosphatase activity reduced subsequent bioprosthetic tissue calcification.

However, this theory is inadequate to explain two other phenomena: the calcification of aging vascular tissue and the calcification of devitalized collagenous and synthetic polymeric implants. In the first case, calcification of aging vascular connective tissue is associated with progressive loss of cells and accumulation of lipids (Schoen et al., 1988). This is manifested through the necrosis of fibroblasts and smooth muscle cells and accumulation of membranous cellular degradation products that subsequently are the loci for calcification. In calcified cells, needle-shaped apatite crystals are noted by electron microscopy in association with cellular degradation products, consistent with the model that influx of calcium caused by mechanical or chemical cellular injury is an important contributing factor to calcification. In vascular connective tissue, large-sized vesicular structures several micrometers in diameter are frequently seen (Kim et al., 1976). Needle-shaped crystals are frequently embedded radially in the thick wall of vesicles that appear to be derived from the plasma membrane.

In the case of devitalized collagenous and synthetic polymer implants, the mode of calcification may be somewhat different. Mineralization during bone formation is associated with high levels of osteocalcin and alkaline phosphatase. In contrast, calcification of glutaraldehyde-cross-linked implants and reconstituted collagen showed only trace amounts of osteocalcin. Alkaline phosphatase activity was undetectable in implants of cross-linked collagenous tissues (Nimni et al., 1988).

Summary

Although most implants are tolerated well by the host organism, some individuals react to any implant placed in their bodies. Qualitatively, about one of every several thousand patients has a response that requires an implant or all implants in their body, including all their dental fillings, to be removed before their symptoms disappear. Most of the time, the reason for the pathobiological response involves natural immunity and a macrophage response. In other cases, although a foreign body response is the initial response to a material, T cells become involved. How this transition occurs is unclear, but ultimately the implant or wear particles derived from the implant trigger an immune response. Whether there is a cause-and-effect relationship between implants and acquired autoimmune disease is unclear; however, it is clear the association between the two is a cause of concern. Although all implants pass current cytotoxicity tests, it is possible that low levels of cell cytotoxicity associated with implants cause calcification. Beyond that, activation of blood clotting and complement are two other ways that implants cause systemic effects.

Our knowledge of implants and their effects on cells and tissues has grown rapidly since the 1950s. It is now possible to learn how to minimize the pathobiological side effects associated with use of these devices.

Suggested Reading

Arsenault A.L., Crystal-Collagen Relationship in Calcified Turkey Leg Tendons Visualized by Selected-Area Dark Field Electron Microscopy, Calcif. Tissue Int. 43, 202, 1988.

Ahlswede K.M and Williams S.K., Microvascular Endothelial Cell Sodding of 1-mm Expanded Polytetrafluorethylene Vascular Grafts, Arterioscler. Thromb. 14, 25, 1994.

Basle M.F., Bertrand G., Guyetant S., Chappard D., and Lesourd M., Migration of Metal and Polyethylene Particles From Articular Prostheses May Generate Lymphadenopathy with Histiocytosis, J. Biomed. Mat. Res. 30, 157, 1996.

Boskey A., Noncollagenous Matrix Proteins and Their Role in Mineralization, Bone Mineral 6, 111, 1989.

Buckwalter J.A., Rosenberg L.C., Ungar R., Changes in Proteoglycan Aggregates During Cartilage Mineralization, Calcif. Tissue Int. 41, 228, 1987.

Dotter C.T., Transluminally Placed Coilspring Endarterial Tube Grafts: Long Term Patency in Canine Popliteal Artery, Invest. Radiol. 4, 329, 1969.

Frisch E.E., Technology of Silicones in Biomedical Applications, in Biomaterials in Reconstructive Surgery, edited by L.R. Rubin, C.V. Mosby Co., St. Louis, MO, chapter 8, 1983.

Gabriel S.E., O'Fallon W.M., Kurkland L.T., Beard C.M., Woods J.E., and Melton L.J., Risk of Connective Tissue Diseases and Other Disorders After Breast Implantation, N. Engl. J. Med. 330, 1697, 1994.

Goldring S.R., Schiller A.L., Roelke M., Rourke C.M., O'Neille D.A., and Harris W.H., The Synovial-Like Membrane at the Bone-Cement Interface in Loose Total Hip Replacement and Its Proposed Role in Bone Lysis, J. Bone Joint Surg. 65A, 575, 1983.

Goldring S.R., Jasty M., Roelke M.S., Rourke C.M., Bringhurst F.R., and Harris W.H., Formation of a Synovial-Like Membrane at the Bone-Cement Interface. Its Role in Bone Resorption and Implant Loosening After Total Hip Replacement. Arthritis Rheum. 29, 836, 1986.

Gonzalez O., Lane Smith R., and Goodman S.B., Effect of Size, Concentration, Surface Area, and Volume of Polymethylmethacrylate Particles on Human Macrophages In Vitro, J. Biomed. Mater. Res. 30, 463, 1996.

Gruntzig A.R., King S.B., Schlumpf M., and Siegenthaler W., Long-Term Follow-up After Percutaneous Transluminal Coronary Angioplasty, N. Engl. J. Med. 316, 1127, 1987.

Hakim R.M., Breillant J., Lazarus J.M., and Port F.K., Complement Activation and Hypersensitivity Reaction to Dialysis Membranes, N. Engl. J. Med. 311, 878, 1984.

Hakim R.M., Complement Activation by Biomaterials, Cardiovasc. Pathol. 2, 187S, 1993.

Haynes D.R., Rogers S.D., Hay S., Pearcy M.J., and Howie D.W., The Differences in Toxicity and Release of Bone-Resorbing Mediators Induced by Titanium and Cobalt-Chromium-Alloy Wear Particles, J. Bone Joint Surg. 75A, 825, 1993.

Hirakawa K., Bauer T.W., Stulberg B.N., and Wilde A.H., Comparison and Quantification of Wear Debris of Failed Total Hip and Total Knee Arthroplasty, J. Biomed. Mater. Res. 31, 257, 1996.

Kim K.M., Valigorsky J.M., Mergner W.J., Jones R.T., Pendergrass R.F., and Trump B.F., Aging Changes in the Human Aortic Valve in Relation to Dystropic Calcification, Hum. Pathol. 7, 47, 1976.

Levy R.J., Schoen F.J., Flowers W.B., and Staelin S.T., Initiation of Mineralization in Bioprosthetic Heart Valves: Studies of Alkaline Phosphatase Activity and Its Inhibition by $AlCl_3$ or $FeCl_3$ Preincubations, J. Biomed. Mater. Res. 25, 905, 1991.

Marosk R., Washburn R., Indorf A., Solomon D., and Sherertz R., Contribution of Vascular Catheter Material to the Pathogenesis of Infection: Depletion of Complement by Silicone Elastomer In Vitro, J. Biomed. Mater. Res. 30, 245, 1996.

Myler R.K., Topol E.J., Shaw R.E., Stertzer S.H., Clark D.A., Fishmen J., and Murphy M.C., Multiple Vessel Coronary Angioplasty: Classification, Results, and Patterns of Restenosis in 494 Consecutive Patients. Cathet. Cardiovasc. Diag. 13, 1, 1987.

Nimni M.E., Bernick S., Cheung D.T., Ertyl D.C., Nishimoto S.K., Paule W.J., Salka C., and Strates B.S., Biochemical Differences Between Dystrophic Calcification of Cross-linked Collagen Implants and Mineralization During Bone Induction, Calcif. Tissue Int. 42, 313, 1988.

Orentreich D.S. and Orentreich N., Injectable Fluid Silicone for Soft Tissue Augmentation, in Applications of Biomaterials in Facial Plastic Surgery, edited by A.I. Glasgold and F.H. Silver, CRC Press Inc., Boca Raton, FL, p. 219, 1991.

Palmatz J.C., Intravascular Stents: Tissue-Stent Interactions and Design Considerations, AJR 160, 613, 1992.

Revell P.A., Weightman B., Freeman M.A.R., and Roberts V.B., The Production and Biology of Polyethylene Wear Debris, Arch. Orthop. Trauma Surg. 91, 167, 1978.

Reynolds L.O., Newren W.H. Jr., Scolio J.F., and Miller I.F., A Model for Thromboembolization on Biomaterials, J. Biomater. Sci. Polym. Ed. 4, 451, 1993.

Rogers C. and Edelman E.R., Endovascular Stent Design Dictates Experimental Restenosis and Thrombosis, Circulation 91, 2995, 1995.

Schatz R.A., Introduction to Intravascular Stents, Intervent. Cardiol. 6, 357, 1988.

Schoen F.J., Harasaki H., Kim K.M., Anderson H.C., and Levy R.J., Biomaterial-Associated Calcification: Pathology, Mechanisms, and Strategies for Prevention, Appl. Biomater. 22, 11, 1988.

Shanklin D.R. and Smalley D.L., Silicone Immunopathology, Sci. Med. September/October 22, 1996.

Shons A.R. and Schubert W., Silicone Breast Implants and Immune Disease, Ann. Plast. Surg. 28, 491, 1992.

Silver F.H., Pins G.D., Rizvi A., Olson R.M., and D'Aguillo A., Silicone Gel-Filled Breast Implants: Is Local Inflammation Associated with Fat Cell Necrosis? Breast J. 1, 17, 1995.

Vanholder R., Ringoir S., Dhondt A., and Hakim R., Phagocytosis in Uremic and Hemodialysis Patients: A Prospective and Cross Sectional Study, Kidney Int. 39, 320, 1991.

Vasey F.B., Havice D.L., Bocanegra T.S., Seleznick M.J., Bridgeford P.H., and Germain B.F., Clinical Manifestations of Fifty Women With Silicone Breast Implants and Connective Tissue Disease, Arthritis Rheum. 35, S212, 1992.

Veith F.J., Gupta S.K., Ascer E., White-Flores S., Samson R.H., Scher L.A., Towne J.B., Bernard V.M., Bonier P., Flinn W.R., Astelford P., Yao J.S.T., and Bergan J.J., Six-Year Prospective Multi-Center Randomized Comparison of Autologous Saphenous Vein and Expanded Polytetrafluoroethylene Grafts in Infrainguinal Arterial Reconstructions, J. Vasc. Surg. 3, 104, 1986.

Warren-Poole L.A., Schindhelm K., Graham A.R., Slowiaczek P.R., and Noble K.R., Performance of Small Diameter Synthetic Vascular Prostheses with Confluent Autologous Endothelial Cell Linings, J. Biomed. Mater. Res. 30, 221, 1996.

11

Tissue Engineering

Introduction to Tissue Engineering

Although medical device development began in the 1950s with the introduction of devices and materials derived from synthetic polymers, the field has progressed to the point where devices now use cells and biopolymers made using cultured or genetically engineered cells (Figure 11.1). Physicians have treated organ or tissue failure by transplanting organs from one individual into another, performing surgical reconstruction using autologous tissues, or using mechanical devices such as blood oxygenators. Although these approaches are used daily to save lives, they fall short of being available for everyone at a reasonable cost. Transplantation is severely limited by the number of donor organs available. Surgical reconstruction requires in some cases multiple procedures, and the recovery period can be extensive. Mechanical devices fail to pro-

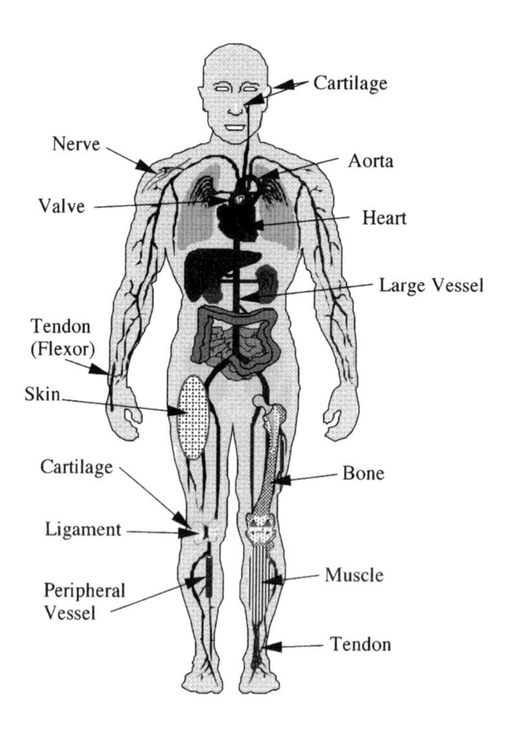

Figure 11.1 Overview of tissue engineering. The diagram illustrates the potential applications of tissue engineering of both soft and hard tissues.

vide all the biological functions required for normal homeostasis. Therefore, the replacement of tissue and organ structure and functions requires a combination of biological components engineered into a device.

The field of tissue engineering involves the application of the principles of engineering and life sciences toward the development of biological substitutes that restore, maintain, or improve tissue function (Nerem, 1992; Langer and Vacanti, 1993). The approaches envisioned to be included in tissue engineering include use of isolated cells to replace specific functions, substances that induce proliferation of cells and tissues, and cells in combination with matrixes that are either in immunological contact with host tissues or are encapsulated to prevent immunological contact (Figures 11.2 and 11.3).

Cells derived from external tissues, skeletal tissues, cardiovascular tissues, and specialized organs have been used in this approach. In the case of external tissues, tissue engineering of skin has progressed rapidly, and repair of skeletal tissues, including bone and cartilage repair using tissue engineered materials, has received much attention. Cardiovascular tissue engineering has focused on un-

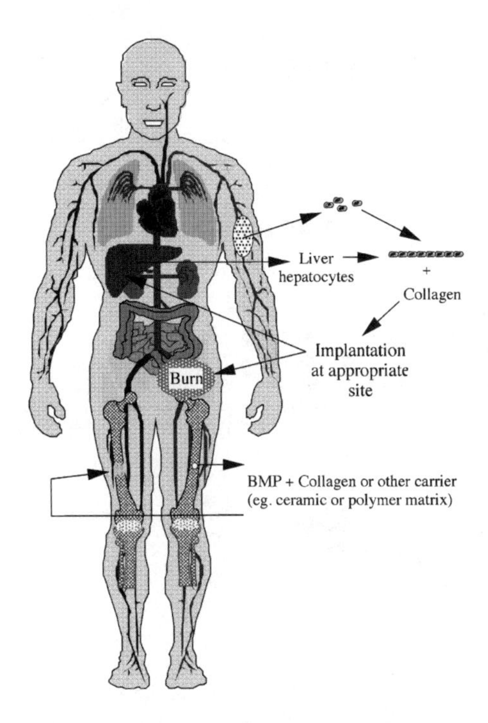

Figure 11.2 Tissue engineering approaches. The diagram illustrates the various approaches used, including use of isolated cells from liver and skin in combination with collagen to form hepatocyte sandwiches or skin replacements and use of BMP or other growth factors to improve healing.

derstanding the behavior of the endothelial cell and creation of small-diameter vessels (see Chapter 10), and specialized organs, including the pancreas and liver, are still engineering challenges. This chapter discusses details of this progress in tissue engineering after examining the immunological problems that must be overcome to allow the transplantation of cells from a donor to an unrelated recipient.

Immunology of Cell and Tissue Transplantation

Transplantation of cells that contain class I and II MHC markers between unrelated individuals normally causes rejection of the transplanted cell, tissue, or organ. The rejection response typically eliminates all but autogenous and tissue-

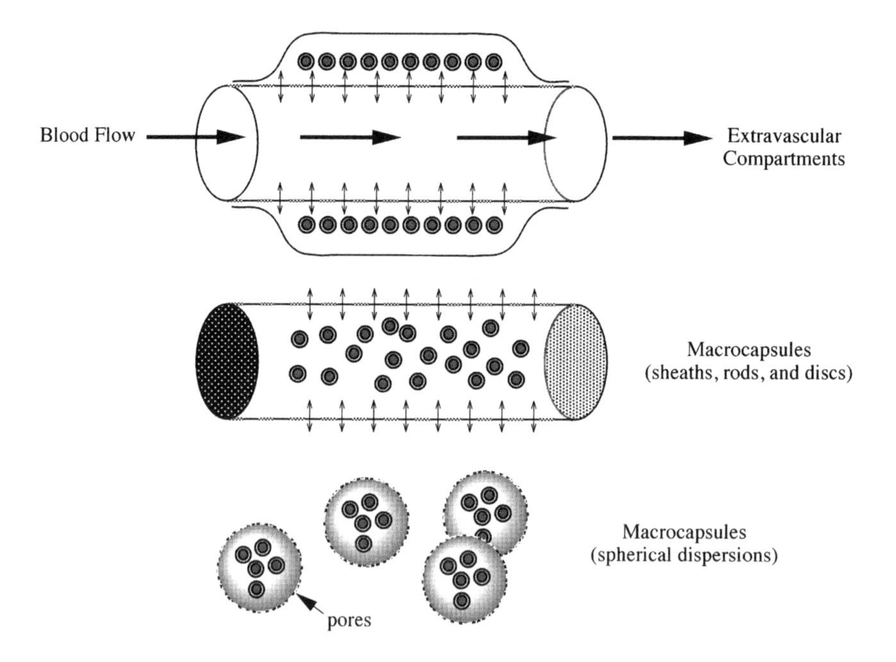

Figure 11.3 Configurations of cell transplant devices. The diagram illustrates the possible arrangements of cell transplant devices that contain cells from hosts that may be immunologically incompatible. In these cases, the contact between blood and cells from nonhost tissue can be achieved through an extravascular shunt. During blood flow through the shunt, molecules synthesized within the cellular elements enter the blood by passive diffusion. In the other approaches, cells are contained in polymer rods or spheres that allow diffusion of molecules from the cells but protect the encapsulated cells from stimulating the hosts' immune system. *Source:* Adapted from Langer and Vacanti (1993).

matched grafts as candidates for long-term transplantation. However, it was recognized by workers in the skin transplantation field during the 1980s that skin transplanted from unrelated donors would not be rejected if the epidermis was stripped away from the dermis before transplantation or if the epidermal cells were passaged in culture before use on an unrelated recipient (Figure 11.4). These observations opened the door to transplantation of a variety of cells and tissues without the need for immunosuppression. However, this also raises questions concerning the mechanism of transplant tolerance, because skin grafts are normally rejected if left on a burn patient more than a few days.

Two additional observations were crucial to understanding the survival of allografted cells and immunotolerance. The first observation was that human

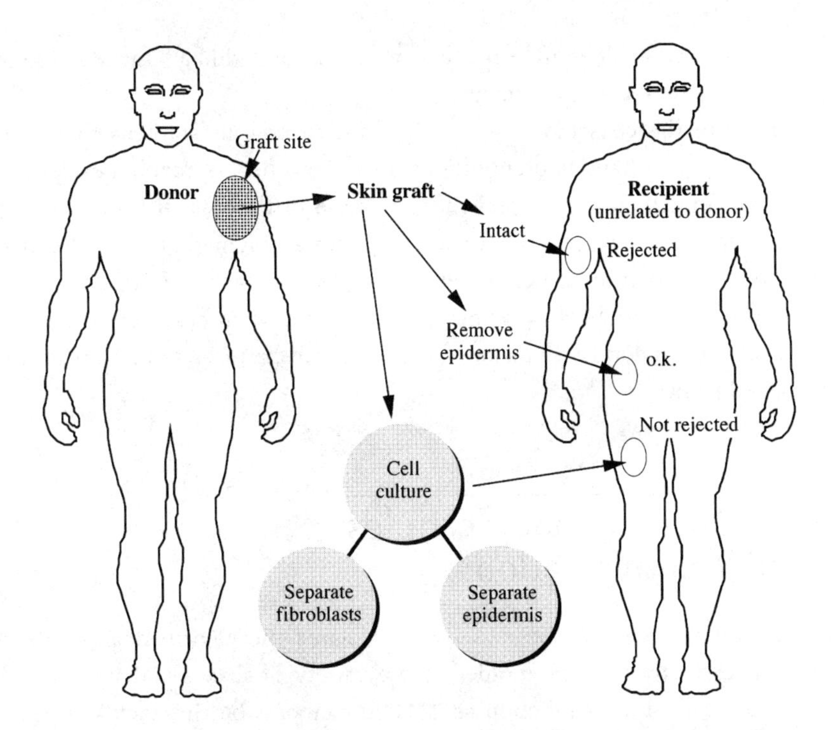

Figure 11.4 Outcomes of transplantation of allogeneic cells. Transplanted skin will be rejected if it is used intact as shown; however, if the epidermis is removed before transplantation or if the epidermal cells are passaged in culture before transplantation, the graft is not rejected.

epidermal cells grown from a small skin specimen in tissue culture formed multilayered sheets. In addition, they were not rejected when grafted onto skin ulcers across major histocompatibility barriers in nonimmunosuppressed adult patients, whereas control noncultured skin was rejected (Thivolet et al., 1986). These workers concluded that cultured epidermal cells did not express class II antigens when passaged in tissue culture. The possibility that passage of many types of cells in cell culture would deplete their cell-surface class II markers and allow for transplantation to unrelated donors without rejection provided much excitement for the possibility of engineering a variety of tissues.

The second observation was that living skin equivalents prepared by combining cultured allogeneic fibroblasts with a collagen matrix and overlaying this with isogeneic keratinocytes were not rejected when transplanted onto skin wounds in an animal of the same species (Sher et al., 1983). These workers concluded that fibroblasts express class I but not class II MHC antigens and therefore are acceptable graft constituents if incorporated in a tissue equivalent, excluding cells with class II antigens.

In skin, the cells most frequently cited to present antigens that cause rejection are Langerhans' and endothelial cells. Another approach, besides passaging them in cell culture to prevent expression of the class II MHC markers causing rejection, is either to coat each cell with a polymeric membrane or to enclose the tissue within a membrane. This is an approach used to prevent rejection of differentiated cells that may lose function when passaged in cell culture. This has been done with pancreatic islet cells to create an artificial pancreas, as discussed later.

Role of Matrixes and Scaffolds in Cell Transplantation

Although isolated cells have been used as transplants, the majority of applications require materials that can be molded into a variety of shapes and sizes. In addition, most cells do not function as isolated elements but interact with other cell types and the extracellular matrixes that they sit on. Therefore, as discussed later in the case of skin cells and hepatocytes, the interaction between cells and matrixes that they associate with is an important aspect of tissue engineering.

Much of the information about optimizing cell attachment to substrates has come from designing surfaces and coatings for cell culture plates. The role of surface morphology has been studied extensively; of primary importance is the observation that cells become oriented in response to underlying surface topography, commonly known as contact guidance. Several physicochemical surface properties, such as surface composition, surface charge, surface energy, surface oxidation, curvature, and morphology, affect cell attachment and behavior (Singhvi et al., 1994).

Many studies report the use of collagen and other extracellular matrix (ECM) macromolecules as substrates for attachment and growth of cells. The interaction between ECM and cells is an important aspect of tissue engineering because cell shape, cytoskeleton, cell migration, control of cell growth, and differentiation all involve the ECM found around cells and in connective tissue.

The ECM is a structural material that interacts with the cell cytoskeleton and binds growth factors. These interactions allow the transduction of chemical and mechanical signals across the cell membrane and result in changes in cell shape, protein synthesis, and other cellular functions (Figure 11.5). Although

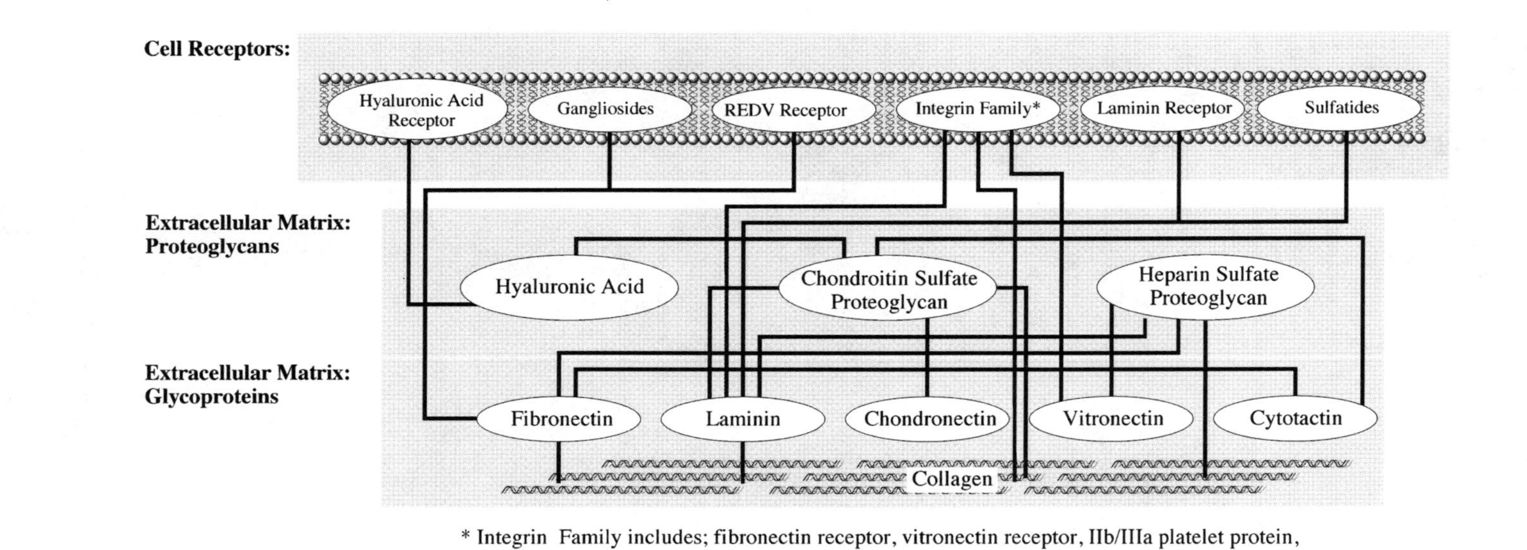

Figure 11.5 Relationship between cell membrane and ECM components. The illustration shows the relationship between the receptors on the cell membrane and ECM components. Via these interactions, signals outside the cell influence events such as rearrangement of the cytoskeleton that occur within the cell.

collagen provides tensile strength and creep resistance to ECM, elastic fibers and proteoglycans help dissipate loads elastically and viscously, respectively. Attachment between the cell membrane and the ECM occurs via cell surface proteoglycans, integrins, and cell attachment glycoproteins. These factors influence adhesion and migration of cells to substrates.

In normal skin, basal epithelium rests on a basement membrane at the dermal–epidermal junction. At this location, the basal epithelia associate with ECM in the basement membrane, including fibronectin, laminin, heparan sulfate proteoglycan, and type IV collagen. Human epithelial cell adherence and growth in culture were shown to be maximized on fibronectin compared to other ECM macromolecules (Kubo et al., 1987). Epithelial cell migration appears to be associated with an increase of β_1 integrin subunits (Guo et al., 1990), and exposure of endothelial cells to shear stress caused fibronectin fibrils to group into thicker tracks of fibrils and to align with the direction of flow (Thiomine et al., 1995). These results suggest that attachment and alignment with ECM substratum are linked to the presence of specific macromolecules and physical interaction between the ECM and the cell membrane. Although phenomenologically it is understood that ECM is essential for the normal adhesion, attachment, migration, and growth of cells, the exact mechanism for these interactions is still unknown.

As discussed later, synthetic and biological macromolecules have been used as matrixes for tissue engineering. Synthetic matrixes have an advantage that they can be produced reproducibly from abundantly available raw materials. The disadvantage is that they must induce cells to make ECM, because without ECM, normal cell function is compromised. Biological substrates are more difficult to produce; however, they can be made to mimic the normal ECM of the tissue or organ to be replaced and therefore have an advantage in minimizing the healing time associated with introduction of the implant. The use of both types of matrixes to replace the function of external surface linings, skeletal structures, and specialized organs is discussed next.

Tissue Engineering of External Lining Structures

As discussed in Chapter 4, external lining structures are composed of layers, including the external epithelia, a dermal/connective tissue layer that sometimes contains muscular tissue, and the subcutaneous tissue that contains blood vessels, fat, and connections to surrounding tissues. Engineering of skin replacements historically led the way to the development of several other tissue replacements. We will focus on tissue engineering of skin in this section.

Evolution of an artificial skin derived from three research activities that originated in the 1970s and 1980s: the formulation of a cell-free collagen matrix, the growth of epidermal cells (also known as keratinocytes) in the form of sheets for transplantation, and the growth of epidermal and connective tissue cells on a collagen matrix. All these products were envisioned to be used to treat patients with severe burns that were life-threatening.

Cell-free collagen matrixes for treatment of burns were produced from cow corium and animal dermis in the 1960s. In 1965, Abbenhaus et al. reported the use of collagen extracted from cow hides for treatment of large excised areas of skin. Oneson et al. (1970) reported the preparation of highly purified insoluble collagen for medical applications. Oliver and co-workers (1976) studied the use of trypsin-treated porcine dermis as a wound dressing. In 1977, Chvapil reviewed the use of purified collagen in the form of a sponge as a wound dressing. Later in 1980, Yannas and co-workers (Yannas and Burke, 1980, Yannas et al., 1980) introduced the use of collagen co-precipitated with glycosaminoglycans covered with a silicone layer as the chemical basis for the design of artificial skin. A subsequent clinical trial on artificial skin for treatment of major burns showed that in combination with a thin epidermal graft, it provided a wound covering that was as effective as thicker autografts (Figure 11.6). This was an advantage because the donor sites healed more rapidly using thinner epidermal grafts. The skin graft was placed on the artificial skin after the silicone layer was removed. The commercialized product was named Integra Artificial Skin (Integra Life-Sciences, Plainsboro, N.J.).

Another approach to the problem of skin replacement involves producing sheets of differentiated epithelia from cell cultures of small biopsies. Eisinger et al. (1980) used this approach to create sheets of differentiated epithelial cells that after 13 weeks showed normal dermal and epidermal structures when implanted on animal wounds. Clinical use of expanded epidermal autografts by Pittlekow and Scott (1986) showed that one month after application, the grafts showed many characteristics of normal epidermis. The primary drawback to such a procedure was the 3- to 4-week lag time between harvest and use of autogenous epithelium. This time lag could be circumvented by use of allogeneic epidermal cells that were available immediately for grafting. Use of allograft epidermal sheets to cover severe burns resulted in wound healing, although no rete pegs or adrenal structures were present after nine months (Hefton et al., 1983). Madden et al. (1986) reported that third-degree burns did not support the growth of

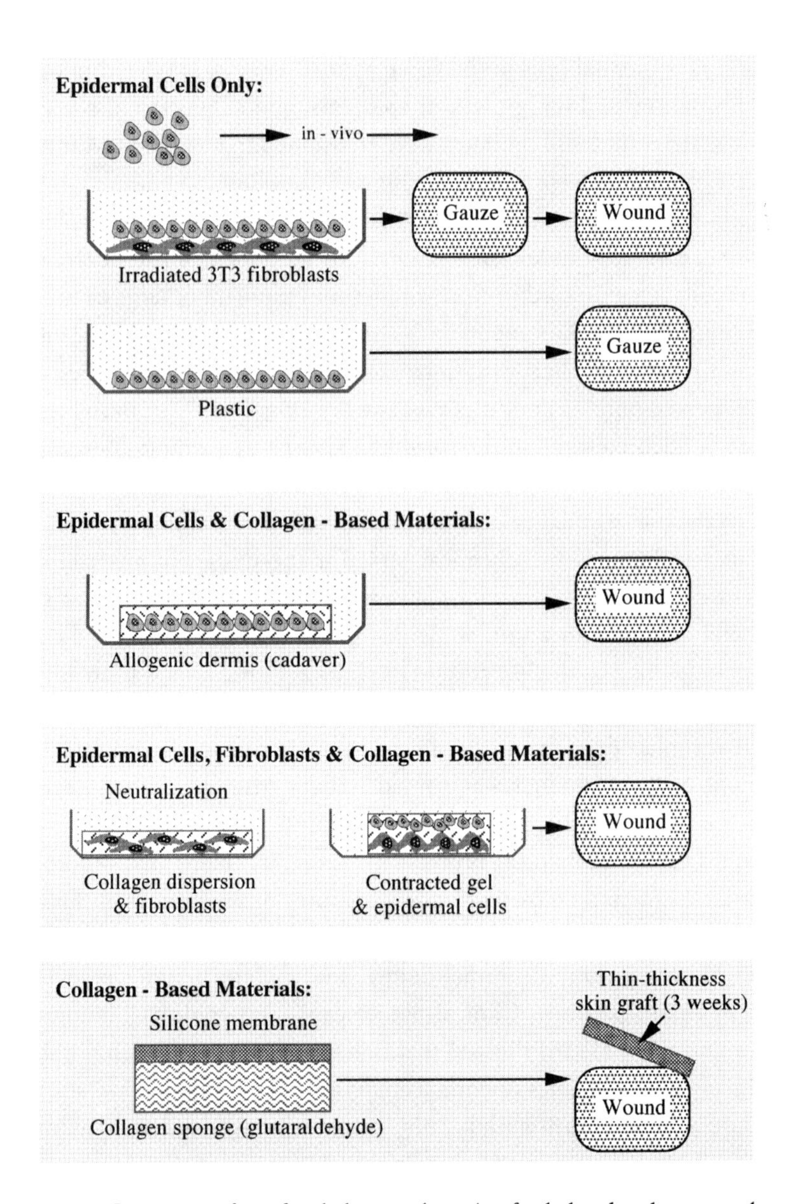

Figure 11.6 Summary of artificial skin studies. Artifical skin has been produced by several different approaches, including use of epidermal cells only (top), epidermal cells or epidermal cells and fibroblasts in combination with collagen (middle), and a collagen matrix with an occlusive layer (bottom).

allograft epidermal sheets. These results suggested that allograft epidermal cells alone do not properly function as an artificial skin.

Green and co-workers pioneered a slightly different approach. Instead of growing epithelia without a substrate, they used a feeder layer of lethally irradiated 3T3 fibroblasts to support keratinocyte growth for the formation of autologous keratinocyte sheets for skin replacement on patients with severe burns (Reinhwald and Green, 1975). Cultured autografts form a differentiated and confluent epidermal layer within one month after transplantation (O'Connor et al., 1981). Although cultured autografts perform satisfactorily as replacements, they have a 3- to 4-week lag period before they are available for grafting. In addition, histological and morphological examinations suggest that the absence of a dermal interface between graft and the wound contributes to graft fragility (Woodley et al., 1988). This proves to be the limiting variable in making autologous keratinocyte sheets a viable method to restore limited skin coverage in patients with severe burns.

To further stabilize the wound bed using the approach of forming keratinocyte sheets on a 3T3 cell feeder layer, two reports suggest that severe burns grafted with cryopreserved cadaver skin that have had their original epidermal layers removed act as a good substrate for grafting of autologous epidermal sheets (Cuono et al., 1986). Long-term follow-up on these patients indicates that a mature dermis forms between 11 and 12 months after grafting (Langdon et al., 1988). The disadvantage of this approach is that grafting involves two surgical procedures rather than a single procedure.

Skin tissue engineering involving growth of skin cells on collagenous substrates dates back to the early 1970s. Karasek and Charlton (1971) reported growth of postembryonic skin epithelia cells on a collagen gel, and Worst and colleagues (1974) used rat tail tendon as a substrate for perinatal mouse skin epidermal cells. Freeman and co-workers (1974) extended this work to show that epithelia could be expanded in culture and then transferred as an autograft to rabbit skin wounds and placed on top of a collagenous matrix. These studies paved the way for further studies by Bell and co-workers (Bell et al., 1981a, 1981; Topol et al., 1986; Bell et al., 1989) who developed a skin equivalent composed of rat-tail tendon collagen that was contracted by fibroblasts and then subsequently covered with epidermal cells. This approach was used commercially by Organogenesis in clinical trials on third-degree burns and skin ulcers. Although the approach appears to have merit for enhancing healing of skin ulcers, it was subsequently abandoned for treatment of third-degree burns because of graft adhesion problems. Other commercial competitors, including Advanced Tissue Sciences, Ortec, and LifeCell, have competitive products under development.

Other engineered materials for creation of artificial skin and skin equivalents are variants of the basic approaches discussed earlier. Boyce et al. (1988) use artificial skin as a support for human keratinocytes and fibroblasts. They prepared human fibroblasts from 10- to 20-cm^2 biopsies of skin for seeding onto the porous surface of a collagen-glycosaminoglycan sponge. One day later, they inoculated human keratinocytes onto the matrix.

Results of studies aimed at tissue engineering of skin have demonstrated that the basic concept—introducing different cell types into a scaffold that mimics that of the tissue to be replaced—is a viable approach. The components of this approach are the engineering of scaffolds that mimic the normal hierarchical structure of the tissue parenchyma or stroma, the isolation and the passage of parenchymal cells in culture until they do not express class II MHC markers, and introduction of the appropriate cell types into the scaffold. This approach assumes that cells that lack class II MHC markers still remain differentiated so that they will perform the function of their counterparts in normal tissues (Figure 11.7).

Skeletal Tissue Engineering

Replacement of cartilage, tendon, and bone has traditionally involved use of transplanted tissue or total joint replacement with metallic implants. In the case of tendons and ligaments, most synthetic materials fail to withstand the cyclic loading required by the application and eventually fail as a result of fatigue. Bone transplants of either autogenous iliac crest or bone from cadavers are used to repair defects that are small. Nonunions traditionally are difficult to heal and are not treated with bone grafts. Although many soft tissue defects are treated successfully by transfer of vascularized flaps of skin, muscle, or fascia, vascularized bone grafts are typically used to promote healing of complex bone defects. This is because of factors such as limited availability of donor tissue, donor site morbidity, difficulty shaping the graft to fit the defect, and the availability of synthetic materials that give less-than-ideal results. Cartilage replacement typically requires total joint removal; however, cartilage repair using tissue autografts and tissue-cultured materials has been reported. The next section examines the different approaches to tissue engineering of cartilage and bone using cartilage- and bone-inducing substances and cellular grafts.

The ability of bone extracts to induce bone formation in muscle was first demonstrated by Lacroix (1945), who showed that acid-alcohol bone extracts

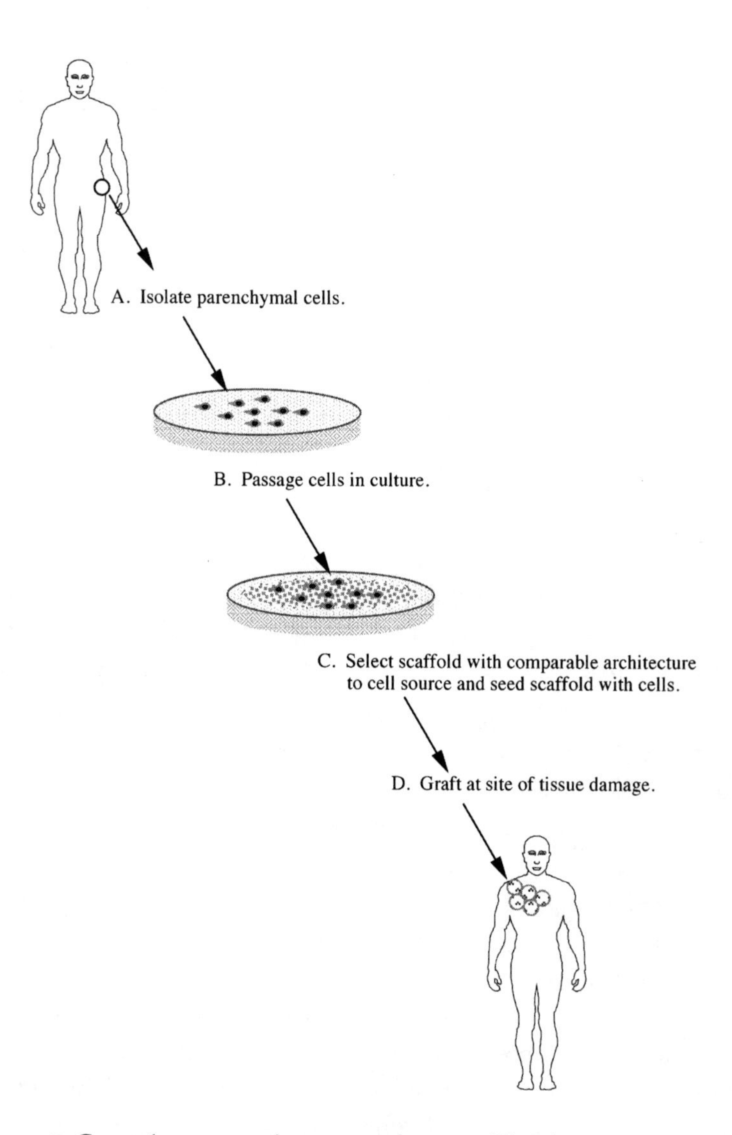

Figure 11.7 General concept of tissue engineering. The diagram illustrates a universal approach to engineering of replacement tissues. The first step is to isolate parenchymal cells of the tissue to be regenerated. These cells are passaged in culture until they lose their class II MHC markers. This may require the passage of more than one cell line if the functional units within the tissue have a number of cell types. Then a scaffold is selected that has the same architecture as does the tissue to be regenerated, and it is seeded with the passaged cells. The seeded scaffold is then grafted into the desired site.

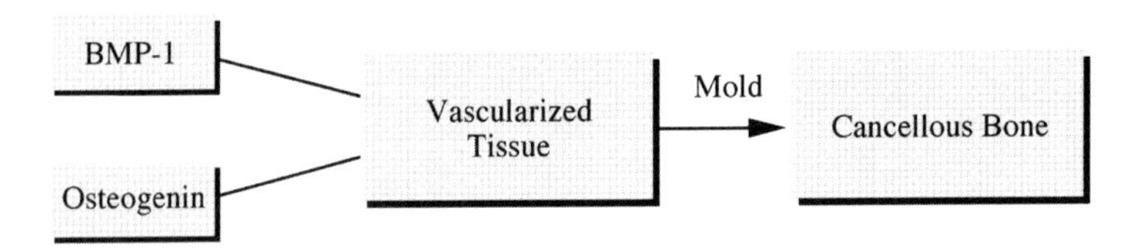

Figure 11.8 Bone tissue engineering using growth factors. Materials for bone repair can be made by combining BMP-1 and osteogenin with vascularized tissue in a mold, which then causes formation of cancellous bone.

produced bone when implanted into rabbit muscle (Khouri et al., 1991). This extract became known as osteogenin. Osteoinduction is the term given to a material that induces formation of bone in the absence of bone, whereas osteoconduction is the term given to denote that a material will support bone formation. Osteogenin was the first bone-induction factor discovered. Later, Urist (1965) reported that a fraction of demineralized bone matrix that he called bone morphogenic protein (BMP) formed bone when implanted subcutaneously. Purified BMP has been isolated and consists of three polypeptides named BMP-1, BMP-2A and BMP-3 (Wozney et al., 1988). BMP-2A and BMP-3 were found to be new members of the TGF-β supergene family (TGF-β is a family of growth factors), but BMP-1 appeared to be a novel regulatory protein. Osteogenin appears to be similar to BMP-3.

One group investigated the transformation of muscle flaps into vascularized bone by taking thigh abductor muscle island flaps in silicone rubber molds coated with BMP and injected with osteogenin (Khouri et al., 1991). The flaps treated with osteogenin and BMP were transformed into cancellous bone in the exact shape of the mold (Figure 11.8).

A number of biological grafts have been used to repair articular cartilage defects, including perichondrium, osteochondral graft, meniscus, epiphyseal growth plate, periosteum, and mandibular cartilage (Silver and Glasgold, 1995). Articular cartilage is a specialized connective tissue that covers the surface of long bones. The poor healing of articular cartilage and the limited amount of autogenous cartilage that can be transplanted make it an excellent candidate for tissue engineering. The approaches used to replace and repair cartilage include use of biological grafts and cell-cultured materials.

Perichondral autografts have been used to completely fill in defects in rabbit medial femoral condyle with neocartilage. Rib perichondrium has also been used to fill in full-thickness cartilage defects. Improved articular cartilage

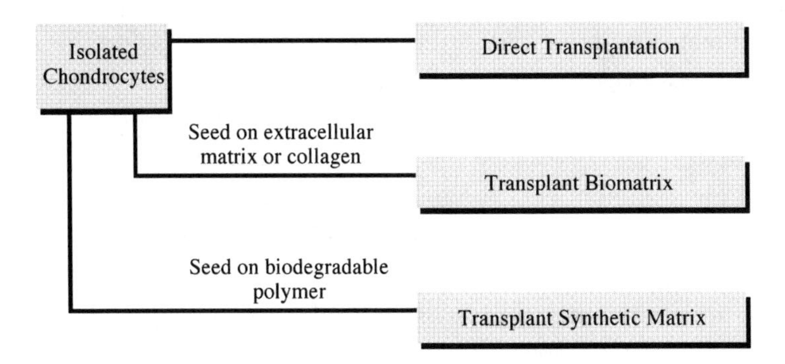

Figure 11.9 Tissue engineering of cartilage. The approaches used to engineer cartilage include use of isolated chondrocytes for direct transplantation and chondrocytes in combination with a collagenous matrix or biodegradable polymers.

healing has also been noted using autogenous meniscus, osteochondral grafts, periosteum, and mandibular condylar cartilage.

Other approaches to cartilage repair include use of isolated chondrocytes with and without substrate matrixes. Aston and Bentley (1986) isolated chondrocytes from epiphyseal growth plate and inserted them into full-thickness rabbit defects (Figure 11.9). Grande et al. (1987) evaluated the effects of autogenous chondrocytes grown in vitro on the healing rates of articular defects. Macroscopic results from grafted specimens showed that defects that received chondrocyte transplants had significantly more reconstituted cartilage than ungrafted controls. Weiss and colleagues (1988) reported organ culture of the zone of progenitor cells of mandibular condyles of neonatal mice. They found that cartilage explants of progenitor cells and their adjacent ECM succeeded in repairing damaged condylar cartilage in vitro. Robinson and colleagues described repair of articular cartilage using fetal allogeneic cells. The cells were grown on biodegradable scaffolds and were later implanted into joint defects. They also described use of rabbit autogenic chondrocyte-enriched cultures derived from mesenchymal stem cells from adult rabbit bone marrow cultured on carbon-fiber mesh impregnated and coated with hyaluronic acid for repair of joint cartilage (Robinson et al., 1993). They reported that the repaired cartilage was more cellular and thicker when the carbon fiber mesh was impregnated with cultured cells.

The use of synthetic polymers as supports for cultured cartilage cells has become popular. Von Schroeder and associates (1991) investigated use of poly(lactic acid) matrix in the presence and absence of a periosteal graft. They concluded that poly(lactic acid) in the absence of a graft allowed for de novo growth of neocartilage at the articular surface. In the presence of a graft, neocartilage was thicker and more closely resembled articular cartilage. Ruuskanen and co-workers (1991) evaluated the role of poly(glycolic acid) rods in regeneration of cartilage from perichondrium. Perichondrium formed a tube around the implant, and a foreign body reaction was observed within the implant. Synthetic biodegradable polymers as cell–polymer delivery devices can be designed to induce cartilage formation in a predetermined shape (Kim et al., 1994), making the product suitable for cosmetic and plastic surgery.

Tissue Engineering of Specialized Organs

Tissue engineering of specialized organs, including liver and pancreas, offers additional challenges to isolation and cell culture of single cells. In addition to a structural role, these organs must perform biochemical functions such as providing proteins necessary to maintain clotting and regulate osmotic pressure. The first step to replace liver function has been to isolate hepatocytes and then place them in suspensions, microcapsules, hollow fibers, or on substrates (Figure 11.10). In animal models, hepatocytes have been shown to produce albumin and other liver markers as well as products of urea and bilirubin metabolism.

The liver is capable of regenerating itself after loss of two-thirds of its mass. After partial hepatectomy, much of the lost liver is replaced by division of

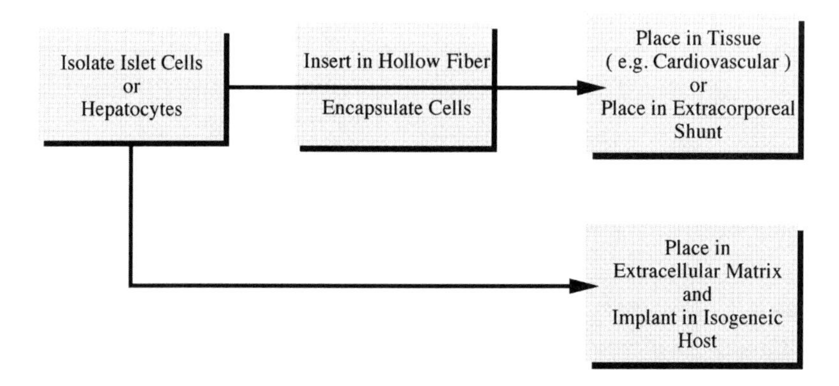

Figure 11.10 Summary of liver tissue engineering. Liver tissue engineering involves use of non-MHC matched cells in a hollow fiber or encapsulated in an extracorporal shunt system or use of hepatocytes in a collagen sandwich in isogeneic animals.

fully differentiated hepatocytes. In conditions of severe liver injury, restoration of liver mass may be accomplished by hepatocyte precursors, also known as stem cells, that subsequently differentiate into hepatocytes. Adult hepatocytes have been cultured in a collagen sandwich system that showed normal morphology and albumin secretion rate for 42 days (Dunn et al., 1989) compared to hepatocytes cultured on a single collagen layer that ceased albumin production in less than one week. In another study it was reported that hepatocyte cultures containing aggregates of cells within collagen gels maintained a twofold higher concentration of functional parameters than singlet cultures (Parsons-Wingerter and Saltzman, 1993). Hepatocyte culture systems have been studied as extracorporeal and implantable devices.

Extracorporeal systems (Langer and Vacanti, 1993), which could be used while a patient's liver is recovering from injury or during the time a patient is unable to get a transplant, are simpler and have the advantage of better regulation and limited potential of immune reactions. Vascular access required for extracorporeal use has proved to be a major problem in dialysis patients. Implantable systems can act as a permanent liver replacement and appear to be the only long-term solution.

In a similar manner to liver, pancreas replacements have been designed as either extracorporeal devices or implantable devices. The purpose of these devices is to provide insulin release from islet cells to promote glucose transport into cells. To prevent or decrease the rejection response, islets cells are physically separated from the blood, which contains antibodies and B and T cells involved in cell-mediated rejection processes. The blood flows through a tube in an extracorporeal circuit and is connected to a membrane surrounded by islet cells. Insulin diffuses across the membrane and into the blood, which is then returned to the patient. In another approach, islets cells are either encapsulated or trapped in a polymer gel that can be introduced into the peritoneal cavity from which insulin can be slowly released and absorbed into the blood. A third approach is to place the islet cells into tiny capillary tubes made of synthetic polymers. The capillaries can be placed in the vascular system or implanted somewhere beneath the skin.

Summary

The rate-limiting step in success of tissue-engineered devices is acquiring enough functional cells or product to produce the required amounts of biochemically active materials necessary for maintaining homeostasis. Whether parenchymal cells lack class II MHC markers after being passaged in cell culture while remaining in their differentiated state will determine if direct allogeneic implantation procedures will be successful. It is important to verify that these cells will maintain a differentiated state indefinitely and divide normally to ensure long-term success of these procedures. In applications where allogeneic cells without these markers cannot be obtained, it is likely that most of the clinical uses of these devices will be as short-term extracorporeal shunts.

Even though many of the techniques developed for skin tissue engineering have been in the literature for three decades, we are only beginning to see the possibility of their widespread applications. Technologically, many of these procedures are possible, and the real challenge will be to scale up the processes so that they can be done economically.

Suggested Reading

Abbenhaus J.L., MacMahon R.A., Rosenkrantz J.G. and Paton B.C., Collagen Sheets as a Dressing for Large Excised Areas, Surg. Forum 16, 477, 1965.

Aston J.E. and Bentley G., Repair of Articular Surfaces by Allografts of Articular Cartilage and Growth-Plate Cartilage. J. Bone Joint Surg 68B, 29, 1986.

Bell E., Ehrlich H.P., Buttle D.J., and Nakatsuji T., Living Tissue Formed In Vitro and Accepted as Skin Equivalent Tissue of Full Thickness, Science 211, 1052, 1981a.

Bell E., Ehrlich H.P., Sher S., Merrill C., Saber R., Hull B., Nakatsuji T., Church D., and Buttle D., Development and Use of a Living Skin Substitute, Plast. Reconstr. Surg., 67, 386, 1981b.

Boyce S.T., Glafkides M.C., Foreman T.J., and Hansbrough J.F., Reduced Wound Contraction After Grafting of Full-Thickness Burns with a Collagen and Chondroitin-6-Sulfate (GAG) Dermal Skin Substitute and Biobrane, J. Burn Care Rehabil. 9, 234, 1988.

Chvapil M., Collagen Sponge Theory and Practice of Medical Applications, J. Biomed. Mater. Res. 11, 721, 1977.

Cuono C., Langdon R., and McGuire J., Use of Cultured Epidermal Autografts and Dermal Allografts as Skin Replacement After Burn Injury, Lancet May 17, p. 1123, 1986.

Dunn J.C.Y., Yarmush M.L., Koebe H.G., and Tomkins R.G., Hepatocyte Function and Extracellular Matrix Geometry: Long-Term Culture in a Sandwich Configuration FASEB J. 3, 174, 1989.

Eisinger M., Monden M., Raaf J.H., and Fortner J.G., Wound Coverage by a Sheet of Epidermal Cells Grown In Vitro From Dispersed Single Cell Populations, Surgery 88, 287, 1980.

Freeman A.E., Igel H.J., Waldman N.L., and Losikoft A.M., A New Method for Covering Large Surface Area Wounds With Autografts. I. In Vitro Multiplication of Rabbit-Skin Epithelial Cells. Arch. Surg. 108, 721, 1974.

Guo M., Toda K.-I., and Grinnell F., Activation of Human Keratinocyte Migration on Type I Collagen and Fibronectin, J. Cell Sci. 96, 197, 1990.

Grande D.A., Singh I.L., and Pugh J., Healing of Experimentally Produced Lesions in Articular Cartilage Following Chondrocyte Transplantation. Anat. Rec. 218, 142, 1981.

Hefton J.M., Madden M.R., Finkelstein J.L., and Shires G.T., Grafting of Burn Patients with Allografts of Cultured Epidermal Cells, Lancet 2, 428, 1983.

Heimbach D., Luterman A., Burke J.F., Cram A., Herndon D., Hunt J., Jordan M., McManus W., Solem L., Warden G., and Zawacki B., Artificial Dermis for Major Burns: A Multi-Center Randomized Clinical Trial, Ann. Surg. 208, 313, 1988.

Karasek M.A. and Charlton M.E., Growth of Post-Embyonic Skin Epithelial Cells on Collagen Gels. J. Invest. Dermatol. 56, 205, 1971.

Khouri R.K., Koudsi B., and Reddi H., Tissue Transformation Into Bone In Vivo, JAMA 266, 1953, 1991.

Kim W.S., Vacanti J.P., Cima L., Mooney D., Upton J., Puelacher W.C., and Vacanti C.A., Cartilage Engineered in Predetermined Shapes Employing Cell Transplantation on Synthetic Biodegradable Polymers, Plast. Reconstr. Surg. 94, 233, 1994.

Kubo M., Kan M., Isemura M., Yamane I., and Tagami H., Effects of Extracellular Matrices on Human Keratinocyte Adhesion and Growth and on Its Secretion and Deposition of Fibronectin in Culture, J. Invest. Dermatol. 88, 594, 1987.

Lacroix P., Recent Investigations on the Growth of Bone, Nature 156, 576, 1945.

Langer R. and Vacanti J.P., Tissue Engineering, Science 260, 920, 1993.

Langdon R.C., Cuono C.B., Birchall N., Madri J.A., Kuklinska E., McGuire J., and Moellmann G.E., Reconstitution of Structure and Cell Function in Human Skin Grafts Derived From Cryopreserved Allogeneic Dermis and Autologous Cultured Keratinocytes, J. Invest. Dermatol. 91, 478, 1988.

Madden M.R., Finkelstein J.L., Staiano-Coico L., Goodwin C.W., Shires G.T., Nolan E.E., and Hefton J.M., Grafting of Cultured Allogeneic Epidermis on Second and Third Degree Burn Wounds on 26 Patients, J. Trauma 26, 955, 1986.

Nerem R.M., Tissue Engineering in the USA, Med. Biol. Eng. Comput. 30, CE8, 1992.

O'Connor N.E., Mulliken J.B., Banks-Schledels S., Kehinds O., and Green H., Grafting of Burns With Cultured Epithelium Prepared From Autologous Epidermal Cells Lancet 1, 75, 1981.

Oliver R.F., Grant R.A., Cox R.W., Hulme M.J., and Mudie A., Histological Studies of Subcutaneous and Intraperitoneal Implants of Trypsin-Prepared Dermal Collagen in the Rat, J. Clin. Orthoped. Rel. Res. 115, 291, 1976.

Oneson I., Fletcher D., Olivo J., Nichols J., and Kronenthal R., The Preparation of Highly Purified Insoluble Collagens, J. Am. Leather Chem. Assoc. LXV, 440, 1970.

Parsons-Wingerter P.A. and Saltzmann W.M., Growth versus Function in the Three-Dimensional Culture of Singlet and Aggregated Hepatocytes within Collagen Gels, Biotechnol. Prog. 9, 600, 1993.

Pittlekow M.R. and Scott R.E., New Techniques for the In Vitro Culture of Human Skin Keratinocytes and Perspectives on Their Use for Grafting of Patients with Extensive Burns, Mayo Clin. Proc. 61, 771, 1986.

Reinwald J.G. and Green H., Serial Cultivation of Strains of Human Epidermal Keratinocytes: The Formation of Keratinizing Colonies From Single Cells, Cell 6, 331, 1975.

Robinson D., Efrat M., Mendes D.G., Halperin N., and Nevo Z., Implants Composed of Carbon Fiber Mesh and Bone- Marrow-Derived Chondrocyte-Enriched Cultures for Joint Surface Reconstruction, Bull. Hosp. J. Dis. 53, 75, 1993.

Ruuskanen M.M., Kallinoinen M.J., Kaarela O.I., Laiho J.A., and Tormala P.O., The Role of Polyglycolic Acid Rods in the Regeneration of Cartilage from Perichondrium in Rabbits, Scand. J. Plast. Reconstr. Surg. Hand Surg. 25, 15, 1991.

Sher S.E., Hull B.E., Rosen S., Church D., Friedman L., and Bell E., Acceptance of Allogeneic Fibroblasts in Skin Equivalent Transplants, Transplantation 36, 552, 1983.

Silver F.H. and Glasgold A.I., Cartilage Wound Healing: An Overview, Otolaryngol. Clin. North Am. 28, 847, 1995.

Singhvi R., Stephanopoulos G., and Wang D.I.C., Review: Effects of Substratum Morphology on Cell Physiology, Biotechnol. Bioeng. 43, 764, 1994.

Thivolet J., Faure M., Demidem A., and Mauduit G., Long-Term Survival and Immunological Tolerance of Humans, Transplantation, 42, 274, 1986.

Thiomine O., Nerem R.M., and Girard P.R., Changes in Organization and Composition of the Extracellular Matrix Underlying Cultured Endothelial Cells Exposed to Laminar Steady State Stress, Lab. Invest. 73, 565, 1995.

Topol B., Haimes H., Dubertret L., and Bell E., Transfer of Melanosomes in a Skin Equivalent Model In Vitro, Invest. Dermatol. 87, 642, 1986.

Urist M.R., Bone: Formation By Autoinduction, Science 150, 893, 1965.

von Schoeder H.P., Kwan M., Amiel D., and Coutts R.D., The Use of Polylactic Acid Matrix and Periosteal Grafts for the Reconstruction of Rabbit Knee Defects, J. Biomed. Mater. Res. 25, 329, 1991.

Weiss A., Livne E., Von der Mark K., Heinegard D., and Silberman M., Growth and Repair Cartilage: Organ Culture System Utilizing Chondroprogenitor Cells of Condylar Cartilage in Newborn Mice, J. Bone Min. Res 3, 93, 1988.

Woodley D.T., Peterson H.D., Herzog S.R., Stricklin G.P., Burgeson R.E., Briggaman R.A., Cronce D.J., and O'Keefe E.J., Burn Wounds Resurfaced By Cultured Epidermal Autografts Show Abnormal Reconstitution of Anchoring Fibrils, JAMA 258, 2566, 1988.

Worst P.K.M., Valentine E.A., and Fusenig N.E., Formation of Epidermis After Reimplantation of Pure Primary Epidermal Cell Cultures From Perinatal Mouse Skin, J. Natl. Cancer Inst. 53, 1061, 1974.

Wozney J.M., Rosen V., Celeste A.J., Mitsock L.M., Whitters M.J., Kriz R.W., Hewick R.M., and Wang E.A., Novel Regulators of Bone Formation: Molecular Clones and Activities, Science 242, 1528, 1988.

Yannas I.V. and Burke J.F., Design of an Artificial Skin. I. Basic Design Principles, J. Biomed. Mat. Res. 14, 65, 1980.

Yannas I.V. Burke J.F., Gordon P.L., Huang C., and Rubenstein, R.H., Design of an Artificial Skin. II. Control of Chemical Composition, J. Biomed. Mat. Res. 14, 107, 1980.

12 Future Considerations

Introduction

The field of biomaterials has expanded greatly in the past decade. Minimally invasive surgery is being performed using instruments and devices that were just dreams a decade ago. Continued development of surgical techniques and engineered tissues will revolutionize health care delivery and contain costs. However, much of this progress will require skillful development so that the cost of new technologies will not swell the already expanded cost of health care. For this to occur, several fundamental questions need to be answered regarding the impact of new technologies in the area of tissue engineering, new polymeric materials, small-diameter vascular grafts, and vascular stents.

New Directions in Tissue Engineering

Although there has been much progress in the development of artificial skin over the past 25 to 30 years, questions continue to be asked concerning the possible commercial impact of cell-cultured materials. Considering that at least $100 million was spent commercializing the Integra artificial skin, it can be projected that the commercialization of tissue culture skin products will cost in excess of several hundred million dollars. At the same time, with the conservative nature of the third-party reimbursement rate for new treatments, there exists a major concern that many of these new technologies may never reach the average consumer because of the potential price tag. Also, the availability of tissue grafts, both autogenous and allogeneic, at very reasonable costs to the patient suggest that much progress in the field of tissue engineering will come from more creative uses of tissue grafts. Much of the progress in use of autogenous transplants has been limited by the time, difficulty, and morbidity associated with historically used harvesting techniques. The development of new minimally invasive harvesting procedures in concert with progress made in understanding tissue engineering uses of graft materials will lead to a new generation of graft materials. Graft materials in conjunction with new biodegradable polymeric implants will also have great utility in both structural and functional roles.

Design of New Polymeric Materials

Although the synthetic polymers used as off-the-shelf materials since the 1950s in medical devices are not new, they function acceptably in many applications. Dacron and Teflon, two polymers developed for other commercial uses, still remain the workhorses in cardiovascular applications. What remains to be developed are synthetic polymers that can reproduce the effects of biological macromolecules in terms of promoting cell attachment, migration, and proliferation and that can be prepared cheaply in large quantities. One question that needs to be answered is whether the biodegradation rate of synthetic materials can be controlled carefully enough to be useful in devices and as scaffolding materials. It is clear that degradation of any polymer, whether it is synthetic or natural, will cause inflammatory and wound healing responses that have a negative impact on the host tissue.

Design of Interfaces

Perhaps the biggest challenge to the biomaterials scientist is the design of the interface between a medical device and host tissue. This proves to be another challenge that will determine whether implants in the musculoskeletal and cardiovascular systems will perform indefinitely in their applications. Failure caused by micromotion or stress concentration at the interface between the tissue and the implant ultimately has negative affects on tissue, leading to tissue necrosis, inflammation, tissue resorption, and scar tissue deposition. The potential solution to this problem involves creating interfaces that are designed to cause less stress concentration or increased interfacial areas. Much attention has focused on development of porous interfaces; however, the use of porous interfaces has the potential of negative side effects if the device has to be separated from tissue because of an infection or poor performance. Perhaps this problem could be eliminated if surfaces were designed to allow tissue deposition without cellular penetration. Solving this problem is important, especially in designing interfaces with cardiovascular devices.

Vascular Device Design

The two limitations to designing small-vessel vascular grafts and stents to keep arteries open include intimal hyperplasia at the attachment to the host vasculature and thrombosis. Intimal hyperplasia continues to be a challenging problem and appears associated with stress concentration on the lumen wall, leading to smooth muscle cell proliferation and obliteration of the vascular lumen. Whether the stress causes direct medial tearing and inflammation or whether the application of stress to the intima leads to thickening of the wall to offset stress increases is unknown. However, it is clear that the aortic wall thickens with increased blood pressure, suggesting that a combination of these factors must cause intimal hyperplasia. Research is needed to understand how normal and shear stresses are transduced by cells into changes in the composition and structure of the wall. Of course, much of our progress will depend on a better understanding of how to control thrombosis induced by the surface.

Summary

The field of biomaterials science has many challenges ahead. We have used the available materials in every conceivable manner. It is now time to develop new graft materials from cells and tissues as well as from synthetic sources. Much of the progress in understanding the role of different macromolecules in growth and development will serve as a basis for design of new scaffolds. Synthetic materials have the distinct advantage of large-scale producibility at low cost. It is time to design new materials with biological activity for tissue engineering and other applications.

Index

D-*N*-Acetyl glucosamine, 33f, 50, 51f, 60t
Acquired immunity, 253
Actin, 30, 81–83, 83f, 142
  actin-myosin interaction, 179–182, 180f
  assembly of cytoskeletal components, 176–179
  cell migration, 228
  cell shape, 229f
  F-actin, 81, 83f, 166, 177–179, 178f, 209
  G-actin, 81, 83f, 163t, 166, 177–179, 178f
  self-assembly, 177–179, 178f
α-Actinin, 217–219, 228, 229f
Actinomyosin, 181
Adaptation, cellular, 228–233, 231t
Addition polymerization, 89–90, 90f, 119
Adhesion, cell, *See* Cell adhesion
Adressin, mucosal, 224
Aggrecan, 12, 67, 69f
Aging, 231t
Albumin, serum, 163t
Alcian blue stain, 125
Alginate, 19
Alkaline phosphatase, 300
Allergic reaction, 19
Allogeneic cells, transplantation, 309f
Alloys, 20, *See also* Metals; *specific alloys*
  annealing, 111
  crystalline lattice structure, 108f
  defined, 109
  deformation, 109, 111
  degradation, 119–120
  formation, 109, 110f
    dendrites, 109
    grains, 109, 110f
    heterogeneous process, 109
    homogeneous process, 109
  hardening, 111
  mechanical properties, 112–114, 112–114t, 188t
  processing, 111
  structure, 106–115
  types used in medical devices, 111–112, 112t
  unit cell, 106–115
α-chain, collagen, 9, 11
α helix, 34, 43–45, 47–48f, 55–56
  mechanical properties, 192–195
Aluminum oxide, 20–21, 116t
Alveoli, 4t, 127, 128t, 129f
Amino acids
  D or L form, 37–40, 40f
  peptide formation, 31f
  stereochemistry of proteins, 33–49
  structure, 36–40, 37f, 38–39t
Aneurysm
  aortic, 205
  vascular graft, 274
Angiogenesis, 227, 268
Annealing, 111
Antibodies, 75–78, 78f, 238, 242
  wound healing, 251, 252f, 253, 254f, 265f, 266
Antigens, 253
Antioxidants, 105
Antithrombotic therapy, 295
Aorta
  aneurysm, 205
  elastic fraction, 205
  mechanical properties, 188t, 202f, 204–205, 206t
  stress-strain curve, 205, 206f
  wound healing, 274
Arachidonic acid metabolites, 259
Arterial wall, 200t
Articular cartilage, 5t, 270–271, 318–319
Assembly
  cytoskeletal components, 176–179
  macromolecules, 165–184, *See also* Self-assembly
Atactic polymer, 91–92, 91f
Atomic force microscopy, 122f
ATP, 217, 218f
Autocorrelation function, 154–155, 157–158f
Autophagy, 228, 231t
Axonal sprouting, 272

Bacterial infection, 236
Balloon angioplasty, 294f
Basement membrane, 7, 65, 67f, 71f, 242, 312
Basophils, 249f, 250, 258t, 259

Index

B cells, 251, 253, 254f, 257
β sheet, 43–45, 47f, 49f, 56
  mechanical properties, 192–195
Biglycan, 12–13, 67, 70f
Biodegradation, 119–120
  polymers, 20
  synthetic materials, 328
Biomaterials science, historical aspects, 1–3
Biomechanics, cellular, 208–209
Biomer, 94t
Bladder, 4t, 137–138, 139f
Blebbing, 227
Blood cells, morphology, 249f
Blood clotting, 79, 182, 183f, 242, 243f, 244, 246, 247f, 266
Blood vessels, 5–6
  angiogenesis, 227, 268
  functions, 4t, 14, 15t
  leukocyte adhesion to vessel walls, 224–226, 226f
  structure, 4t, 133–135, 135t, 136f
BMP, *See* Bone morphogenic protein
Body-centered tetragonal structure, 108, 108f
Bone
  cancellous, 140, 141t, 142f, 208, 318f
  compact, 5t, 140–141, 141t, 143f
  cortical, 208
  endochondral, 273
  lamellar, 16
  long, 16
  mechanical properties, 207–208, 208f
  membranous, 273
  osteonic, 16, 273
  spongy, 5t
  stress-strain curve, 207–208, 208f
  structure, 16
  tissue engineering, 316–320, 318f
  wound healing, 273–274
Bone cement, 288–289, 288f
Bone graft, 316
Bone morphogenic protein (BMP), 273, 307f, 318, 318f
Bone pin, 20
Bradykinin, 242, 246f, 248, 252
Breast implants, 280–286
Bronchi, 4t, 127, 129f
Bronchiole, 4t
Bubble oxygenator, 291, 292f
Burns, skin replacement, 17, 312–316, 314f

Cadherins, 223–224
Calcification, ectopic, 18–19
Calmodulin, 218
Cancellous bone, 140, 141t, 142f, 208, 318f
Carbons, 20–21
  glassy, 116t, 117
  mechanical properties, 116, 116t
  pyrolytic, 114, 116, 116t
Carbon-to-carbon bonds, 28, 28f
Carboxymethylcellulose, 19, 52–53
Cardiac muscle, 15–16
Cardiopulmonary bypass, 291
Cardiovascular tissue
  mechanical properties, 204–205, 206t
  tissue engineering, 306–307
  wound healing, 274
Carotid artery, 206t
Cartilage
  articular, 5t, 270–271, 318–319
  elastic, 142, 144f, 270
  hyaline, 141–142, 141t, 143f, 200t, 269–272
  mechanical properties, 200t, 207
  stress-strain curve, 118, 118f, 207
  structure, 141–142, 141t, 143–144f
  tissue engineering, 316–320, 319f
  wound healing, 269–272
Cartilage-derived growth factor (CDGF), 271–272
Cast metal, 111
Catheters, 291
CDGF, *See* Cartilage-derived growth factor
Cell(s)
  components, 214–223, 215f, 216t
  methods for structural analysis, 121–125
  properties, 22–24
  responses to materials, 24–25
  structure, 22–24
  in wound healing, 249f, 250–251
Cell adhesion, 223–224
  leukocytes, 224–226, 226f
Cell adhesion molecules, 225f, 227, 229f, 261–263
Cell culture, 310
Cell cytoskeleton, 9
  assembly of components, 176–179
  response to tissue injury, 215f, 217–219
  shear-induced structural changes, 209
Cell death, 213, 228, 230f, 232
Cell degeneration, 228, 230f
Cell differentiation, 227–228
Cell division, 221–222, 222f
Cell injury, 213–238, *See also* Immune response; Inflammation
  chemicals, 236–238, 237f
  genetic factors, 234
  infectious agents, 236
  nutritional factors, 234
  physical factors, 235–236
  stimuli leading to, 231t
Cell lysis, 230f
Cell-mediated immune response, 253
Cell membrane, 22, 31, 214–217, 215f, 216t
  macromolecules, 71–78
Cell migration, 227–228
Cell number, increases, 232
Cell polarity, 217
Cell proliferation, 232
Cell shape, 209, 217–219, 229f
Cell size, 219

changes, 232
Cell surface proteins, 13–14
Cell swelling, 228, 230f
Cell-to-cell attachment, 217
Cell transplantation
allogeneic cells, 309f
matrixes and scaffolds in, 310–312, 311f
Cell transplant device, 306, 308f
Cell types, changes, 233
Cellular adaptation, 228–233, 231t
Cellular biomechanics, 208–209
Cellulose derivatives, 19
Cellulose membrane, 291
Ceramics, 20–21
degradation, 119–120
mechanical properties, 116–117, 116t, 188t
properties, 21–22
structure, 88–89, 114–117
unit cell, 115
Cesium chloride, 115, 115f
Chemical injury, 236–238, 237f
Chitin, 19
Chondroblasts, 270
Chondrocytes, 142, 270–272, 298, 319, 319f
Chondroitin sulfate, 12, 12t, 57, 59, 60t, 65–68, 311f
Chondronectin, 311f
Chromatin, 215f, 221
Chromosomes, 221–222
Ciliated epithelium, 7
Cobalt-based alloys, 20–21, 23f, 88, 111
mechanical properties, 113t, 114
Cobalt-chromium alloy, 111
Cochlear implant, 20
Collagen, 2, 9–11, 30, 118–119, 227
α-chain, 9, 11
amino acid content, 36
composite networks, mechanical properties, 205–207
fibrillar, 9, 10–11f
fibrous, 62–63, 63f
mechanical properties of collagenous tissue, 196–207
nonfibrillar, 9–11, 10f
novel, 9–10
orientable networks, mechanical properties, 201–205, 202f
oriented networks, mechanical properties, 198–201, 200t
physical constants, 163t
self-assembly, 171–173f, 174–176, 175f
synthesis, 219–220
triple helix, 9, 10f, 43, 45, 47f, 56–57, 58–59f
mechanical properties, 192–195
type I, 11, 16, 62–63, 63f, 163t, 196, 227, 268, 270, 299
type II, 11, 62–63, 63f, 142, 163t, 270, 299
type III, 11, 62–63, 63f
type IV, 10–11, 67f, 163t, 227, 312
type V, 11
type VI, 11
type VII, 11
type VIII, 11
type X, 11, 299
type XI, 11
Colony-stimulating factors, 264
Compact bone, 5t, 140–141, 141t, 143f
Complement fragments, 252
Complement system
activation by biomaterial surfaces, 289–291
wound healing, 242, 244–249, 245f, 247–248f, 255, 264, 265f, 266
Composites, 21
degradation, 119–120
mechanical properties, 117–119, 118f, 188t
structure, 117–119
Compounding agents, 104–105
Condensation polymerization, 89, 90f, 119
Conductive leads, 112t
Conduit structures, 3, 4t, 5
structure, 133–138, 135t
Connective tissue, 6, 121–122
elastic fraction, 202f
mechanical properties of collagenous tissue, 196–207
silicone implant-associated, 284–286, 287t
stress-strain curve, 197, 197f
structure, 9–14
Connective tissue disease, 284
Contact guidance, 310
Contracture, 17
Copper, *See* Silver-tin-copper alloy
Cornea, 4t, 5, 6f
mechanical and optical properties, 17
structure, 127–130, 128t, 129f
Coronary artery, 292–296, *See also* Vascular stent
Coronary bypass, 292
Corrosion, 119
Cortical bone, 208
Crawling, *See* Cell migration
Creep test, 100, 103
Cross-linking agents, 105
Cruciate ligament, 5t
Cubic unit cell, 107f
Cuprophane membrane, 291
Cytokines, 226, 244, 248f, 253, 259, 260t, 264
Cytoplasm, 214, 215f, 216t, 217
Cytoskeleton, *See* Cell cytoskeleton
Cytosol, *See* Cytoplasm
Cytotactin, 311f

Dacron, *See* Poly(ethylene terephthalate)
Decalcification, processing tissue for microscopy, 124
Decorin, 12–13, 67, 70f, 199
Dehydration, processing tissue for microscopy, 122–124, 123f
Dendrites, 109
Dental amalgam, 88, 111–112, 112–113t, 114

Index

Dental appliance, 111, 112t
Dental implant, 114
Dental porcelains, 89, 115–116, 116t
Dental structures, 5, 5t
Dermatan sulfate, 12, 12t, 57, 59, 60t, 65–68, 199
Desmosome, 215f, 217
Dictyosome, 219–220
Diffusion coefficient, 154–156, 157f, 159f, 172, 173f
Digestive system, 6
Dipeptide, 3-D structure, 40, 41f
Dislocation, 109
DNA, 221–222, 222f
  stereochemistry, 53
  structure, 32–33, 34–35f, 60–61
Double helix, 60–61
Dura mater, 200–201, 202f
Dynamic testing, 103
Dysplasia, 231t, 233

Ectoderm, 5–6, 6f
Ectopic mineralization, implants, 298–300, 299f
EGF, *See* Epidermal growth factor
Eicosanoids, 259
ELAM, *See* Endothelial cell leukocyte adherence molecule
Elastic behavior, tissues, 189–190, 189–193f
Elastic cartilage, 142, 144f, 270
Elastic fibers, 9, 13
  aorta, 205
  skin, 204
  structure, 63–65, 64f
Elastic fraction, 190, 193f, 201, 202f, 205
Elasticity, defined, 189, 189f
Elastic tissue, 196–198
Elastin, 34, 57, 61, 63–65, 195
Electron microscopy, 121–122, 122f
  of individual molecules, 160, 161f
  optics, 126f
  processing tissue for, 122–124, 123f
  protein molecules, 33
  scanning, 125, 126f
  studying self-assembly processes, 174
  transmission, 125, 126f
Electrostatic bonds, proteins, 46–47
Embryonic development, 5–6, 6f
Endochondral bone, 273
Endocytosis, 220f, 250, 251f
Endoderm, 5–6, 6f
Endoplasmic reticulum, 214, 215f, 216t, 218–220
  induction, 230–231, 231t
Endothelial cell(s)
  coating surface of vascular graft, 298
  shear-induced structural changes, 209
  wound healing, 250, 258t, 260t, 266, 267f
Endothelial cell leukocyte adherence molecule (ELAM), 262
Endothelium, 6, 224
Endotoxin, 236, 245f
Endurance limit testing, 104, 104f
Engineering, tissue, *See* Tissue engineering
Engineering stress, 201
Entactin, 13–14
Eosinophils, 249f, 250, 259
Epidermal growth factor (EGF), 263–264, 266, 271–272
Epidermis, 5, 6f
  autograft, 313
Epiphyseal plate, 273
Epithelial cells, 266
Epithelialization, 268
Epithelium, 3, 5–6
  types, 6–9, 8f
Esophagus, 4t, 14, 15t, 130, 131f
Euchromatin, 221
Exocytosis, 220f
Extracellular matrix, 9, 14, 209
  cells, 250
  in cell transplantation, 310–312, 311f
  macromolecules, 62–71
  response to tissue injury, 220, 223, 227–228, 229f
  wound healing, 262, 266–268, 267f
Extracorporeal systems, 321
Eye, 14, 15t

Face-centered cubic structure, 106, 108f, 115
Facial implants, 89, 117, 280–286
Fatigue tests, 103–104
FGF, *See* Fibroblast growth factor
Fiberglass, 117
Fibrillin, 63, 64f
Fibrin, 182, 183f, 243f, 244, 246, 295
Fibrin degradation products, 244f, 247
Fibrinogen, 74f, 79, 80f, 163t, 182, 183f, 225f, 227, 296
Fibrinolysis, 242, 243f, 244, 244f, 246–247, 247f
Fibrinopeptide A, 182, 183f
Fibrinopeptide B, 182, 183f
Fibroblast(s), 23–24
  tissue engineering, 309–310
  wound healing, 250, 260t, 266, 267f, 268
Fibroblast growth factor (FGF), 266, 271–272
  acidic fibroblast growth factor, 263
  basic fibroblast growth factor, 263
Fibrocartilage, 142, 270–271
Fibroglycan, 12
Fibromodulin, 12–13, 70f
Fibronectin, 10, 13–14, 74, 74f
  response to tissue injury, 225f, 227
  structure, 65, 68f
  tissue engineering, 311f, 312
  wound healing, 262, 268
Filamin, 177, 228
Finger implants, 280–286
Fixation, processing tissue for microscopy, 123–124, 123f

Flexibility
  lipids, 50–52
  polymers, 92
  polysaccharides, 50
  proteins, 42–44
Flex temperature, 105, 106f
Flory's mathematical relationships, 167–169, 168f
Fluid transport tissues, *See* Conduit structures
Fracture energy determination, 104, 105f
Fracture plate, 111, 112t
Free energy change
  mixing of solute and solvent, 167–169, 168f
  protein transition from flexible to folded chain, 45–47
Free radicals, 231
Fungal infection, 236
Fused alumina, 116t

Galactose, 60t
Gamma rays, cell injury, 235
Gap junction, 215f, 217
Gasoline, 52–53
Gelsolin, 177
Genetic factors, cell injury, 234
Glass-transition temperature, 28, 95t
Glucose repeat units, 49–50, 51f
Glucuronic acid, 33f, 50, 51f, 60t
  boat and chair forms, 32f
Glutaraldehyde-treated implants, 299–300
Glycogen, 30, 49
Glycoprotein IIb/IIa, 74
Glycosaminoglycan, 12, 12t, 57, 118, 266
Glypican, 12, 72, 73f
GMP-140, 262
Gold foil, 114t
Golgi apparatus, 214, 215f, 216t, 219–220
Grain, 109, 110f
Grain boundary, 109, 110f
Granulation tissue, 266, 267f, 268–269
Graphite, 116, 116t
Growth factors, 13, 19, 217
  tissue engineering, 307f, 310, 318f
  wound healing, 242–244, 248f, 252–253, 263–266, 267f, 271–272
GTP-binding protein, 227
Guide wires, 111, 112t

Hageman factor, 244, 247–248, 296
Hardening of metals, 111
Hard tissue, 2
  mechanical properties, 188t, 207–208, 208f
  repair, 273–274
  structure, 16
Healing, wound, *See* Wound healing
Heart valve, 21, 89, 111, 112t, 114, 291
  ectopic mineralization of implants, 298–300, 299f
Helical aggregates, 166
HEMA-hydrogel, *See* Poly(hydroxyethyl methacrylate)
Hematoma, 274
Hemodialysis membrane, 291, 293f
Heparan sulfate, 12, 12t, 57, 60t, 65–68, 71–72, 268, 311f, 312
Heparin, 12, 12t, 57, 60t, 268
Hepatocytes, 320, 320f
Hepatocyte sandwich, 307f, 321
H & E stain, 124, 124f
Heterochromatin, 221–222
Hexagonal close-packed structure, 108, 108f
Hexagonal unit cell, 107f
Hip replacement, 286–289, 288f, 290f
Histamine, 249, 258, 258t, 264
Histocompatibility markers, *See* Major histocompatibility complex markers
Histones, 221, 222f
Holding structures, 3, 4t, 5
  structure, 133–138, 135t
Hormone receptors, 217
Human leukocyte antigens, 217
Humoral immune response, 253
Hyaline cartilage, 141–142, 141t, 143f, 200t, 269–272
Hyaluronan, 12, 12t, 30, 49–50, 51f, 57, 59, 68–71, 69f
  interaction with aggrecan, 69f
  physical constants, 163t
  repeat disaccharide, 33f
  wound healing, 266–268
Hyaluronic acid, 19, 31, 59, 60t, 311f
Hydrocarbon chains
  mobility, 28, 28f
  packing, 52–53, 52f
  potential energy function, 53, 54f
Hydrogen bonds, proteins, 46–47, 49f
Hydroxyapatite, 16, 298
Hydroxyproline, 10f, 36, 56–57
Hydroxypropylmethylcellulose, 19
Hyperplasia, 231t, 232, 294, 329
Hypertrophic scar tissue, 268–269
Hypertrophy, 231t, 232

ICAM, *See* Intercellular adhesion molecules
Iduronic acid, 60t
IFN, *See* Interferons
IGF, *See* Insulin-like growth factor
IL, *See* Interleukins
Immune reactions, 238
Immune response, 238
  cell-mediated, 253
  humoral, 253
  wound healing and, 252–261
Immunity
  acquired, 253
  natural, 253
Immunoglobulin, 252f
Immunoglobulin domain, 76, 78f

Index

Immunoglobulin-like protein, 224, 225f, 226
Immunoglobulin superfamily, 223
Immunology, transplantation, 307–310
Immunotolerance, 308–309
Impact strength, 104
Impact testing, 104
Implants
  capsule around, 221, 238, 269
  and cell changes, irreversible, 219
  cell replication around, 223
  cellular and tissue response, 24–25
  chemicals in device manufacture, 237
  complement activation by biomaterial surfaces, 289–291
  cytotoxicity study, 231
  degradation products, 120
  ectopic mineralization, 298–300, 299f
  hypertrophy in presence of, 232
  inflammation induced by wear particles, 286–289, 288f, 290f
  loosening, 288–289
  metallic, 20–21, 111
  pathobiological responses, 213–238, 279–301
  tests required, 24–25
  silicone, 280–286
  sterilization, 236
  stimulation of inflammation, 221
Infectious agents, 236
Infiltration, processing tissue for microscopy, 122–124, 123f
Inflammation, 19–20, 220–224, 238
  induced by wear particles, 286–289, 288f, 290f
  leukocyte involvement, 224
  wound healing, 244, 248f, 252–261
Injury, tissue response, 213–238
Insulin, 321
Insulin-like growth factor (IGF), 263, 271–272
Integral proteins, 215f, 217
Integrins, 14, 22, 30–31, 71–73, 73f, 209, 262, 312
  response to tissue injury, 223–224, 225f, 226–227, 229f
  wound healing, 262, 268
Intercellular adhesion molecules (ICAM)
  ICAM-1, 74, 74f, 224, 225f, 262
  ICAM-2, 74, 225f, 262
  ICAM-3, 262
Interfaces, design, 328–329
Interferons (IFN), 259–260, 260–261t
Interleukins (IL), 238
  IL-1, 224, 253, 259, 260t, 266
  IL-2, 260, 261t
  IL-4, 261t
  IL-5, 260–261
  IL-6, 259, 260t
  IL-8, 259, 260t
Intermediate filaments, 79, 218
Internal lining structures, 3, 4t
  cellular and noncellular composition, 128t
  function, 125–126
  structure, 5–9, 125–133
Intervertebral disc, 5t, 141t, 142
Intestine, 4t, 14, 15t, 135–137, 135t, 138f
Ionic bonds, 115, 115f
Ionizing radiation, cell injury, 235
Isotactic polymer, 91–92, 91f

Joint implant, 111, 112t, *See also* Orthopedic implants

Karyotype, 222
Keratan sulfate, 12, 12t, 57
Keratin, 7, 34, 56, 61, 79–81, 82f, 163t, 194
  protofilament, 81, 82f
Keratinocyte sheets, 315
Kinin system, 242, 244, 246–248f, 264
Knee replacement, 286–289, 288f, 290f
Krebs cycle, 217, 218f, 219

LAM, *See* Leukocyte adhesion molecule
Lamellar bone, 16
Laminin, 13–14, 227, 262
  structure, 65, 66–67f
  tissue engineering, 311f, 312
Langer's lines, 204
Latex allergy, 213
LEA-1, 74
Lectin-cell adhesion molecules, 262
Leukocyte(s), 250
  adhesion, 224–226, 226f
  binding to endothelium, 224
Leukocyte adhesion molecule (LAM), 262
Leukocyte-function-associated antigen (LFA), 74, 262
LFA, *See* Leukocyte-function-associated antigen
Leukotrienes, 258t, 259, 264, 265f, 266
Ligament
  function, 5t
  mechanical properties, 200t
  mechanical properties of oriented collagen networks, 198–201, 200t
  replacement, 114, 117
  stress-strain curve, 198
  structure, 5t, 141t, 144
  wound healing, 272
Light microscopy, 121, 122f, 123
  optics, 126f
  processing tissue for, 122–124, 123f
Light scattering, 152–154, 158f
  light intensity scattered at fixed angle, 169–170, 170f
  quasi-elastic, 154–157
  studying self-assembly processes, 169–172, 171–173f
Lipid(s), 27
  flexibility, 50–52
  stereochemistry, 50–53, 52f
  structure, 31
    packing of hydrocarbon chains, 52–53, 52f

Lipid bilayer, 214–217, 215f
Liver, 14, 15t
  tissue engineering, 320–321, 320f
Long bone, 16
Low temperature injury, 235
Lumican, 70f
Lupus erythematosus, systemic, 284, 287t
Lymphatic system, 6, 14, 15t, 133–135, 136f
Lymphocytes, 249f, 250–251, 253, 255
Lymphokines, 259–261, 261t, 264, 265f, 266
Lymphotoxin, 260, 261t
Lysosomes
  response to tissue injury, 214, 215f, 216t, 220, 220f, 228, 230f
  wound healing, 250, 251f, 258

Mac-1, 74, 262
Macromolecules (biological materials), *See also* Carbohydrates; Lipids; Nucleic acids; Proteins
  assembly, 165–184, *See also* Self-assembly
  axial ratios, 167–169
  backbone chemistry, 28–30, 28–29f
  basic building blocks, 30–33
  bead models, 160, 162f
  cell and tissue analysis, 122
  cell membrane, 71–78
  determination of physical parameters, 161–162
  determination of physical structure, 147–163
  extracellular matrix, 62–71
  modeling, 147–163
  packing, 2
  physical constants, 163t
  stereochemistry, 27–30, 33–53
  structure
    higher-order structures found in tissues, 61–62
    primary and secondary, 53–61
Macrophage inflammatory protein (MIP), 260, 261t
Macrophages, 251, 257, 258t, 259, 272
  response to metal particles, 289, 290f
Major histocompatibility complex (MHC) markers, 22, 30–31
  class I, 71–72, 75–78, 76–77f, 255, 256f, 257–258, 307, 309–310
  class II, 71–72, 75–78, 76–77f, 253, 254f, 255–258, 262, 307, 309–310, 316, 317f
  class III, 255
  MHC genes, 255, 257f
Mammary gland, 14, 15t
MAP, *See* Microtubule-associated proteins
Masson's trichrome stain, 124, 124f
Mast cells, 258t, 259
Materials degradation, 119–120
Mechanical tests, polymer characterization, 102–105, 103–106f
Membranous bone, 273
Mesothelium, 3, 6
Metals, 88, *See also* Alloys
  corrosion, 119
  deformation, 109
  degradation, 119–120
  mechanical properties, 112–114, 112–114t, 188t
  properties, 21
  stress-strain curve, 23f
  structure, 106–114
  unit cell, 106, 107f
Metaplasia, 231t, 233
MHC markers, *See* Major histocompatibility complex markers
α_2-Microglobulin, 255
β_2-Microglobulin, 255
Microscopy, *See* Electron microscopy; Light microscopy
Microtubule(s), formation, 179, 180f
Microtubule-associated proteins (MAP), 179
Mineralization, 274
  ectopic, 298–300, 299f
MIP, *See* Macrophage inflammatory protein
Mitochondria, 214, 215f, 216t, 219
Modeling, 147–163
Modulus, 18, 18f, 194–195
  defined, 191f
  polymers, 97–98, 98–99f, 101, 102f
  tissues and synthetic materials, 187, 188t
Molecular weight
  from light scattering data, 152–154, 158f
  from sedimentation coefficient, 160
Monoclinic unit cell, 107f
Monocytes, 249f, 250–251, 253, 259
Mooney-Rivlin equation, 100
Mouth, 4t, 128t, 130, 131f
mRNA, *See* RNA, messenger
Muscle, 15–16
  cardiac, 15–16
  function, 5t
  smooth, 15–16
  striated, 15–16
  structure, 5t, 141t, 142–144
Muscle fiber, 15
Mutation, 234
Myofibroblasts, 268
Myosin, 30, 81–83, 84f, 142, 163t, 217–219
  actin-myosin interaction, 179–182, 180f

Nails, 112t
Natural immunity, 253
Natural killer (NK) cells, 253, 255, 257f, 261t
NCAM, 224
Nerve repair, 272–273
Neurocan, 12
Neutrophils, 249f, 250, 253, 259
Nitinol, 294
NK cells, *See* Natural killer cells
Nuclear membrane, 214, 215f, 221

Nucleic acids, 27
  stereochemistry, 53
  structure, 32–33, 34f
    primary and secondary, 60–61
Nucleolus, 215f, 221
Nucleus, 214, 216t, 221
Null cells, 253
Nutritional factors, 234

Oral histology, 130, 131f
Organelles, 122, 214–223, 215f, 216t
Organs, *See also specific organs*
  functions, 15t
  specialized, 14, 15t
    tissue engineering, 320–321, 320f
Organ-supporting structures, *See* Parenchyma
Orthopedic implants, 20–21, 114, 117
Orthorhombic unit cell, 107f
Osmotic pressure regulation, 242
Osteoblasts, 273, 298
Osteocalcin, 300
Osteoclasts, 274
Osteoconduction, 318
Osteocytes, 140–141
Osteogenin, 318, 318f
Osteoid, 298
Osteoinduction, 318
Osteoinductive proteins, 273
Osteon, 141, 141t
Osteonic bone, 16, 273
Osteopontin, 227
Oxidative phosphorylation, 218f, 219
Oxygenator, blood, 291, 292f

p150/95, 74
Pacemaker, 112t
Pancreas, 14, 15t
  tissue engineering, 320–321
Parenchyma, 3, 139
Particle dissipation factor, 170
Particle-scattering factor, 153–154, 156f
Pathobiology
  responses to implants, 279–301
  responses to tissue injury, 213–238
Paxillin, 228, 229f
PDGF, *See* Platelet-derived growth factor
PECAM, *See* Platelet-endothelial cell adhesion molecule
Pellethane, 94t
Peptide unit, 40, 41t, 42
Percutaneous transluminal coronary angioplasty (PTCA), 292, 294f, 296
Pericardium, 200–201, 202f
Perichondral autograft, 318–319
Periodontal ligament, 5t
Peripheral nerve repair, 272–273
Peripheral proteins, 215f, 217
Peritoneum, 4t, 128t, 130, 132f
Perlecan, 68, 71f, 74
Phagocytosis
  of metal particles by macrophages, 289, 290f
  response to tissue injury, 220, 220f
  wound healing, 250, 251f, 264–266, 265f
Phagolysosome, 251f
Phagosome, 250
Phospholipids, 31
  metabolites, 258–259
Physical structure, determination, 147–163
Plasma cells, 249f, 251, 254f, 266
Plasma membrane, *See* Cell membrane
Plasmin, 244f, 247
Plasminogen, 246, 248f
Plasminogen activator, 258t
Plasminogen inhibitor, 258t
Plastic behavior, 190
Plasticizer, 105, 106f, 238
Platelet(s), 244, 249f, 250, 260t
Platelet-activating factor, 258t, 266
Platelet-derived growth factor (PDGF), 263–264, 266, 271–272, 296
Platelet-endothelial cell adhesion molecule (PECAM), 224
Pleura, 4t, 128t, 130, 132f
Plexiglas, *See* Poly(methyl methacrylate)
Poisson's ratio, 190
Polyamide, 93t
Poly(dimethylsiloxane), *See* Silicone
Polyester, 93t
Polyethylene, 19, 90, 92, 93t, 95t, 221, 288–289, 288f
Poly(ethylene terephthalate), 19, 88–89, 93t, 95t, 103–104, 213
Poly(glycolic acid), 20, 320
Poly(hydroxyethyl methacrylate), 92, 93t
cis-Polyisoprene, 95t
Poly(lactic acid), 20, 319–320
Polymer(s) (nonbiological material), 19
  amorphous, 95, 96f, 97, 98–99f
  atactic, 91–92, 91f
  biodegradation, 20
  crystalline, 95, 96f, 97, 98f
  degradation, 119–120
  design of new materials, 328
  flexibility, 92
  glass-transition temperature, 95t
  hydrophobicity, 92
  isotactic, 91–92, 91f
  mechanical properties, 95–106, 188t
    mechanical tests, 102–105, 103–106f
    modulus, 97–98, 98–99f, 101, 102f
    rubber elasticity theory, 99–100
    viscoelasticity, 100–101, 101–102f
  melting temperatures, 95t
  random-coil, 95, 96f
  repeat structure of medical polymers, 93t
  semicrystalline, 95, 96f
  stress-strain curve, 97, 98f
  structure, 89–95
    physical structure of polymer chains, 92–95, 96f

syndiotactic, 91–92, 91f
synthesis, 89–92
addition, 89–90, 90f, 119
condensation, 89, 90f, 119
Polymerization, 89–92
Poly(methyl methacrylate), 88, 92, 93t, 95t, 103–104, 289
Polymyositis-dermatomyositis, 284, 287t
Polypeptides, 30
model, mechanical properties, 192–195, 194–196f
Polypropylene, 95t
Polysaccharides, 27
flexibility, 50
functions, 30
stereochemistry, 49–50, 51f
structure, 30–31, 32–33f
α linkage, 50
β linkage, 50
glucose repeat units, 49–50, 51f
primary and secondary, 57–60, 60t
thixotropic properties, 50
Polystyrene, 95t
Polysulfone, 93t
Poly(tetrafluoroethylene), 19, 93t, 213
Polyurethane, 19, 88, 93–94t
Poly(vinyl acetate), 95t
Poly(vinyl chloride), 91–92, 91f, 93t, 95t
Poly(vinyl fluoride), 93t
Porcelains, dental, *See* Dental porcelains
Prekallikrein, 247–248
Pressure, excessive, cell injury, 235
Procollagen, 11f
Proline, 10f, 36, 37f, 43, 46f, 56–57
Properdin, 249
Prostaglandins, 258t, 259, 264, 265f, 266
Proteins, 27
containing proline, 36, 43
flexibility, 42–44
free energy change with transition from flexible chain to folded chain, 45–47
functions, 30
interatomic distances for nonbonded atoms, 42, 42t
membrane, 215f, 217
stereochemistry, 33–49
structure, 30, 31f
α helix, 34, 43–45, 47–48f, 55–56
amino acid sequence, 34
backbone, 40–41, 41f, 41t
β sheet, 43–45, 47f, 49f, 56
conformational map, 42–44, 44–46f
electrostatic and hydrogen bonds, 46–47, 49f
higher-order, 61
peptide unit, 40, 41t, 42
primary and secondary, 54–57
synthesis, 30, 31f, 217, 218f, 219, 221
Proteoglycans, 9, 11–13, 49, 65–68, 70f, 142, 196, 198, 207, 215f, 299
tissue engineering, 311f, 312
wound healing, 266, 270
Pseudostratified columnar epithelium, 7–9, 8f
PTCA, *See* Percutaneous transluminal coronary angioplasty
Purine, 32–33, 34f, 60
Pyrimidine, 32–33, 34f, 60

Quasi-elastic light scattering, 154–157

Radiation injury, 235–236, 237f
Ramachandran plot, 29f, 30
Rayleigh factor, 153–154
Receptors, 311f
Rectum, 4t
Red blood cells, 249f, 250
Reepithelialization, 268
Remodeling phase, wound healing, 268–269
Reproductive system, 14, 15t
Residual body, 220f, 250, 251f
Respiratory chain, 218f, 219
Respiratory system, 14, 15t
Restenosis, after vascular stenting, 292–296
Reticuloendothelial system, 251
RGD sequence, 227
Rheumatoid arthritis, 284, 287t
Rhombohedral unit cell, 107f
Ribosome, 215f, 216t, 219
RNA, 215f
messenger, 219, 221
ribosomal, 221
stereochemistry, 53
structure, 32–33, 34f
primary and secondary, 60–61
Rubber(s), 87–88, 97, 98f
Rubber elasticity theory, 99–100
Ruffling, 227–228

Scaffold, in cell transplantation, 310–312, 311f
Scanning electron microscope, 125, 126f
Scar tissue, 17, 223
hypertrophic, 268–269
wound healing, 242, 244, 248f, 264, 267f, 268, 271, 274
Schwann cells, 272
Screws, 111, 112t
Sectioning, processing tissue for microscopy, 122–124, 123f
Sedimentation coefficient, 160
Selectins, 223–224, 226
Self-assembly, macromolecules, 165–184
actin, 177–179, 178f
collagen, 171–173f, 174–176, 175f
methods of studying, 169–174, 171–173f
Semicrystalline materials, 88
Serotonin, 258, 264
Shape factor, 149–150, 150f
Shear modulus, 190
Silica glass, 20–21, 116t

Silicon dioxide, 114
Silicone, 19, 88, 93t, 95t, 213
  preparation, 282f
Silicone catheters, 291
Silicone implants
  chemistry, 280–281, 281f
  connective tissue disease and, 284–286, 287t
  local reactions, 281–286, 283–286f, 287t
  macrophage response, 284, 285f
  pathobiological complications, 280–286
  rupture, 284, 286f
  vacuoles in tissue surrounding, 283–284, 283–284f, 286f
Silk, 56, 194, 195f
Silver-tin-copper alloy, 111–112, 113t, 114
Silver-tin-mercury-copper alloy, 111–112
Simple columnar epithelium, 7, 8f
Simple cuboidal epithelium, 7, 8f
Simple epithelium, 6
Simple squamous epithelium, 7, 8f
Sjögren syndrome, 284, 287t
Skeletal muscle
  actin-myosin interaction, 179–182, 180f
  contraction cycle, 181–182, 181f
Skeletal structures, 3
  functions, 5, 5t
  structure, 5, 5t, 139–144, 141t
  tissue engineering, 306, 316–320, 318–319f
Skin
  cellular and noncellular composition, 128t
  elastic fraction, 201, 202f, 204
  functions, 4t
  mechanical properties, 17, 200t, 203t
    directional dependence, 204
  mechanical properties of orientable collagen networks, 201–205, 202f
  optical properties, 17
  reepithelialization, 268
  stress-strain curve, 201, 203f, 204
  structure, 4t, 5, 130–133, 132f
  tissue engineering, 306, 307f, 312–316, 314f, 328
  transplantation, 308–310, 309f
  wound healing, 242
Slow-reacting substances of anaphylaxis, 259
Small intestine, 4t, 138f
Smooth muscle, 15–16
Soft tissue, 2
  mechanical properties, 188t
  structure, 3–16, 4–5t
Solvents, 234, 238
Spectrin, 218
Spleen, 14, 15t
Spongy bone, 5t
Staining, processing tissue for microscopy, 122–124, 123–124f
Stainless steel, 20–21, 23f, 88, 111, 119, 288
Starch, 30, 49
Steel, 119
Stereochemistry, 27–30
  lipids, 50–53, 52f
  macromolecules, 33–53
  nucleic acids, 53
  polysaccharides, 49–50, 51f
  proteins, 33–49
Stomach, 4t, 14, 15t, 135–137, 135t, 137t
Strain, 190
  defined, 191f
Strain at failure, 190
Stratified columnar epithelium, 7, 8f
Stratified epithelium, 6
Stratum corneum, 194, 196f, 204
Stress, 190
  defined, 191f
  engineering, 201
Stress fibers, 209
Stress-strain curve, 21, 22f, 190
  aorta, 205, 206f
  bone, 207–208, 208f
  cartilage, 118, 118f, 207
  connective tissue, 197, 197f
  incremental, 103f, 190, 193f, 200, 202f
  ligament, 198
  metals, 23f
  polymers, 97, 98f
  silk, 194, 195f
  skin, 201, 203f, 204
  stratum corneum, 194, 196f
  tendon, 192, 192f, 194f, 198–199
  tissue, 17–18, 18f
Striated muscle, 15–16
Sugar, boat and chair forms, 32f, 49–50
Surface lining structures, 3, 4t
  cellular and noncellular composition, 128t
  function, 125–126
  structure, 5–9, 125–133
  tissue engineering, 312–316, 314f
Sutures, 20, 89
Syndecan, 12–13, 22, 31, 72, 73f, 217
Syndiotactic polymer, 91–92, 91f
Synthetic materials, *See also* Alloys; Ceramics; Composites; Polymers (nonbiological)
  biodegradation, 328
  properties, 21–22
Systemic lupus erythematosus, 284, 287t
Systemic sclerosis, 284, 287t

Talin, 228, 229f
T cells
  cytolytic, 253, 255, 256f
  T-helper, 253, 254f
  T-suppressor, 253, 254f
  wound healing, 251, 253, 254f, 255, 260–261t, 266
Tecoflex, 94t
Teflon, *See* Poly(tetrafluoroethylene)
Telopeptides, self-assembly of collagen, 176
Temperature extremes, cell injury, 235
Tenascin, 13–14

Tendon
elastic fraction, 202f
mechanical properties, 200t
mechanical properties of oriented collagen networks, 198–201, 200t
replacement, 117
stress-strain curve, 192, 192f, 194f, 198–199
structure, 141t, 144, 145f
wound healing, 272
Tensile testing, 101
Tensile testing device, 96, 97f
Tetragonal unit cell, 107f
TGF, *See* Transforming growth factor
Thermal injury, 235
Thermoplastics, 88
Thick filament, 179
Thin filament, 179
Thoracic aorta, 206t
Thrombin, 264
Thromboembolism, 291
Thrombosis, 291, 329
small-diameter vascular grafts, 296–298, 297f
stent-associated, 294–295
Thrombospondin, 13–14, 227
Thymus, 14, 15t
Tight junction, 215f, 217
Tin, *See* Silver-tin-copper alloy
Tissue
elastic and viscous behavior, 189–190, 189–193f
force required to deform, 17
macromolecular structures in, 61–62
mechanical properties, 16–19, 187–210
methods for structural analysis, 121–125
modulus, 18, 18f
optical properties, 16–19
response to injury, 213–238
response to materials, 24–25
stress-strain curve, 17–18, 18f
Tissue engineering, 305–322
approaches, 306, 307f
cardiovascular, 306–307
external lining structures, 312–316, 314f
general concept, 317f
introduction, 305–307, 306f
liver, 320–321, 320f
matrixes, 312
new directions, 328
pancreas, 320–321
skeletal structures, 306, 316–320, 318–319f
skin, 306, 307f, 312–316, 314f, 328
specialized organs, 320–321, 320f
Tissue factors, 244
Titanium-based alloys, 20–21, 23f, 88, 111, 287–288
mechanical properties, 113t, 114
TNF, *See* Tumor necrosis factor
Trachea, 4t
Transforming growth factor-α (TGF-α), 263
Transforming growth factor-β (TGF-β), 260, 261t, 263–264, 273
Transforming growth factor-β (TGF-β) supergene family, 318
Transitional epithelium, 7, 8f
Translational diffusion coefficient, 155–156, 157f, 159f, 172, 173f
Transmission electron microscope, 125, 126f
Transparency, 16–17
Transplantation, 305
allogeneic cells, 309f
immunology, 307–310
matrixes and scaffolds in, 310–312, 311f
skin, 308–310, 309f
Trichloroethylene, 234
Triclinic unit cell, 107f
Triple helix, collagen, 9, 10f, 43, 45, 47f, 56–57, 58–59f
mechanical properties, 192–195
Tropoelastin, 63
Trypan blue uptake test, 217
Tubulin, 30, 163t, 218
assembly of cytoskeletal components, 176–179
microtubule formation, 179, 180f
Tumor necrosis factor (TNF), 224, 259, 260t
Turbidity-time curve, 169–172, 171f

Ulcer, skin, 315
Ultimate strain, 18f
Ultimate tensile strength (UTS), 17–18, 18f, 190, 192f, 194–195
tissues and synthetic materials, 187, 188t
Ultracentrifugation, 157–160
equilibrium, studying self-assembly processes, 173–174
Unit cell
ceramics, 115
metals, 106, 107f
Ureter, 4t, 137–138, 140f
Urinary system, 14, 15t
Uterus, 4t, 128t, 133, 134f, 233
UTS, *See* Ultimate tensile strength

Vagina, 4t
van Gieson's solution, 125
Vascular cell adhesion molecule, VCAM-1, 74, 224, 225f
Vascular graft, 89
coating surface with endothelial cells, 298
compliance mismatch, 298
small-diameter
device design, 329
thrombosis, 296–298, 297f
wound healing, 274
Vascular stent, 111, 112t, 329
balloon-expandable, 294–295
placement, 294f
restenosis after stenting, 292–296
spring-loaded, 294, 295f

Vasculitis, silicone implant-associated, 284, 285f
Vasoactive mediators, 258
Vasodilation, 246f, 252, 264
VCAM, *See* Vascular cell adhesion molecule
Versican, 12
Very late activation antigens (VLA), 224, 262
Vinculin, 228, 229f
Viral infection, 236
Viscoelasticity, 192f
  defined, 189, 189f
  polymers, 100–101, 101–102f
Viscosity, 148–152
  defined, 148
  determination of shape factor, 150, 150f
  intrinsic, determination, 150–152, 151f
  measurement, 148–149
Viscous behavior, tissues, 189–190, 189–193f
Vitamin C, 234, 266
Vitronectin, 13–14, 227, 311f
VLA, *See* Very late activation antigens
von Willebrand factor, 227

Wear particles, inflammation induced by, 286–289, 288f, 290f
White blood cells, 250
Wound dressing, 313
Wound healing, 220, 224, 241–275
  biological cascades, 242–249
  capsule formation around implant, 269
  cardiovascular, 274
  cartilage, 269–272
  cell adhesion molecules, 261–263
  cell replication, 222
  cells in, 249f, 250–251
  growth factors, 263–264
  hard tissue repair, 273–274
  inflammatory and immunological aspects, 252–261
  ligament, 272
  mechanical properties of wound tissue, 118–119, 119f
  peripheral nerve, 272–273
  process
    inflammatory phase, 264–266, 265f
    proliferative phase, 266–268, 267f
    remodeling phase, 268–269
  skin, 242
  tendon, 272
  vitamin C, 234
Wrought metal, 111

X-rays
  cell injury, 235
  X-ray diffraction, 33

Yield strength at 2% offset, 21
Young's modulus, 190

Zimm plot, 153